AF361711

# The Systematics, Morphological, and Molecular Characterization of Economically Important Plant–Parasitic Nematodes: A Themed Issue in Honor of Dr. Gary Bauchan

# The Systematics, Morphological, and Molecular Characterization of Economically Important Plant–Parasitic Nematodes: A Themed Issue in Honor of Dr. Gary Bauchan

Editors

**Zafar Handoo**
**Mihail Kantor**

MDPI • Basel • Beijing • Wuhan • Barcelona • Belgrade • Manchester • Tokyo • Cluj • Tianjin

*Editors*
Zafar Handoo
Mycology & Nematology
Genetic Diversity & Biology
Laboratory
USDA, ARS
Beltsville
United States

Mihail Kantor
Mycology & Nematology
Genetic Diversity & Biology
Laboratory
USDA, ARS
Beltsville
United States

*Editorial Office*
MDPI
St. Alban-Anlage 66
4052 Basel, Switzerland

This is a reprint of articles from the Special Issue published online in the open access journal *Plants* (ISSN 2223-7747) (available at: www.mdpi.com/journal/plants/special_issues/plant_nematodes).

For citation purposes, cite each article independently as indicated on the article page online and as indicated below:

LastName, A.A.; LastName, B.B.; LastName, C.C. Article Title. *Journal Name* **Year**, *Volume Number*, Page Range.

ISBN 978-3-0365-5462-4 (Hbk)
ISBN 978-3-0365-5461-7 (PDF)

# Contents

# About the Editors

**Zafar Handoo**

Since 1990, Dr. Zafar Handoo has served as a Microbiologist at the United States Department of Agriculture Agricultural Research Service (USDA ARS) and the National Specialist for the US, providing accurate nematode identifications. He conducts research on the systematics and morphology of plant-parasitic nematodes. He identifies nematode samples received from federal, state, and foreign agencies and scientists for research, control, and regulatory purposes. He is also the curator of the USDA Nematode Collection, managing and expanding the collection with thousands of valuable slides and vials. He oversaw the development of a computerized internet-accessible database for the collection with thousands of sample records on hosts, occurrence, and distribution. He loaned hundreds of slides to scientists around the world to enable them to perform accurate identifications of nematodes. His research interests include plant-parasitic nematodes of the order Tylenchida and some forms of Dorylaimida. Dr. Handoo has served as Editor of the *Journal of Nematology* over the last ten years and on the Editorial Board of the *Int. J. of Nematology*, the Pakistan *J. of Nematology* and Turkish *J. of Agri. and Nat. Sciences*, served as Chair, and is now a member of the Systematic Resources Committee and Regulatory Committee of the Society of Nematologists.

**Mihail Kantor**

Dr. Mihail Kantor is an Established Researcher at the United States Department of Agriculture Agricultural Research Service (USDA ARS). He graduated with an M.S. in Plant Biotechnology from Claflin University and obtained a Ph.D. from University of Agricultural Sciences and Veterinary Medicine of Cluj-Napoca. After his Ph.D, he continued his career in academia, having held several teaching and research positions until 2017 when he joined the Mycology and Nematology Genetic Diversity and Biology Laboratory USDA, ARS. His main research interest focuses on the identification of plant-parasitic nematodes by integrating classical systematics and morphology with molecular techniques. In addition to his primary research, he is studying the effects of plant metabolites, plant extracts, and cover crop amendments on plant-parasitic nematodes. Other ongoing research includes studying the interactions between plant-parasitic nematodes and vegetables or microbes. Dr. Kantor manages the curation of plant-parasitic nematodes, as well as the maintenance and expansion of the USDA Nematode Collection.

# Preface to "The Systematics, Morphological, and Molecular Characterization of Economically Important Plant–Parasitic Nematodes: A Themed Issue in Honor of Dr. Gary Bauchan"

This is a compilation of articles published in the Special Issue "The Systematics, Morphological, and Molecular Characterization of Economically Important Plant–Parasitic Nematodes: A Themed Issue in Honor of Dr. Gary Bauchan" in *Plants*. It includes a series of original research (seven) and review articles (four) focused on plant-parasitic nematodes, including two new species descriptions, *Pratylenchus dakotaensis* n.sp. and *Xiphinema malaka* n. sp. Several original articles present integrative taxonomy and molecular phylogeny methods to identify plant-parasitic nematodes of the genera *Paratylenchus*, *Xiphinema*, *Pratylenchus*, and *Rotylenchulus*. A few articles present reports of plant-parasitic nematodes found in major economically important crops such as soybean and potato, while some other articles describe new sensitive and rapid detection methods for *Meloidogyne hapla*, *Globodera pallida* and *Globodera rostochiensis*. We would like to express our gratitude to all the authors who submitted their work to be included in this Special Issue.

**Zafar Handoo and Mihail Kantor**
*Editors*

Article

# Morphometric and Molecular Diversity among Seven European Isolates of *Pratylenchus penetrans*

Mesfin Bogale [1,†], Betre Tadesse [2,†], Rasha Haj Nuaima [3], Bernd Honermeier [2], Johannes Hallmann [2] and Peter DiGennaro [1,*]

1  Department of Entomology and Nematology, University of Florida, Gainesville, FL 32611, USA; mazene@ufl.edu
2  Justus Liebig University, Schubertstraße 81, 35392 Gießen, Germany; betretadesse@yahoo.com (B.T.); Bernd.Honermeier@agrar.uni-giessen.de (B.H.); johannes.hallmann@julius-kuehn.de (J.H.)
3  Julius Kühn-Institut, Federal Research Centre for Cultivated Plants, Institute for Epidemiology and Pathogen Diagnostics, Toppheideweg 88, 48161 Münster, Germany; rasha.haj-nuaima@julius-kuehn.de
*  Correspondence: pdigennaro@ufl.edu
†  These authors contributed equally to this work.

**Abstract:** *Pratylenchus penetrans* is an economically important root-lesion nematode species that affects agronomic and ornamental plants. Understanding its diversity is of paramount importance to develop effective control and management strategies. This study aimed to characterize the morphological and genetic diversity among seven European isolates. An isolate from the USA was included in the molecular analyses for comparative purposes. Morphometrics of the European *P. penetrans* isolates generally were within the range of the original descriptions for this species. However, multiple morphometric characteristics, including body length, maximum body width, tail length and length of the post-vulval uterine sac showed discrepancies when compared to other populations. Nucleotide sequence-based analyses revealed a high level of intraspecific diversity among the isolates. We observed no correlation between D2-D3 rDNA- and COXI-based phylogenetic similarities and geographic origin. Our phylogenetic analyses including selected GenBank sequences also suggest that the controversy surrounding the distinction between *P. penetrans* and *P. fallax* remains.

**Keywords:** *Pratylenchus penetrans*; *Pratylenchus fallax*; root-lesion nematode; genetic diversity; morphometrics; COXI; D2-D3 rDNA; PP5; $\beta$-1,4-endoglucanase

## 1. Introduction

With a global distribution and significant economic impact [1], sometimes requiring quarantine measures [2], species within the plant parasitic nematode genus *Pratylenchus* are some of the most agriculturally important pests. Species identification within the genus is traditionally based on morphological and morphometric characterization [1,2]. The main diagnostic characteristics are presence/absence of males, body length, head shape, stylet length, and other cuticular characters including the number of lip annules, the number of lateral field lines, the presence/absence of areolated bands on the lateral fields within the vulval region, the length and structure of the post-vulval uterine sac and shape of the spermatheca, the shape of the female tail and tail tip, and de Man's indices [3–6].

Identification and delineation of *Pratylenchus* species using these anatomical and morphometric features alone can pose many issues due to interspecific similarity and intraspecific variability of some of these characters [1,7,8]. For example, the high intraspecific morphological variations that exist within *P. penetrans* and *P. fallax* have contributed to the taxonomic confusion of these species. *P. fallax* was separated from *P. penetrans* by Seinhorst [7], only to be considered conspecific later by Tarte and Mai [8], who attributed the variations to environmental factors. The separation of the two species was confirmed using breeding experiments [9], isozyme [10] and PCR Restriction Fragment Length Polymorphism (PCR-RFLP; [11]) analyses. The presence/absence of males also does not appear

**Citation:** Bogale, M.; Tadesse, B.; Nuaima, R.H.; Honermeier, B.; Hallmann, J.; DiGennaro, P. Morphometric and Molecular Diversity among Seven European Isolates of *Pratylenchus penetrans*. *Plants* **2021**, *10*, 674. https://doi.org/10.3390/plants10040674

Academic Editors: Zafar Handoo and Mihail Kantor

Received: 31 January 2021
Accepted: 25 March 2021
Published: 31 March 2021

**Publisher's Note:** MDPI stays neutral with regard to jurisdictional claims in published maps and institutional affiliations.

to be a robust taxonomic characteristic as some asexual species such as *P. thornei*, *P. neglectus* and *P. hippeastri* have been reported to occasionally have males though these males may not play a role in reproduction [12]. The large number of species (110 species) described within the genus [13] is also a contributing factor owing to the limited number of distinguishing morphological features that are available. Consequently, different molecular methods have been developed for species identification and assessment of genetic variation within and between species of *Pratylenchus*. Commonly used molecular methods include quantitative PCR (qPCR; [14]), Amplified Fragment Length Polymorphism (AFLP; [15,16]), RFLP [11,17], Random Amplified Polymorphic DNA (RAPD; [18–20]), Sequence Characterized Amplified Region (SCAR; [16,21]), Single Nucleotide Polymorphism (SNP; [22]) and Simple Sequence Repeats or Variable Number Tandem Repeats (SSR or VNTR, respectively; [23,24]).

One of the most economically important species within this genus is *P. penetrans*, which affects a wide range of agronomic and ornamental plants, and has the potential to parasitize over 400 plant species [1,25]. *P. penetrans* is cosmopolitan though more significant in temperate regions, harbours high morphological variation, and it is considered to represent a species complex [26]. The objective of this work was to determine the diversity among seven populations of *P. penetrans* that were collected from different geographical regions in Europe based on morphometric and molecular analyses. An isolate (VA) obtained from Virginia, USA, was also included in nucleotide sequence analyses for comparative purposes.

## 2. Results

### *2.1. Morphometrical Observations*

Significant similarities and differences in morphometric characters were observed amongst the seven *P. penetrans* isolates (Table 1). The ratio (b′) of body length (L) to length of pharynx (from anterior end to posterior end of pharyngeal gland) was the largest for NL, FR and UK, and the smallest for MN, WZ and BL. The ratio (c) of body length to tail length (tail) ranged from 14.10 in BN to 23.30 in FR. These isolates were significantly different from each other in terms of this ratio. The excretory pore (EP) was most anterior in MN, WZ and some UK isolates, and most posterior in BL, FR, NL and some BN isolates. Ovary length (Ovary) was significantly different between MN and BL isolates. MN and WZ isolates had shorter tails than BN, BL, FR and NL. Some morphological characters varied among the seven populations, but no distinct groupings were observed in terms of these characters. Such characters included stylet length (Stylet), pharynx length (Ph-L; anterior end to end of pharyngeal gland) and length of pharyngeal overlap (Ph-O). The distance of vulva from anterior end divided by body length (V) did not vary significantly among the seven populations.

Coefficient of variation (CV) for the various morphometric characters ranged from 2.40% to 14.92% (Table 1). CV was the lowest for Stylet length (2.40%) and a value (2.85%); and the highest for ovary length (14.92%) and length of post-vulval uterine sac (PUS; 14.59%).

### *2.2. Nucleotide Sequence Analysis*

For each of the eight isolates, we sequenced the partial β-1,4-endoglucanase gene, the D2-D3 expansion of rDNA and the partial mitochondrial COXI gene region. The rDNA amplicon for each isolate was cloned (see below) and two transformed bacterial colonies were sequenced to check for the presence/absence of gene variants and/or intrapopulation variants. Both colonies that were sequenced for each isolate's rDNA fragment had identical D2-D3 sequences. We included in our sequence alignments selected GenBank sequences spanning the D2-D3 rDNA expansion and the mitochondrial COXI sequences for which our sequences found the highest hits during nucleotide Basic Local Alignment Search Tool (BLASTn) analysis (Table 2). We also included *P. neglectus* sequences for outgroup purposes (Table 2). The aligned D2-D3 and COXI sequences (each consisting of 23 taxa, including our eight isolates; Supplementary Data S1) were analyzed as a combined dataset. The β-1,4-endoglucanase sequences were not included in the phylogenetic analyses for lack of related sequences in the public databases for use as references.

**Table 1.** Morphometry of the seven European *Pratylenchus penetrans* female isolates and their geographical origins.

| Char. | P. penetrans Isolates | | | | | | | CV [4] (%) |
|---|---|---|---|---|---|---|---|---|
| | MN | WZ | BN | BL | UK | FR | NL | |
| L | 449 ± 9.70 [1] a [2] (431–462) [3] | 437 ± 9.60 a (381–492) | 506 ± 10.30 bc (465–578) | 525 ± 10.50 c (443 ± 594) | 470 ± 10.00 ab (428–517) | 544 ± 10.70 c (505–625) | 527 ± 10.50 c (465–572) | 6.23 |
| a | 26.00 ± 0.25 ab (24.80–28.60) | 25.10 ± 0.25 a (22.20–28.90) | 27.50 ± 0.25 d (24.40–31.20) | 26.30 ± 0.25 bc (22.10–29.70) | 25.10 ± 0.25 a (22.20–27.30) | 27.70 ± 0.25 d (25.90–30.20) | 27.10 ± 0.25 cd (24.60–30.80) | 2.85 |
| b′ | 4.34 ± 0.08 a (5.84–9.22) | 4.38 ± 0.08 a (6.63–9.55) | 4.52 ± 0.08 ab (6.32–10.30) | 4.33 ± 0.08 a (5.62–8.44) | 4.85 ± 0.08 bc (6.45–9.23) | 4.98 ± 0.08 c (5.33 ± 9.17) | 4.87 ± 0.08 c (6.92–8.50) | 4.59 |
| c | 19.30 ± 0.35 bc (17.10–20.50) | 19.10 ± 0.35 bc (16.40 -21.10) | 17.20 ± 0.33 a (14.10–20.30) | 18.20 ± 0.34 ab (14.40–20.70) | 18.20 ± 0.34 ab (14.60–21.70) | 20.00 ± 0.35 c (16.90–23.30) | 18.40 ± 0.34 ab (16.00–21.00) | 5.06 |
| V | 79.2 ± 0.81 a (77.90–80.80) | 79.9 ± 0.82 a (78.60–81.90) | 79.70 ± 0.81 a (73.30–82.90) | 80.90 ± 0.82 a (76.80–85.90) | 78.80 ± 0.81 a (76.70–81.60) | 79.80 ± 0.81 a (77.30–82.70) | 79.60 ± 0.81 a (76.00–82.30) | 6.68 |
| Stylet | 15.10 ± 0.12 a (14.60–15.60) | 15.40 ± 0.12 abc (14.50–16.00) | 15.30 ± 0.12 ab (15.00–15.80) | 15.80 ± 0.12 c (15.10–16.80) | 15.70 ± 0.12 bc (15.20–16.40) | 15.30 ± 0.12 abc (15.00–16.20) | 15.20 ± 0.12 ab (14.60–15.60) | 2.40 |
| Ph-L | 104.00 ± 3.50 ab (89–115) | 100.00 ± 3.50 ab (90–112) | 112.00 ± 3.50 bc (97–133) | 121.00 ± 3.50 c (100–136) | 97.00 ± 3.50 a (95–111) | 111.00 ± 3.50 abc (90–161) | 108.00 ± 3.50 abc (98–119) | 9.20 |
| Ph-O | 37.60 ± 1.54 a (30.90–40.20) | 44.10 ± 1.67 bc (37.60–50.00) | 45.40 ± 1.69 bc (32.60–58.70) | 46.30 ± 1.71 c (35.90–52.00) | 37.30 ± 1.53 a (33.10–41.50) | 40.40 ± 1.60 abc (30.50–45.80) | 42.10 ± 1.63 abc (34.80–50.40) | 11.33 |
| EP | 70.60 ± 1.26 a (67.10–72.40) | 67.70 ± 1.23 a (58.60–72.60) | 76.40 ± 1.31 bc (70.50–83.30) | 81.80 ± 1.35 c (74.30–94.70) | 71.60 ± 1.27 ab (69.50–74.10) | 79.30 ± 1.33 c (68.20–84.30) | 78.00 ± 1.32 c (73.50–82.00) | 4.98 |
| MBW | 17.30 ± 0.38 a (16.00–18.00) | 17.40 ± 0.38 a (16.20–19.40) | 18.40 ± 0.39 ab (16.10–20.20) | 19.90 ± 0.41 b (17.40–23.40) | 18.70 ± 0.39 ab (17.80–20.00) | 19.70 ± 0.40 b (18.00–24.00) | 19.40 ± 0.40 b (16.60–21.20) | 6.30 |
| Ovary | 152 ± 8.40 a (134–174) | 182 ± 8.40 ab (155–218) | 172 ± 8.40 ab (137–242) | 191 ± 9.40 b (114–244) | 163 ± 8.40 ab (131–221) | 155 ± 8.80 ab (122–184) | 177 ± 8.40 ab (142–220) | 14.92 |
| PUS | 23.60 ± 1.04 a (18.50–29.30) | 20.50 ± 1.04 a (17.40–28.50) | 19.60 ± 1.04 a (15.60–26.90) | 23.10 ± 1.04 a (21.30–24.40) | 19.70 ± 1.04 a (17.20–23.70) | 20.70 ± 1.04 a (15.70–29.30) | 22.60 ± 1.04 a (18.60–26.70) | 14.59 |
| P | 14.70 ± 0.40 a (1.18–2.47) | 15.70 ± 0.40 a (1.06–1.72) | 17.80 ± 0.40 b (0.88–1.42) | 18.20 ± 0.40 b (1.14–1.55) | 17.90 ± 0.40 b (0.94–1.41) | 18.10 ± 0.40 b (0.83–1.45) | 17.90 ± 0.40 b (1.00–1.50) | 7.06 |
| V-A | 71.00 ± 2.29 ab (68.30–74.50) | 66.00 ± 2.21 a (59.80–73.30) | 71.30 ± 2.30 ab (64.70–80.60) | 75.20 ± 2.36 abc (64.90–93.00) | 77.40 ± 2.40 bc (65.50–96.5) | 85.20 ± 2.51 c (70.20–107.0) | 78.60 ± 2.41 bc (68.20–87.80) | 9.33 |
| Tail | 23.30 ± 0.74 a (22.00–25.40) | 22.80 ± 0.73 a (21.90–24.70) | 29.30 ± 0.83 b (26.40–33.4) | 29.00 ± 0.83 b (25.60–36.2) | 26.00 ± 0.78 ab (20.30–30.40) | 27.60 ± 0.81 b (23.70–31.60) | 28.70 ± 0.82 b (24.70–32.10) | 8.50 |

[1] Average and standard error ($n = 10$), [2] Different letters between columns in the same row indicate significant differences according to generalized linear models and estimated marginal means with Sidak corrections for multiple comparison of means at $p \leq 0.05$, [3] Range, [4] Coefficient of variation.

**Table 2.** Sequences used/generated in this study.

| Species | Strain/Voucher | Accession Number | | | Reference |
|---|---|---|---|---|---|
| | | D2-D3 | COXI | β-1,4-endoglucanase | |
| *P. penetrans* | MN | MW720686 | MW742327 | MW737621 | This study |
| *P. penetrans* | WZ | MW720687 | MW742328 | MW737622 | This study |
| *P. penetrans* | BN | MW720688 | MW742329 | MW737623 | This study |
| *P. penetrans* | BL | MW720689 | MW742330 | MW737624 | This study |
| *P. penetrans* | UK | MW720690 | MW742331 | MW737625 | This study |
| *P. penetrans* | FR | MW720691 | MW742332 | MW737626 | This study |
| *P. penetrans* | NL | MW720692 | MW742333 | MW737627 | This study |
| *P. penetrans* | VA | MW720693 | MW742334 | MW737628 | This study |
| *P. penetrans* | T666 | KY828351 | KY816982 | – | [13] |
| *P. penetrans* | T295 | KY828352 | KY816991 | – | [13] |
| *P. penetrans* | CA82 | EU130859 | KY817022 | – | [27] |
| *P. penetrans* | T132 | KY828358 | KY817015 | – | [13] |
| *P. penetrans* | V3F | KY828346 | KY816940 | – | [13] |
| *P. penetrans* | V1B | KY828348 | KY816942 | – | [13] |
| *P. fallax* | V5C | KY828361 | KY816937 | – | [13] |
| *P. fallax* | T85 | KY828367 | KY817017 | – | [13] |
| *P. fallax* | T283 | KY828364 | KY816996 | – | [13] |
| *P. fallax* | T272 | KY828365 | KY816998 | – | [13] |
| *P. fallax* | T353 | KY828363 | KY816988 | – | [13] |
| *P. fallax* | V4C | KY828362 | KY816938 | – | [13] |
| *P. neglectus* | GSY24S | KY424315 | KX349423 | – | Unpublished |
| *P. neglectus* | CA94 | EU130854 | KU198941 | – | [27] |
| *P. neglectus* | CD1735 | KU198962 | KU198940 | – | [12] |

Aligned sequences were trimmed at the 5′- and 3′-ends such that nucleotide sequences including the primer sequences, or their complimentary nucleotides were excluded. This was to match the regions that we sequenced for our isolates. In the case of COXI sequences, this was also to exclude the two nucleotide differences that we observed in the middle of the JB3 binding sites (see below; indicated by boldface letters) in some GenBank sequences. In some (accession numbers MK877993–MK877996, MK877985–MK877987) the JB3 binding site had the sequence 5′-TTT TTT G**G**T CAT CC**G** GAG GTT TAT-3′, while in others (accession numbers MN453207–MN453217) this sequence was 5′-TTT TTT G**GG** CAT CC**T** GAG GTT TAT-3′. A third group of sequences (accession numbers MK877989–MK877992) had a JB3 site 5′-TTT TTT G**G**T CAT CC**A** GAG GTT TAT-3′. The D2-D3 and COXI datasets incorporated 692-and 321-characters including alignment gaps, respectively.

Maximum Likelihood and Maximum Parsimony analysis of the concatenated D2-D3 rDNA and COXI dataset resulted in the trees presented in Figures 1 and 2. The MP and ML trees had the same general topology though the level of bootstrap support for the two lineages and branches in these lineages differed. Both ML and MP analyses resolved the ingroup into two well-supported lineages, one of which (Lineage 2) exclusively consisted of three of our eight *P. penetrans* isolates (UK, MN and WZ) and *P. fallax* sequences from GenBank. The remaining five of our isolates fell in Lineage 1 either within well-supported groups or scattered throughout this branch. Both analyses used the General Time Reversible model [28] and all nucleotide positions were included.

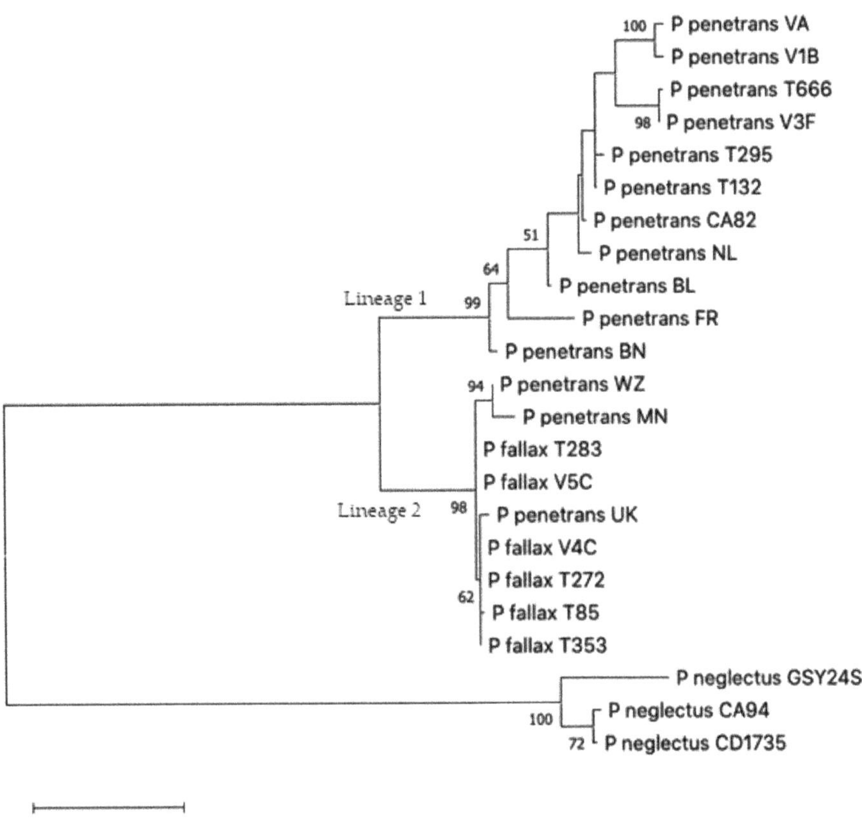


**Figure 1.** ML tree based on the combined D2-D3 rDNA and COXI dataset. Bootstrap values > 50 are shown. Scale bar indicates number of substitutions per site.

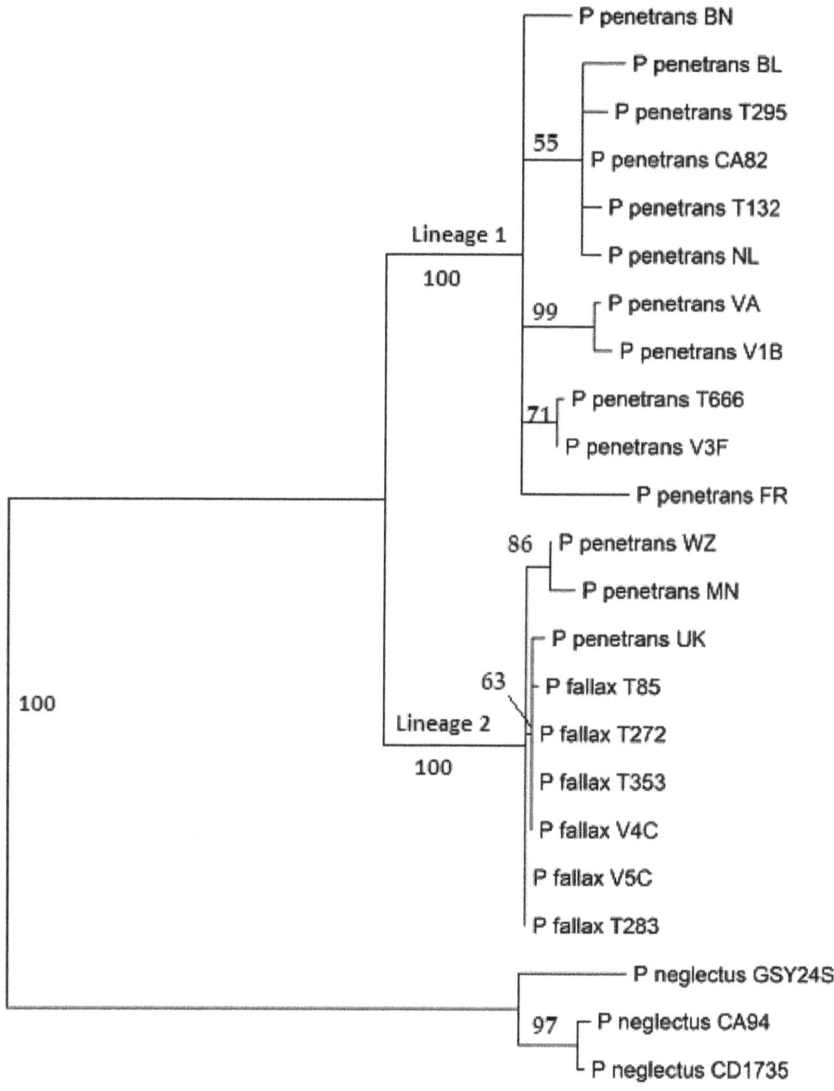


**Figure 2.** Maximum Parsimony tree generated using the combined D2-D3 and COXI dataset. Bootstrap values > 50 are indicated above nodes. Scale bar indicates number of changes.

### 3. Discussion

#### 3.1. Morphometrical Observations

Morphometric measurements of the seven *P. penetrans* populations studied here were within the range of the original descriptions [29,30]. Most of these measurements also largely corresponded with those described for populations from China [31,32]; Colombia, Ethiopia, France, Japan, Rwanda, The Netherlands, and USA [15]; and Morocco [33]. However, remarkable differences were also observed for some characters.

Average ratios of body length to maximum body width (a) observed in the isolates examined here (25.10–27.70) were comparable to those described by Janssen et al. [15] (24.00–27.00), but lower than those reported by Chen et al. [31] (29.90–32.00) and Mokrini et al. [33] (29.20–33.00). The range of ratios of body length to pharynx length from anterior end to posterior end of pharyngeal gland (b') in our isolates (4.33–4.98) was comparable to those described by Mokrini et al. [33] (4.40–5.00). Average body length to tail length ratios (c) ranged from 17.00 to 19.90 among our isolates. Most of these values were lower than those measured for population(s) of Wu et al. [32] (21.40), Chen et al. [31] (20.20–22.10) and Janssen et al. [13] (20.00–25.00). The *P. penetrans* isolates we studied were shorter (437–545 μm) than those described by Wu et al. [32] (666 μm), Chen et al. [31] (540–610 μm) and Janssen et al. [13] (593–684 μm). Position of the vulva relative to body length (V) in our isolates was comparable to those described by Chen et al. [31], Wu et al. [32], Mokrini et al. [33] and Janssen et al. [13]. Similarly, positions of the excretory pore (EP), maximum body width (MBW; Table 1) and tail length in the isolates we studied were comparable to those reported for other populations by Mokrini et al. [33]. Except for MBW, which was considerably higher in our isolates, EP and tail length among our isolates were also comparable to those studied by Chen et al. [31] (69.00–80.00 μm, 9.40–10.40 μm and 25.00–28.00 μm, respectively). However, measurements for these three morphometrical features were shorter in populations described by Wu et al. [32] (91.90 μm, 25.40 μm and 31.40 μm, respectively) and Janssen et al. [13] (97–120 μm, 21–28 μm and 29–32 μm, respectively). The isolates we studied had a shorter post-vulval uterine sac (PUS; 19.60–23.60 μm) than those of Mokrini et al. [33] (26.20–30.90 μm) and Wu et al. [32] (24.90 μm).

Stylet length was the least variable character among our isolates. Previous studies on *P. penetrans* [5,32] and other *Pratylenchus* species [34,35] also reported the same. This suggests that stylet length is a stable characteristic that may allow for clear demarcations among different populations of *P. penetrans* and species of *Pratylenchus*. On the contrary, ovary length and length of the post-vulval uterine sac (PUS) showed high CV among our isolates, confirming previous studies by Román and Hirschmann [5], Tarjan and Frederick [34] and Wu et al. [32]. Ph-L and Ph-O were also among the morphometric characters with high variability that we observed (Table 1). These characteristics with high CVs would be of less value in the morphological taxonomy of *P. penetrans* owing to this high variability.

### 3.2. Sequence Analysis

Mekete et al. [36] designed primer set PP5F/PP5R based on aligned β-1,4-endoglucanase sequences from GenBank for the purpose of identifying *P. penetrans* isolates via amplification of a species-specific 520-bp-fragment. The authors tested the primer set using isolates representing *P. penetrans*, *P. crenatus*, *P. scribneri*, *Helicotylenchus pseudorobustus*, *Hoplolaimus galeatus*, *Xiphinema americanum* and *X. rivesi*, where it resulted in amplification of the expected 520-bp-product only in *P. penetrans* isolates, indicating specificity of the primer set. Similarly, the authors developed a second set of primers (PSC3) that was specific to *P. scribneri* and amplified a 280-bp-fragment only in isolates of this species. In our study, PP5 amplified a PCR product in all the eight isolates. However, the size of the PP5 product among our isolates was only ~346 bp, as opposed to the expected 520 bp. BLASTn analysis of PP5-sequenes of our isolates returned *P. penetrans* β-1,4-endoglucanase as the only one or two significant match(es) from among the eight *Pratylenchus* β-1,4-endoglucanase sequences currently available in GenBank; unfortunately, Mekete et al. [36] did not sequence their PP5 PCR products. To rule out the possibility that Mekete et al. [36] confused amplicon sizes of PP5 and PSC3 in their report, we tested primer set PSC3 in our isolates. PSC3 did not produce amplification products at any of the annealing temperatures reported for this primer set [36]. While we cannot discount the usefulness of PP5 for the identification of *P. penetrans* isolates based on amplification of a PCR product, we can, however, confirm that the size of the amplicon may not always be 520 bp.

Three of our eight isolates which are grouped in Lineage 2 (UK, WZ and MN) shared several morphological characteristics apart from the remaining five isolates. The three isolates had the most anterior excretory pores, 71.60 ± 1.27 μm, 67.70 ± 1.23 μm, and

70.60 ± 1.26 µm, respectively. This was in sharp contrast to that described for *P. fallax* by Janssen et al. [13]. This measurement for *P. fallax* isolates by Janssen et al. [13] were 87 ± 8.3 µm (Ysbrechitum F2455), 91 ± 11 µm (Uddel F0689) and 108 ± 14 µm (Doornenburg–Type locality). Body and tail length in UK, WZ and MN isolates were in the short end of the spectrum for our seven isolates and matched that reported by Janssen et al. [13] for two of their *P. fallax* populations. The third *P. fallax* population (Ysbrechtum F2455), however, had much longer bodies (527 ± 32 µm). The range of pharynx length (Ph-L) reported for *P. fallax* [13] was much wider than what we found among our seven isolates. Stylet length, which showed the least variation among isolates of *P. penetrans* [this study; 6,31] and other *Pratylenchus* species [34,35], did not correlate with phylogenetic groupings. Janssen et al. [13] have attempted to resolve the controversy surrounding the separation of *P. fallax* from *P. penetrans* using morphology and sequence information. However, our findings suggest that *P. fallax* may remain to be a cryptic species along several others in the *P. penetrans* species complex [26].

Phylogenetic resolution of the seven European isolates we studied did not correspond with the geographical origins of these isolates. For example, the three German isolates that were collected not more than 40 km away from each other, grouped in two different lineages. Isolate BN grouped in Lineage 1, while isolates WZ and MN grouped in Lineage 2. On the other hand, isolates UK and WZ, which had the largest distance between their geographical origins (861 km), grouped together in Lineage 2. The isolate from the USA also grouped in Lineage 1, together with some of the European isolates, confirming that geographical origin did not correspond with phylogenetic grouping. The *P. penetrans* group [13] is known to include several more cryptic species than that represented by the two lineages here.

The separation of *P. fallax* from *P. penetrans* was based on breeding experiments that produced infertile interspecific offspring [9], and distinctive isozyme [10] and ITS-RFLP [37] patterns. We have not done any of these analyses using our isolates and cannot confirm or refute the validity of these techniques for the separation of the two species. However, the morphological variations that we observed among our Lineage 2 isolates, and the variation that Janssen et al. [13] reported among their *P. fallax* populations, taken together with the fact that MN, WZ and UK isolates grouped with *P. fallax* isolates in a strongly-supported-Lineage 2, indicates that neither morphology nor D2-D3 rDNA- and COXI-based phylogenetic analyses are sufficient to separate the two species.

## 4. Materials and Methods

### 4.1. Nematode Isolates and Microscopy

Seven of the isolates were collected from soils in different regions in Europe, multiplied from single females on carrot disc cultures for two–three generations (Table 3; [38]) and used in morphometric and molecular analyses. The eighth isolate (VA) obtained from Virginia, USA, was used in the nucleotide sequence analyses for comparative purposes.

**Table 3.** Isolate designation, geographical origin, and distance (km) between geographical origins of the seven European isolates.

| Geographical Origin | Isolates | MN | WZ | BN | BL | UK | FR | NL |
|---|---|---|---|---|---|---|---|---|
| Germany (Münster) | MN | – | | | | | | |
| Germany (Witzenhausen) | WZ | 169 | – | | | | | |
| Germany (Bonn) | BN | 143 | 206 | – | | | | |
| Belgium | BL | 288 | 428 | 237 | – | | | |
| United Kingdom | UK | 693 | 861 | 712 | 493 | – | | |
| France | FR | 616 | 704 | 501 | 366 | 650 | – | |
| The Netherlands | NL | 129 | 128 | 127 | 159 | 594 | 499 | – |

Killing, fixing, and mounting of nematode specimens was done following Hooper et al. [39]. For each isolate, nematode suspensions were transferred into 10 mL glass vials in ~2 mL of water. A double-strength TAF fixative stock solution consisting of 10 mL formalin (35% formaldehyde), 1 mL triethanolamine and 56 mL aqua dest was prepared and heated to 70 °C in a water bath. Two mL of the hot fixative was then dispensed into each of the vials containing nematode suspensions, which were then left at room temperature for 24 h. The TAF fixative was removed from the vials leaving ~1 mL nematode suspension, which were then transferred onto 5 cm sterile plastic Petri dishes. The Petri dishes were filled with a solution consisting of 30% ethanol, 67% aqua dest and 3% glycerine, and placed in a wooden cabinet at room temperature for 5–7 weeks, covered only partially to allow evaporation. Specimens were permanently mounted in anhydrous glycerol.

The selection of morphometric characters studied was in accordance with Decraemer and Hunt [40] and Castillo and Vovlas [1]. Ten females were evaluated for each nematode sample. Measurements were performed using a Nikon ECLIPSE Ni-U microscope at 100X magnification with the aid of a Nikon DS Fi-2 camera and exclusive NIS-Elements image analysis software (Nikon, Tokyo, Japan). Morphometric data were analysed using generalized linear models using Gaussian (for homogeneous) or quasipoisson (inhomogeneous variances) families. Estimated marginal means (R version 4.0.2; [41]) were used to generate means and standard errors as well as for separation of treatments at $p \leq 0.05$.

### 4.2. DNA Extraction

For each isolate, DNA was extracted following Holterman et al. [42] from ten nematodes (4-stage juveniles and adults). Nematodes were transferred individually into 0.2 mL PCR tubes using micropipette in a total of 25 µL. An equal volume of lysis buffer (25 µL) consisting of 0.2 M NaCl, 0.2 M Tris-HCl (pH 8.0), 1% *v/v* β-Mercaptoethanol, 0.8 mg/mL Proteinase K was then added to each sample. The tubes were briefly centrifuged at 16,000 rpm and incubated at 65 °C and 750 rpm for 2 h followed by 10 min at 100 °C in a Thermomixer (Eppendorf, Hamburg, Deutschland). Nematode lysates were used immediately or stored at −20 °C till used.

### 4.3. Nucelotide Sequence Analysis

Amplicons of ~2000 base pair (bp), ~350 bp and ~286 bp of the genes encoding for the 28S rDNA, the mitochondrial COXI gene and "PP5 region" were amplified using primer pairs 18S CL-F2 [43] and D3B [44], JB3 and JB4.5 [45], and PP5F and PP5R [36], respectively. The reaction and cycling conditions for the COXI and PP5 gene regions were as described by Bowles et al. [45] and Mekete et al. [36], respectively. These fragments were sequenced using the same primers as for the respective PCRs. The PCR cycles for the 28S rDNA consisted of an initial denaturation at 95 °C for 4 min followed by 35 cycles of denaturation at 95 °C for 45 s, annealing at 64 °C for 30 s and extension at 72 °C for 2 min; and a final extension at 72 °C for 10 min. The resulting fragments were cloned using a NEB PCR Cloning Kit (New England Biolabs Inc., Ipswich, MA, USA) following the manufacturer's recommendations. For each isolate, two colonies were PCR-amplified using the primers supplied with the kit and sequenced using the D3B primer [44]. All amplification reactions were performed on a GeneAmp PCR System 2700 (Applied Biosystems, Thermo Fisher Scientific, Waltham, MA, USA). PCR products were purified using QIAquick PCR purification kit (QIAGEN, Germantown, MD, USA), and sequenced at Eurofins USA (https://www.eurofinsgenomics.com (accessed on 1 February 2021).

For COXI and D2-D3 rDNA gene regions sequenced in this study, selected sequences were obtained from GenBank and included here for reference and outgroup purposes (Table 2). DNA sequences generated in this study have been deposited in GenBank (Table 2). Nucleotide sequences were assembled using Geneious (Version 11.1.5, Biomatters Ltd., Auckland, New Zealand), and aligned using Clustal Omega [46], after which the alignments were manually corrected where needed using Phylogenetic Analysis Using Parsimony (PAUP, Version 4.0b 10; [47]). Maximum Parsimony (MP) and Maximum Likelihood (ML) analyses were done on the concatenated D2-D3 and COXI dataset using MEGA-X [48]. Heuristic searches based on 1000 random addition sequences and tree bisection-reconnection were used for this purpose, with the branch swapping option set on 'best trees' only. Bootstrap analysis [49] was based on 1000 replications.

**Supplementary Materials:** The following are available online at https://www.mdpi.com/article/10.3390/plants10040674/s1, Data S1: D3 rDNA and COXI sequences of 37 taxa (including our eight isolates) still groups MN, WZ and UK isolates in a strongly supported branch together with *P. fallax* isolates.

**Author Contributions:** Conceptualization, M.B., B.T., P.D. and J.H.; methodology, M.B., B.T., R.H.N. and B.H.; formal analysis, M.B., B.T. and R.H.N.; investigation, M.B., B.T. and R.H.N.; resources, B.H., J.H. and P.D.; data curation, M.B., B.T. and R.H.N.; writing—original draft preparation, M.B. and B.T.; writing—review and editing, M.B., B.T., R.H.N., B.H., J.H. and P.D.; supervision, B.H., J.H. and P.D.; project administration, B.H., J.H. and P.D. All authors have read and agreed to the published version of the manuscript.

**Funding:** This research received no external funding.

**Institutional Review Board Statement:** Not applicable.

**Informed Consent Statement:** Not applicable.

**Data Availability Statement:** Data is contained within the article or Supplementary Materials. The data presented in this study are available in Supplementary Data S1 and sequences produced can be found in GenBank, accessions listed in Table 2.

**Acknowledgments:** We would like to acknowledge the contributions of Peter-Jan Jongenelen from Joordans Zaden in acquiring nematode samples throughout Europe that significantly increased the breadth of this work, as well as Jan Henrik Schmidt from Julius Kühn-Institut for his support in statistical analysis.

**Conflicts of Interest:** The authors declare no conflict of interest.

# References

1. Castillo, P.; Vovlas, N. *Pratylenchus* (Nematoda: Pratylenchidae): Diagnosis, biology, pathogenicity and management. *Nematol. Monogr. Perspect.* **2007**, *6*, 1–543.
2. Ryss, A. Genus *Pratylenchus* Filipjev (Nematoda: Tylenchida: Pratylenchidae): Multientry and monoentry keys and diagnostic relationships. *Zoosyst. Ross.* **2002**, *10*, 11–25.
3. Hernández, M.; Jordana, R.; Goldaracena, A.; Pinochet, J. SEM observations of nine species of the genus *Pratylenchus* Filipjev, 1936 (Nematoda: Pratylenchidae). *J. Nematode Morphol. Syst.* **2001**, *3*, 165–174.
4. Hunt, D.; Luc, M.; Manzanilla-Lopez, R. Identification, morphology and biology of plant-parasitic nematodes. In *Plant-Parasitic Nematodes in Tropical Agriculture*; Luc, M., Sikora, R., Bridge, J., Eds.; CAB International: Wallingford, UK, 2005; pp. 11–52.
5. Roman, J.; Hirschmann, H. Morphology and morphometrics of six species of *Pratylenchus* I. *J. Nematol.* **1969**, *1*, 363–385. [PubMed]
6. Handoo, Z.A.; Carta, L.K.; Skantar, A.M. Taxonomy, morphology and phylogenetics of coffee-associated root-lesion nematodes, *Pratylenchus* spp. In *Plant-Parasitic Nematodes of Coffee*; Springer: Dordrecht, The Netherlands, 2008; pp. 29–50.
7. Seinhorst, J.W. Three new *Pratylenchus* species with a discussion of the structure of the cephalic framework and of the spermatheca in this genus. *Nematologica* **1968**, *14*, 497–510. [CrossRef]
8. Tarte, R.; Mai, W.F. Morphological variation in *Pratylenchus penetrans*. *J. Nematol.* **1976**, *8*, 185–195. [PubMed]
9. Perry, R.; Plowright, R.; Webb, R. Mating between *Pratylenchus penetrans* and *P. fallax* in sterile culture. *Nematologica* **1980**, *26*, 125–129.
10. Ibrahim, S.K.; Perry, R.N.; Webb, R.M. Use of isoenzyme and protein phenotypes to discriminate between six *Pratylenchus* species from Great Britain. *Ann. Appl. Biol.* **1995**, *126*, 317–327. [CrossRef]
11. Waeyenberge, L.; Ryss, A.; Moens, M.; Pinochet, J.; Vrain, T. Molecular characterisation of 18 *Pratylenchus species* using rDNA restriction fragment length polymorphism. *Nematology* **2000**, *2*, 135–142. [CrossRef]
12. Troccoli, A.; Subbotin, S.; Chitambar, J.; Janssen, T.; Waeyenberge, L.; Stanley, J.; Duncan, L.; Agudelo, P.; Uribe, G.; Franco, J.; et al. Characterisation of amphimictic and parthenogenetic populations of *Pratylenchus bolivianus* Corbett, 1983 (Nematoda: Pratylenchidae) and their phylogenetic relationships with closely related species. *Nematology* **2016**, *18*, 651–678. [CrossRef]
13. Janssen, T.; Karssen, G.; Orlando, V.; Subbotin, S.; Bert, W. Molecular characterization and species delimiting of plant-parasitic nematodes of the genus *Pratylenchus* from the penetrans group (Nematoda: Pratylenchidae). *Mol. Phylogenet. Evol.* **2017**, *117*, 30–48. [CrossRef]

14.  Orlando, V.; Grove, I.G.; Edwards, S.G.; Prior, T.; Roberts, D.; Neilson, R.; Back, M. Root-lesion nematodes of potato: Current status of diagnostics, pathogenicity and management. *Plant Pathol.* **2020**, *69*, 405–417. [CrossRef]
15.  Jung, J.; Han, H.; Ryu, S.; Kim, W. Amplified fragment length polymorphism analysis and genetic variation of the pinewood nematode *Bursaphelenchus xylophilus* in South Korea. *Animal Cells Syst.* **2010**, *14*, 31–36. [CrossRef]
16.  Correa, V.R.; Mattos, V.S.; Almeida, M.R.A.; Santos, M.F.A.; Tigano, M.S.; Castagnone-Sereno, P.; Carneiro, R.M.D.G. Genetic diversity of the root-knot nematode *Meloidogyne ethiopica* and development of a species-specific SCAR marker for its diagnosis. *Plant Pathol.* **2014**, *63*, 476–483. [CrossRef]
17.  Širca, S.; Stare, B.; Strajnar, P.; Urek, G. PCR-RFLP diagnostic method for identifying *Globodera* species in Slovenia. *Phytopathol. Mediterr.* **2010**, *49*, 361–369.
18.  Pinochet, P.; Ceres, J.L.; Fern, C.; Doucet, M.; Marull, A.J. Reproductive fitness and random amplified polymorphic DNA variation among isolates of *Pratylenchus vulnus*. *J. Nematol.* **1994**, *26*, 271–277.
19.  Hahn, M.; Burrows, P.; Wright, D. Genomic diversity between *Radopholus similis* populations from around the world detected by RAPD-PCR analysis. *Nematologica* **1996**, *42*, 537–545.
20.  Fallas, G.A.; Hahn, M.L.; Fargette, M.; Burrows, P.R.; Sarah, J.L. Molecular and biochemical diversity among isolates of *Radopholus* spp. from different areas of the world. *J. Nematol.* **1996**, *28*, 422–430.
21.  Tigano, M.; De Siqueira, K.; Castagnone-Sereno, P.; Mulet, K.; Queiroz, P.; Dos Santos, M.; Teixeira, C.; Almeida, M.; Silva, J.; Carneiro, R. Genetic diversity of the root-knot nematode *Meloidogyne enterolobii* and development of a SCAR marker for this guava-damaging species. *Plant Pathol.* **2010**, *59*, 1054–1061. [CrossRef]
22.  Figueiredo, J.; Simões, M.J.; Gomes, P.; Barroso, C.; Pinho, D.; Conceição, L.; Fonseca, L.; Abrantes, I.; Pinheiro, M.; Egas, C. Assessment of the geographic origins of pinewood nematode isolates via single nucleotide polymorphism in effector genes. *PLoS ONE* **2013**, *8*, e83542. [CrossRef]
23.  De Luca, F.; Reyes, A.; Veronico, P.; Di Vito, M.; Lamberti, F.; De Giorgi, C. Characterization of the (GAAA) microsatellite region in the plant parasitic nematode *Meloidogyne artiellia*. *Gene* **2002**, *293*, 191–198. [CrossRef]
24.  Arias, R.S.; Stetina, S.R.; Tonos, J.L.; Scheffler, J.A.; Scheffler, B.E. Microsatellites reveal genetic diversity in *Rotylenchulus reniformis* populations. *J. Nematol.* **2009**, *41*, 146–156. [PubMed]
25.  Davis, E.L.; MacGuidwin, A.E. Lesion nematode disease. Available online: https://www.apsnet.org/edcenter/disandpath/nematode/pdlessons/Pages/LesionNematode.aspx (accessed on 31 January 2021).
26.  Fanelli, E.; Troccoli, A.; Capriglia, F.; Lucarelli, G.; Vovlas, N.; Greco, N.; De Luca, F. Sequence variation in ribosomal DNA and in the nuclear hsp90 gene of *Pratylenchus penetrans* (Nematoda: Pratylenchidae) populations and phylogenetic analysis. *Eur. J. Plant Pathol.* **2018**, *152*, 355–365. [CrossRef]
27.  Subbotin, S.A.; Ragsdale, E.J.; Mullens, T.; Roberts, P.A.; Mundo-Ocampo, M.; Baldwin, J.G. A phylogenetic framework for root lesion nematodes of the genus *Pratylenchus* (Nematoda): Evidence from 18S and D2-D3 expansion segments of 28S ribosomal RNA genes and morphological characters. *Mol. Phylogenet. Evol.* **2008**, *48*, 491–505. [CrossRef] [PubMed]
28.  Nei, M.; Kumar, S. *Molecular Evolution and Phylogenetics*; Oxford University Press: New York, NY, USA, 2000.
29.  Loof, P. Taxonomic studies on the genus *Pratylenchus* (Nematoda). *Tijdschr. Plantenziekten* **1960**, *66*, 29–30.
30.  Corbett, M.; Clark, S.A. Surface features in the taxonomy of *Pratylenchus* species. *Rev. Nematol.* **1983**, *6*, 85–98.
31.  Chen, D.Y.; Ni, H.F.; Yen, J.H.; Wu, W.S.; Tsay, T.T. Identification of root-lesion nematode *Pratylenchus penetrans* and *P. loosi* (Nematoda: Pratylenchidae) from strawberry and tea plantations in Taiwan. *Plant Pathol. Bull.* **2009**, *18*, 247–262.
32.  Wu, H.Y.; Tsay, T.T.; Lin, Y.Y. Identification and biological study of *Pratylenchus* spp. isolated from the crops in Taiwan. *Plant Pathol. Bull.* **2002**, *11*, 123–136.
33.  Mokrini, F.; Waeyenberge, L.; Viaene, N.; Andaloussi, F.A.; Moens, M. Diversity of root-lesion nematodes (*Pratylenchus* spp.) associated with wheat (*Triticum aestivum* and *T. durum*) in Morocco. *Nematology* **2016**, *18*, 781–801. [CrossRef]
34.  Tarjan, A.; Frederick, J. Intraspecific morphological variation among populations of *Pratylenchus brachyurus* and *P. coffeae*. *J. Nematol.* **1978**, *10*, 152–160.
35.  Tuyet, N.; Elsen, A.; Nhi, H.; De Waele, D. Morphological and morphometrical characterisation of ten *Pratylenchus coffeae* populations from Vietnam. *Russ. J. Nematol.* **2012**, *20*, 75–93.
36.  Mekete, T.; Reynolds, K.; Lopez-Nicora, H.; Gray, M.; Niblack, T. Distribution and diversity of root-lession nematode (*Pratylenchus* spp.) associated with *Miscanthus* × *giganteus* and *Panicum virgatum* used for biofuels, and species identification in multiplex polymerase chain reaction. *Nematology* **2011**, *13*, 673–686.
37.  Waeyenberge, L.; Viaene, N.; Moens, M. Species-specific duplex PCR for the detection of *Pratylenchus penetrans*. *Nematology* **2009**, *11*, 847–857. [CrossRef]
38.  Betre, T. *Studies on the Biological and Molecular Variation among Seven Isolates of Pratylenchus Penetrans from Different Geographical Locations in Europe*; Justus Liebig University: Giessen, Germany, 2019.
39.  Hooper, D.; Hallmann, J.; Subbotin, S. Extraction, processing and detection of plant and soil nematodes. In *Plant-Parasitic Nematodes in Tropical Agriculture*; Luc, M., Sikora, R., Bridge, J., Eds.; CAB International: Wallingford, UK, 2005; pp. 53–86.
40.  Decraemer, W.; Hunt, D.J. Structure and classification. In *Plant Nematology*; Perry, R.N., Moens, M., Eds.; CAB International: Wallingford, UK, 2013; pp. 3–39. ISBN 9781780641515.
41.  Core R Team. A Language and Environment for Statistical Computing. *R Found. Stat. Comput.* **2019**, *2*. Available online: https://www.R-project.org (accessed on 31 January 2021).

42. Holterman, M.; Karssen, G.; Van Den Elsen, S.; Van Megen, H.; Bakker, J.; Helder, J. Small subunit rDNA-based phylogeny of the Tylenchida sheds light on relationships among some high-impact plant-parasitic nematodes and the evolution of plant feeding. *Am. Phytopath. Soc.* **2009**, *99*, 227. [CrossRef]

43. Carta, L.K.; Li, S. Improved 18S small subunit rDNA primers for problematic nematode amplification. *J. Nematol.* **2018**, *50*, 533–542. [CrossRef]

44. Nunn, G.B. *Nematode Molecular Evolution*; University of Nottingham: Nottingham, UK, 1992.

45. Bowles, J.; Blair, D.; McManus, D.P. Genetic variants within the genus *Echinococcus* identified by mitochondrial DNA sequencing. *Mol. Biochem. Parasitol.* **1992**, *54*, 165–173. [CrossRef]

46. Madeira, F.; Park, Y.; Lee, J.; Buso, N. The EMBL-EBI search and sequence analysis tools. *Nucleic* **2019**, *47*, W636–W641.

47. Swofford, D.L. *PAUP*: Phylogenetic Analysis Using Parsimony (*and Other Methods)*, version 4.0b4a; Sinauer Associates: Sunderland, MA, USA, 2002.

48. Kumar, S.; Stecher, G.; Li, M.; Knyaz, C.; Tamura, K. MEGA-X: Molecular evolutionary genetics analysis accross computing platforms. *Mol. Biol. Evol.* **2018**, *35*, 1547–1549. [CrossRef]

49. Hillis, D.; Bull, J. An empirical test of bootstrapping as a method for assessing confidence in phylogenetic analysis. *Syst. Biol.* **1993**, *42*, 182–192. [CrossRef]

*Article*

# Integrative Taxonomy and Molecular Phylogeny of the Plant-Parasitic Nematode Genus *Paratylenchus* (Nematoda: Paratylenchinae): Linking Species with Molecular Barcodes

Phougeishangbam Rolish Singh [1,*], Gerrit Karssen [1,2], Marjolein Couvreur [1], Sergei A. Subbotin [3,4] and Wim Bert [1]

[1] Nematology Research Unit, Department of Biology, Ghent University, K.L. Ledeganckstraat 35, 9000 Ghent, Belgium; g.karssen@nvwa.nl (G.K.); marjolein.couvreur@ugent.be (M.C.); wim.bert@ugent.be (W.B.)
[2] National Plant Protection Organization, Wageningen Nematode Collection, P.O. Box 9102, 6700 HC Wageningen, The Netherlands
[3] Plant Pest Diagnostic Center, California Department of Food and Agriculture, 3294 Meadowview Road, Sacramento, CA 95832, USA; sergei.a.subbotin@gmail.com
[4] Center of Parasitology of A.N. Severtsov Institute of Ecology and Evolution of the Russian, Academy of Sciences, Leninskii Prospect 33, 117071 Moscow, Russia
* Correspondence: PhougeishangbamRolish.Singh@Ugent.be

**Citation:** Singh, P.R.; Karssen, G.; Couvreur, M.; Subbotin, S.A.; Bert, W. Integrative Taxonomy and Molecular Phylogeny of the Plant-Parasitic Nematode Genus *Paratylenchus* (Nematoda: Paratylenchinae): Linking Species with Molecular Barcodes. *Plants* **2021**, *10*, 408. https://doi.org/10.3390/plants10020408

Academic Editor: Francesc Xavier Sorribas Royo

Received: 15 January 2021
Accepted: 15 February 2021
Published: 22 February 2021

**Publisher's Note:** MDPI stays neutral with regard to jurisdictional claims in published maps and institutional affiliations.



**Abstract:** Pin nematodes of the genus *Paratylenchus* are obligate ectoparasites of a wide variety of plants that are distributed worldwide. In this study, individual morphologically vouchered nematode specimens of fourteen *Paratylenchus* species, including *P. aculentus*, *P. elachistus*, *P. goodeyi*, *P. holdemani*, *P. idalimus*, *P. microdorus*, *P. nanus*, *P. neoamblycephalus*, *P. straeleni* and *P. veruculatus*, are unequivocally linked to the D2-D3 of 28S, ITS, 18S rRNA and *COI* gene sequences. Combined with scanning electron microscopy and a molecular analysis of an additional nine known and thirteen unknown species originating from diverse geographic regions, a total of 92 D2-D3 of 28S, 41 ITS, 57 18S rRNA and 111 *COI* new gene sequences are presented. *Paratylenchus elachistus*, *P. holdemani* and *P. neoamblycephalus* are recorded for the first time in Belgium and *P. idalimus* for the first time in Europe. *Paratylenchus* is an excellent example of an incredibly diverse yet morphologically minimalistic plant-parasitic genus, and this study provides an integrated analysis of all available data, including coalescence-based molecular species delimitation, resulting in an updated *Paratylenchus* phylogeny and the corrective reassignment of 18 D2-D3 of 28S, 3 ITS, 3 18S rRNA and 25 *COI* gene sequences that were previously unidentified or incorrectly classified.

**Keywords:** D2-D3 of 28S; ITS; 18S; *COI*; morphology; morphometrics; *Paratylenchus*; plant-parasitic nematodes; phylogeny; taxonomy

## 1. Introduction

The plant-parasitic nematode (PPN) genus *Paratylenchus* Micoletzky, 1922, commonly known as pin nematodes, are obligate ectoparasites of a wide variety of plants, including herbs, shrubs and trees, that are distributed worldwide and cause various symptoms in their host plants [1–5]. This genus was reviewed by Tarjan [6], who provided the first key to the species. In subsequent years, several attempts were made to split the genus and group its representatives into new genera. The genus *Gracilacus* Raski, 1962, was proposed for members of the *Paratylenchus* species with stylet lengths longer than 48 μm [7]. The validity of *Gracilacus* was supported by Thorne and Malek [8], Raski and Luc [9], Maggenti et al. [10], Raski [11], Esser [12], Andrássy [13] and Yu et al. [14], while Siddiqi [15] treated it as a subgenus of *Paratylenchus*. *Gracilacus* was synonymised with *Paratylenchus* by Brzeski [16], and it was recognized in further works of Siddiqi and Goodey [17], Geraert [18], Brzeski [19], Nguyen et al. [20], Decraemer and Hunt [21], Van den Berg et al. [22], Ghaderi et al. [23],

Hesar et al. [24] and Maria et al. [25]. The genus *Paratylenchoides* Raski, 1973 was assigned to *Paratylenchus* species with stronger cephalic sclerotisations, dorso-ventrally narrower heads and small narrow rounded protrusions on the anterior surface of conoid lip region [26]. Siddiqi [15] subsequently lowered *Paratylenchoides* to a sub-generic level for *Paratylenchus*, while Raski and Luc [9] synonymised the two genera owing to the apparent lack of morphological differences between them and Siddiqi [2] accepted this. It was proposed that another genus, *Gracilpaurus* Ganguly and Khan, 1990, included four species displaying long stylets and tubercles on the surface of the cuticle [27]. However, Brzeski [19] did not consider cuticular ornamentation as a generic characteristic, a decision that led to the synonymising of *Gracilpaurus*. The monotypic genus *Cacopaurus* Thorne, 1943 was also proposed and distinguished from *Paratylenchus* by the obese female body, tubercles on annuli of the female cuticle and sessile parasitism [28]. Although Goodey [29] synonymised *Cacopaurus* with *Paratylenchus* due to the lack of consistent differential traits—apart from the female of the former sometimes being sessile and slightly swollen—*Cacopaurus* has been, nevertheless, accepted by Raski [7], Raski and Luc [9], Ebsary [30], Raski [11], Brzeski [19], Siddiqi [2], Andrassy [13] and Ghaderi et al. [23,31].

Nematodes of the genus *Paratylenchus* in a broad sense or *sensu lato* are characterised by: small size (<0.7 mm); females being vermiform to obese; C, J or 6 shapes when heat relaxed; two to four lateral lines; cuticle with or without ornamentations; often continuous cephalic regions of rounded to conoid, truncate or trapezoid shapes; protruding or non-protruding submedian lobes; stylet lengths ranging between 10 and 120 µm; well-developed valves of median bulb, slender isthmuses and rounded to pyriform end bulbs in female pharynges; secretory-excretory pores are often at the level between median bulb and end bulb; spermathecae with or without sperm cells; commonly swollen prevulval region; vulvae with or without lateral flaps; presence or absence of a short post-vulval uterine sac; tails ranging from conoid to hemispherical with variable tail termini. The diagnostic traits of juveniles and males are less frequently used for identification, except for looking for the presence of a stylet and looking at the length of the spicules of males.

Recently, Ghaderi et al. [23] recognized 117 species of *Paratylenchus sensu lato* (excluding *Cacopaurus*), six species of *inquirendae* and four of *nomina nuda*. The nominal species were pragmatically divided into eleven groups based on stylet length, number of lateral lines and absence vs. presence of vulval flaps in females. Since then, seven more species of *Paratylenchus* have been described and linked to DNA sequences [14,25,32–36]. Molecular work on this nematode group is gaining momentum and provides an attractive solution to difficulties encountered in species identification, as well as phylogenetic relationships among species. Subbotin et al. [37], Chen et al. [38,39] and van Megen et al. [40] started to molecularly characterise some *Paratylenchus* spp. using the D2-D3 of 28S rRNA, ITS rRNA and 18S rRNA gene sequences, respectively. Lopez et al. [41] used ITS rRNA gene to examine phylogenetic relationships among four nematode genera; two of the included genera were *Paratylenchus* and *Gracilacus*. Van den Berg et al. [22] conducted the first comprehensive phylogenetic study including several *Paratylenchus* spp. by using 58 28S rRNA and 40 ITS rRNA gene sequences. Several other studies provided additional molecular characterisations, phylogenetic analyses and descriptions of new *Paratylenchus* species [14,25,32–36,42–50]. In a study by Hesar et al. [24], 28S rRNA and ITS rRNA gene sequences of several *Paratylenchus* spp. as well as *Cacopaurus pestis* Thorne, 1943, were updated. In addition to providing the first molecular characterisation of *C. pestis*, their phylogenetic analyses based on the two partial gene sequences did not support the monophyly of the genera *Cacopaurus*, *Gracilacus* and *Gracilpaurus* that were all found embedded within the clade of *Paratylenchus*.

Despite these recent efforts to integrate and include molecular information in species descriptions and species delineations of *Paratylenchus*, several taxonomic challenges still remain. This is often the case in the field of nematology in general, but the genus *Paraty-lenchus* is a perfect case in point. Most of the traditionally described species are not yet linked to molecular data, numerous sequences that are currently available are not linked to established species and/or morphological information, sequences are often misplaced and the existence of cryptic species within the genus is common.

Species boundaries in *Paratylenchus* are sometimes difficult to delimit based solely on morphology because of the limited diagnostic features and morphological plasticity. As of December 2020, only 40 *Paratylenchus sensu lato* species have been linked to molecular data in the GenBank and this database also includes several putative, new, unidentified and incorrectly classified sequences. These misidentified sequences may result in a cascade of erroneous interpretations, including incorrect morphological identification [51] and flawed interpretations of species identity based on relationships in phylogenetic trees. Cryptic species are also likely to represent a component of *Paratylenchus* diversity [22]. It is important to note that correct differentiation of species belonging to agricultural nematode pests from its sibling species has gained importance for a number of reasons, including food security, quarantine regulations and nonchemical pest management strategies [52].

The aims of this study are to: (1) provide and update molecular barcodes of several known and unknown *Paratylenchus* species using four partial sequences—D2-D3 of 28S, ITS and 18S rRNA gene and *COI* gene of mtDNA; (2) link these molecular data to comprehensive morphological information, including light microscopy (LM) and scanning electron microscopy (SEM) images and morphometrics; (3) reconstruct an updated *Paratylenchus* phylogeny; (4) provide a molecular species delimitation for all four markers; (5) reassign unidentified and/or incorrectly classified GenBank sequences to the appropriate species.

## 2. Results

### 2.1. Species Identification, Characterisation and Delimitation

Ten identified and four unidentified *Paratylenchus* species, recovered from soil samples collected in Belgium, were morphologically and molecularly characterised. The identified species were *Paratylenchus aculentus* Brown, 1959, *Paratylenchus elachistus* Steiner, 1949, *Paratylenchus goodeyi* Oostenbrink, 1953, *Paratylenchus holdemani* Raski, 1975, *Paratylenchus idalimus* (Raski, 1962) Siddiqi and Goodey, 1964, *Paratylenchus microdorus* Andrássy, 1959, *Paratylenchus nanus* Cobb, 1923, *Paratylenchus neoamblycephalus* Geraert, 1965, *Paratylenchus straeleni* (De Coninck, 1931) Oostenbrink, 1960 and *Paratylenchus veruculatus* Wu, 1962. The unidentified *Paratylenchus* spp. were *Paratylenchus* sp.2, *Paratylenchus* sp.BE11, *Paratylenchus* sp.D, and *Paratylenchus* sp.F. *Paratylenchus elachistus*, *P. holdemani*, *P. idalimus* and *P. neoamblycephalus* were reported for the first time in Belgium and *P. idalimus* was recorded for the first time in Europe. Additional sequences of *Paratylenchus aquaticus* Merny, 1966, *Paratylenchus dianthus* Jenkins and Tylor, 1956, *Paratylenchus hamatus* Thorne and Allen, 1950, *Paratylenchus leptos* Raski 1975, *P. nanus*, *Paratylenchus projectus* Jenkins 1956, *Paratylenchus shenzhenensis* Wang, Xie, Li, Xu, Yu and Wang, 2013, *P. straeleni* and *Paratylenchus tenuicaudatus* Wu, 1961 and thirteen unidentified *Paratylenchus* species that originated from diverse geographic regions are also provided (Table 1). In total, 68 D2-D3 of 28S, 38 ITS, 57 18S rRNA and 84 *COI* gene sequences were linked to morphological data of the abovementioned ten known and four unknown species collected from Belgium, and 24 D2-D3 of 28S, 3 ITS rRNA and 27 *COI* gene sequences were added to the other nine known and thirteen unidentified species.

**Table 1.** List of *Paratylenchus* populations included in this study. Accession numbers of three ribosomal RNA genes (D2-D3 of 28S, ITS and 18S) and a mitochondrial gene (*COI*) fragments are provided for 18 identified and 14 unidentified *Paratylenchus* species. Accession numbers in italics are ones generated in this study.

| Species | Locality | Associated Plant Host | Sample Code | GenBank Accession Numbers | | | | Source |
|---|---|---|---|---|---|---|---|---|
| | | | | 28S rRNA | ITS rRNA | 18S rRNA | *COI* of mtDNA | |
| *P. aculentus* | Belgium, Ghent, Citadel Park; 51°02′05″ N; 3°43′10″ E | Grasses under a tree | BE9 | *MW413626–MW41328* | *MW413588–MW413589* | *MW413692–MW413693* | *MW421639–MW421642* | C.M. Etongwe |
| *P. aquaticus* type A | USA, Florida, Princeton | *Aechmea* sp. | CD3375 | *MW413557* | - | - | - | S.A. Subbotin |
| *P. aquaticus* type A | USA, Hawaii, Waimanalo | Bromeliad (*Neoregelia* sp.) | CD619 | KF242239, KF242240 | KF242277, KF242278 | - | *MW411845* | S.A. Subbotin, Van den Berg et al. [22] |
| *P. aquaticus* type B | USA, Kansas, Manhattan, Washinton-Marlatt park | Grasses | CD868 | KF242241 | - | - | *MW411838* | S.A. Subbotin, Van den Berg et al. [22] |
| *P. dianthus* | South Africa, Gauteng, Tarlton | Chrysanthemum | CD552 | KF242226–KF242229 | KF242271, KF242272 | - | *MW411837* | Van den Berg, Van den Berg et al. [22] |
| *P. enigmaticus* | Belgium, Ghent, Ghent University Botanical Garden; 51°2′7.53″ N; 3°43′20.07″ E | Leek | BE2 | - | *MW413621–MW413622* | *MW413735, MW413737–MW413739* | *MW421686* | C.M. Etongwe |
| *P. enigmaticus* | Belgium, Ghent, Ghent University Botanical Garden; 51°2′7.10″ N; 3°43′19.28″ E | Wild oregano | BE4 | - | - | *MW413732–MW413734* | - | C.M. Etongwe |
| *P. enigmaticus* | USA, Idaho | Unknown plant | CD2485 | *MW413568* | *MW413583* | - | *MW411828* | S.A. Subbotin |
| *P. elachistus* | Belgium, Kortrijk; 50°47′58″ N; 3°11′37″ E | Grasses under a thorny tree | BE15 | *MW413629–MW413630* | *MW413590–MW413593* | *MW413694–MW413697* | *MW421643–MW421646* | C.M. Etongwe |
| *P. goodeyi* | Belgium, Merendree; 51°04′12″ N; 3°34′37″ E | Grasses around a beech tree | BE22 | *MW413631–MW413633* | *MW413594* | *MW413698–MW413699* | *MW421647–MW421649* | C.M. Etongwe |
| *P. hamatus* | USA, California, Merced County, Planada | Fig tree (*Ficus carica*) | CD1155 | KF242212 | - | - | *MW411821* | S.A. Subbotin, Van den Berg et al. [22] |
| *P. hamatus* | USA, California | Trees | CD1914 | *MW413564* | *MW413585* | - | *MW411823* | S.A. Subbotin |

**Table 1.** *Cont.*

| Species | Locality | Associated Plant Host | Sample Code | GenBank Accession Numbers | | | | Source |
|---|---|---|---|---|---|---|---|---|
| | | | | 28S rRNA | ITS rRNA | 18S rRNA | COI of mtDNA | |
| *P. hamatus* | USA, California, Kern county | Grape (*Vitis* sp.) | CD2534a, b | *MW413565, MW413566* | - | - | - | A. Westphal |
| *P. hamatus* | USA, California, Kern county, Delano | Grape, Cherry | CD3372 | *MW413558* | - | - | - | S.A. Subbotin |
| *P. hamatus* | USA, California, Kern county, Maricopa | Apricot (*Prunus* sp.) | CD454 | KF242206, KF242216, KF242217 | KF242247, KF242256 | - | *MW411822* | S.A. Subbotin, Van den Berg et al. [22] |
| *P. holdemani* | Belgium, Gouvy, Rogery; 50°14′39.8″ N; 5°57′21.9″ E | Grasses under tree *Fraxinus* sp. | AR3 | *MW413636– MW413638, MW413640, MW413642* | - | *MW413701* | *MW421650– MW421652* | P.R. Singh |
| *P. holdemani* | Belgium, Ghent, Blaarmeersen; 51°02′18.9″ N; 3°41′17.2″ E | Grasses under a thorny tree next to a stream | BE19 | *MW413634– MW413635* | *MW413595* | *MW413700* | *MW421658* | C.M. Etongwe |
| *P. holdemani* | Belgium, Ghent, Blaarmeersen; 51°02′14″ N; 3°41′23″ E | Grasses under a tree | BE20 | *MW413639, MW413641* | *MW413596* | *MW413702* | *MW421653– MW421657* | C.M. Etongwe |
| *P. idalimus* | Belgium, Ghent, Blaarmeersen; 51°02′18.9″ N; 3°41′17.2″ E | Grasses under a thorny tree next to a stream | BE19 | *MW413644* | - | - | - | C.M. Etongwe |
| *P. idalimus* | Belgium, Ghent, Blaarmeersen; 51°02′14″ N; 3°41′23″ E | Grasses under a tree | BE20 | *MW413643* | - | *MW413703– MW413704* | - | C.M. Etongwe |
| *P. idalimus* | USA, California, Napa county, Napa | Grape (*Vitis* sp.) | CD106 | KF242237, KF242238 | KF242275, KF242276 | - | *MW411839* | S.A. Subbotin, Van den Berg et al. [22] |
| *P. leptos* | Ethiopia, Jimma Zone, Gera district | Coffee | Ge16c | *MW413645– MW413653* | - | - | *MW421659– MW421665* | A.W. Aseffa |
| *P. microdorus* | Belgium, Zwijnaarde; 51°00′19″ N; 3°42′11″ E | Grasses | BE11 | - | *MW413597* | - | - | C.M. Etongwe |
| *P. microdorus* | Belgium, Ghent, Citadel Park; 51°02′05″ N; 3°43′10″ E | Grasses under a tree | BE9 | *MW413654– MW413655* | *MW413598– MW413600* | *MW413705– MW413706* | *MW421666– MW421667* | C.M. Etongwe |

Table 1. *Cont.*

| Species | Locality | Associated Plant Host | Sample Code | GenBank Accession Numbers | | | | Source |
|---|---|---|---|---|---|---|---|---|
| | | | | 28S rRNA | ITS rRNA | 18S rRNA | *COI* of mtDNA | |
| *P. nanus* | USA, California, Humboldt county, Trinidad, sample 5B; 41°02′40.6″ N; 124°07′18.1″ W | Unknown plant | CD3141 | *MW413576* | - | - | - | S.A. Subbotin |
| *P. nanus* | USA, Washington, Mason County, Skokomish, sample 32B; 47°18′07.0″ N; 123°10′95.6″ W | Unknown plant | CD3217 | *MW413575* | - | - | - | S.A. Subbotin |
| *P. nanus* | USA, California, Riverside | Grasses | CD728 | KF242194, KF242197 | KF242267, KF242268 | - | *MW411835* | S.A. Subbotin, Van den Berg et al. [22] |
| *P. nanus* | USA, California, Marin county | *Festuca* sp. | CD850 | KF242192, KF242193 | - | - | *MW411834* | S.A. Subbotin, Van den Berg et al. [22] |
| *P. nanus* | USA, California, Marin county | Grasses | CD860 | KF242191, KF242195 | - | - | *MW411836* | S.A. Subbotin, Van den Berg et al. [22] |
| *P. nanus* | Belgium, Gouvy, Rogery;50°14′39.8″ N; 5°57′21.9″ E | Grasses under tree *Fraxinus* sp. | AR3 | - | - | - | *MW421673* | P.R. Singh |
| *P. nanus* | Belgium, Zwijnaarde; 51°00′19″ N; 3°42′11″ E | Grasses under a tree | BE11 | *MW413658– MW413659* | *MW413601– MW413603* | *MW413707, MW413711– MW413712* | *MW421668– MW421671, MW421674* | C.M. Etongwe |
| *P. nanus* | Belgium, Ghent, Blaarmeersen; 51°07′14″ N; 2°39′29″ E | Grasses under a tree | BE18 | *MW413657* | *MW413604* | *MW413708* | *MW421672* | C.M. Etongwe |
| *P. neoambly-cephalus* | Belgium, Ghent, Citadel Park; 51°02′09″ N; 3°43′06″ E | Cypress tree | BE10 | *MW413660– MW413663* | *MW413606– MW413610* | *MW413713– MW413718* | *MW421675– MW421682* | C.M. Etongwe |
| *P. neoambly-cephalus* | USA, California, Madera county, Madera | Grasses | CD1223 | KF242190 | - | - | *MW411843* | S.A. Subbotin, Van den Berg et al. [22] |

**Table 1.** *Cont.*

| Species | Locality | Associated Plant Host | Sample Code | 28S rRNA | ITS rRNA | 18S rRNA | *COI* of mtDNA | Source |
|---|---|---|---|---|---|---|---|---|
| | | | | **GenBank Accession Numbers** | | | | |
| *P. projectus* | Belgium, Ghent, Ghent University Botanical Garden; 51°2′7.53″ N; 3°43′20.07″ E | Leek | BE2 | *MW413656* | *MW413605* | *MW413709– MW413710* | - | C.M. Etongwe |
| *P. projectus* | USA, California, Butte county, Gridley | Walnut (*Juglans* sp.) | CD137 | KF242199 | KF242265, KF242266 | - | *MW411840* | S.A. Subbotin, Van den Berg et al. [22] |
| *P. projectus* | South Africa, Western Cape, George | Bent grass | CD587 | KF242198, KF242200 | KF242263, KF242264 | - | *MW411842* | Van den Berg, Van den Berg et al. [22] |
| *P. shenzhenensis* | USA, Florida, Apopka | Unknown plant | CD2728 | *MW413579* | - | - | - | S.A. Subbotin |
| *P. straeleni* | Belgium, Zwijnaarde; 51°00′19″ N; 3°42′11″ E | Grasses under a tree | BE11 | *MW413686* | *MW413623* | - | *MW421713– MW421715* | C.M. Etongwe |
| *P. straeleni* | Belgium, Kortrijk; 50°47′58″ N; 3°11′37″ E | Grasses under a thorny tree | BE15 | *MW413685* | *MW413624– MW413625* | *MW413743– MW413746* | *MW421708– MW421712* | C.M. Etongwe |
| *P. straeleni* | Belgium, Ghent, Blaarmeersen; 51°02′18.9″ N; 3°41′17.2″ E | Grasses under a thorny tree next to a stream | BE19 | - | - | - | *MW421716* | C.M. Etongwe |
| *P. straeleni* | USA, North Carolina | Unknown plant | CD1433 | *MW413577* | - | - | - | W. Ye |
| *P. straeleni* | USA, California, Monterey county | Oak | CD1775 | *MW413578* | - | - | *MW411831* | S.A. Subbotin |
| *P. straeleni* | USA, California, Napa county | Tree | CD899 | KF242236 | - | - | *MW411832* | S.A. Subbotin, Van den Berg et al. [22] |
| *P. tenuicaudatus* | USA, California, Glenn county, Orland | Prune (*Prunus* sp.) | CD57 | KF242223, KF242225 | KF242261, KF242262 | - | *MW411826* | S.A. Subbotin, Van den Berg et al. [22] |
| *P. tenuicaudatus* | USA, California, Glenn county, Butte City | Prune (*Prunus* sp.) | CD61 | KF242224 | KF242259, KF242260 | - | *MW411827* | S.A. Subbotin, Van den Berg et al. [22] |

**Table 1.** *Cont.*

| Species | Locality | Associated Plant Host | Sample Code | GenBank Accession Numbers | | | | Source |
|---|---|---|---|---|---|---|---|---|
| | | | | 28S rRNA | ITS rRNA | 18S rRNA | COI of mtDNA | |
| *P. veruculatus* | Belgium, Ghent, Blaarmeersen; 51°02′14″ N; 3°41′23″ E | Grasses under a tree | BE20 | *MW413687–MW413691* | - | *MW413747–MW413748* | *MW421717–MW421722* | C.M. Etongwe |
| *Paratylenchus* sp.2 | Belgium, Kortrijk; 50°47′58″ N; 3°11′37″ E | Grasses under a thorny tree | BE15 | *MW413670–MW413671* | *MW413615–MW413616* | *MW413724–MW413725* | *MW421683–MW421684* | C.M. Etongwe |
| *Paratylenchus* sp.2 | Belgium, Ghent, Blaarmeersen; 51°02′14″ N; 3°41′23″ E | Grasses under a tree | BE20 | - | - | *MW413726* | *MW421685* | C.M. Etongwe |
| *Paratylenchus* sp.2 | Kyrgyzstan | Trees and grasses | CD2139 | *MW413567* | - | - | - | S.A. Subbotin |
| *Paratylenchus* sp.2 | USA, California, Yolo county, Davis | Grasses under a willow tree | CD604 | KF242220–KF242222 | KF242243 | - | *MW411825* | S.A. Subbotin, Van den Berg et al. [22] |
| *Paratylenchus* sp.3 | USA, California, Santa Barbara county, Goleta | Lemon (*Citrus* sp.) | CD232 | KF242231, KF242232 | KF242273, KF242274 | - | *MW411819* | S.A. Subbotin, Van den Berg et al. [22] |
| *Paratylenchus* sp.3 | USA, Florida | Unknown plant | CD2726 | *MW413573* | - | - | *MW411820* | S.A. Subbotin |
| *Paratylenchus* sp.4 | USA, Oregon | Trees | CD986 | KF242203 | - | - | *MW411829* | S.A. Subbotin, Van den Berg et al. [22] |
| *Paratylenchus* sp.7 | USA, California, Riverside, UCR campus | Unknown plant | CD1004 | KF242242 | - | - | *MW411830* | S.A. Subbotin |
| *Paratylenchus* sp.AH | USA, California, El Dorado county, Placerville, Apple hills, sample N12 | Unknown plant | CD1692 | *MW420921* | - | - | *MW411844* | S.A. Subbotin |
| *Paratylenchus* sp.BE11 | Belgium, Zwijnaarde; 51°00′19″ N; 3°42′11″ E | Grasses under a tree | BE11 | *MW413672–MW413674* | MW413617 | - | *MW421687–MW421688* | C.M. Etongwe |
| *Paratylenchus* sp.CaD | USA, California, El Dorado county, Placerville, Apple hills, sample, N7 | Unknown plant | CD1686 | *MW413561* | - | - | *MW411841* | S.A. Subbotin |

**Table 1.** *Cont.*

| Species | Locality | Associated Plant Host | Sample Code | 28S rRNA | ITS rRNA | 18S rRNA | *COI* of mtDNA | Source |
|---|---|---|---|---|---|---|---|---|
| *Paratylenchus* sp.CaD | USA, California, El Dorado county, Placerville, Apple hills, sample N16 | Unknown plant | CD1695a, b | *MW413560, MW413562* | *MW413584* | - | *MW411824* | S.A. Subbotin |
| *Paratylenchus* sp.CaD | USA, California, El Dorado county, Placerville, Apple hills, sample N20 | Unknown plant | CD1696 | *MW413563* | - | - | - | S.A. Subbotin |
| *Paratylenchus* sp.CaD | USA, California, Yolo county, Putah Creek | *Rubus* sp. | CD1791 | *MW413559* | - | - | - | S.A. Subbotin |
| *Paratylenchus* sp.D | Belgium, Ghent, Blaarmeersen; 51°02′14″ N; 3°41′23″ E | Grasses under a tree | BE20 | *MW413664– MW413669* | *MW413611– MW413614* | *MW413719– MW413723* | *MW421689– MW421699* | C.M. Etongwe |
| *Paratylenchus* sp.Dia | USA, California, Contra Costa County, Mount Diablo State Park | Unknown plant | CD1776 | *MW413574* | - | - | *MW411833* | S.A. Subbotin |
| *Paratylenchus* sp.F | Belgium, Ghent, Blaarmeersen; 51°07′14″ N; 2°39′29″ E | Grasses under a tree | BE18 | - | - | *MW413728* | - | C.M. Etongwe |
| *Paratylenchus* sp.F | Belgium, Merendree; 51°04′12″ N; 3°34′37″ E | Grasses around a beech tree | BE22 | *MW413675– MW413679* | *MW413618– MW413620* | *MW413727, MW413729– MW413731* | *MW421700– MW421702* | C.M. Etongwe |
| *Paratylenchus* sp.F | Russia, Primorsky Krai, Olginsky district | Unknown plant | CD1842, CD1844 | *MW413571, MW413572* | - | | - | J. Zograf |
| *Paratylenchus* sp.Ge16 | Ethiopia, Jimma Zone, Gera district | Coffee | Ge16c | *MW413680– MW413682* | - | *MW413732– MW413734* | *MW421703– MW421705* | C.M. Etongwe |
| *Paratylenchus* sp.J | USA, Washington, Mason County, Skokomish, sample 32E; 47°18′07.0″ N 123°10′95.6″ W | Unknown plant | CD3216 | *MW413570* | - | - | - | S.A. Subbotin |

**Table 1.** *Cont.*

| Species | Locality | Associated Plant Host | Sample Code | GenBank Accession Numbers | | | | Source |
|---|---|---|---|---|---|---|---|---|
| | | | | 28S rRNA | ITS rRNA | 18S rRNA | *COI* of mtDNA | |
| *Paratylenchus* sp.J | USA, Oregon, Douglas County, Oakland, sample 35; 43°28′59.9″ N 123°19′24.5″ W | Unknown plant | CD3220 | *MW413569* | - | - | - | S.A. Subbotin |
| *Paratylenchus* sp.NL | The Netherlands, Hilversum | Holly | NL | *MW413683– MW413684* | - | *MW413740– MW413742* | *MW421706– MW421707* | G. Karssen |
| *Paratylenchus* sp.SK | South Korea | *Pinus* sp. | CD1384 | *MW413580* | - | | - | S.A. Subbotin |

### 2.1.1. *Paratylenchus aculentus*

*Females* (Sample BE9; Figure 1, Table 2): Heat relaxed specimens open C- to J-shape. Lateral field with three lateral lines. Deirids not observed (not necessarily an indication that they are absent). Cephalic region rounded, low, sometimes appearing slightly truncated, submedian lobes not protruded. Stylet 52–61 µm long, cone 80–91% of stylet length, knobs 2–4 µm across. Pharynx well developed, about one-third of body length. Secretory-excretory pore between median bulb and isthmus level. Spermatheca rounded to slightly oval and filled with sperm cells. Prevulval swelling not prominent. Vulval flaps very small and can be visible under LM. Vulval located at 71–76% of body length from anterior end. Vagina straight to slightly oblique, reaching to almost half of body width. Anus obscure. Tail 18–25 µm long, tapers gradually to a finely or bluntly rounded terminus.

*Molecular characterisation:* Three D2-D3 of 28S, two ITS, two 18S rRNA and four *COI* gene sequences were generated without intraspecific sequence variations. The D2-D3 of 28S and the 18S sequences, respectively, were found to be similar to KP966492 (99% similarity; 4 out of 544 bp difference) and KP966494 (100% similarity; 800 bp) of *P. colinus* from Iran after Hesar et al. [24].

*Remarks:* Males were not found. Female morphology and morphometrics matched very well with *P. aculentus*. This species has been reported earlier in Belgium [53]. Although the D2-D3 and 18S sequences pointed towards *P. colinus*, the current population had no cuticular ornamentations present in the anterior part of the body and female bodies were not swollen and submedian lobe protrusions were not seen, which are important characteristics for *P. colinus*. According to Ghaderi et al. [23], *P. aculentus* is part of Group 9 of the *Paratylenchus* species with stylet lengths longer than 40 µm, three lateral lines and absence of vulval flaps. Here, we confirm the presence of small vulval flaps in *P. aculentus*, clearly supported by SEM. This was also an observation originally made by Brzeski [19]. *Paratylenchus aculentus* should, therefore, be placed in Group 8 with *P. colinus* and *P. idalimus*; furthermore, the close affinity of our *P. aculentus* population with *P. colinus* is also molecularly supported by the very conserved 18S rRNA gene fragment.

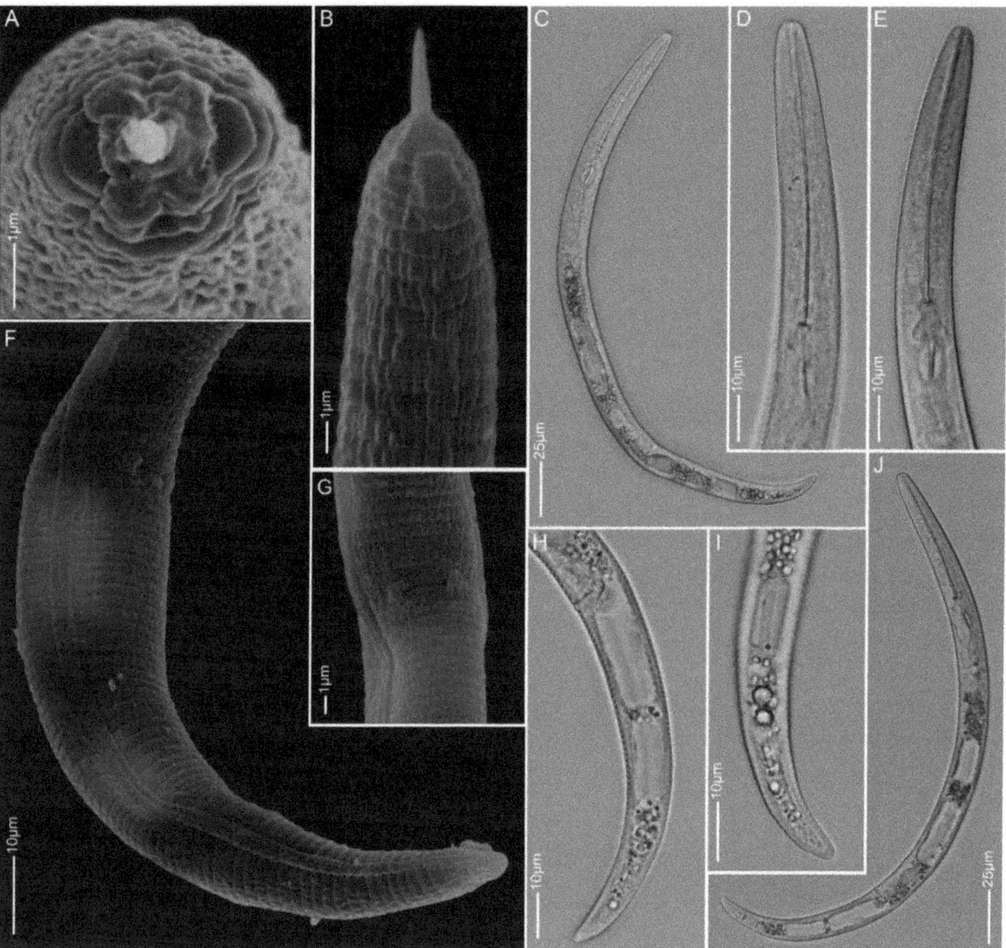

**Figure 1.** Light and scanning electron microscopy images of *Paratylenchus aculentus* females: (**A**) face view; (**B,D,E**) anterior region; (**C,J**) total body; (**G**) vulva region; (**F,H,I**) tail region.

**Table 2.** Female morphometrics of *Paratylenchus aculentus*, *Paratylenchus goodeyi*, *Paratylenchus idalimus* and *Paratylenchus straeleni* from fixed specimens mounted in glycerine. All measurements except for ratios and percentages are given in μm and in the form mean ± stdev (range).

| Population | *P. aculentus* (BE9) | *P. goodeyi* (BE22) | *P. idalimus* (BE19 and BE20) | *P. straeleni* (BE15) |
|---|---|---|---|---|
| n | 12 | 17 | 7 | 11 |
| L | 266 ± 20.1 (233–03) | 348 ± 42.5 (266–452) | 299 ± 20.7 (278–332) | 358 ± 13.1 (330–379) |
| a | 19.6 ± 2.0 (16.3–23.2) | 20.7 ± 1.7 (16.7–23.2) | 21.0 ± 1.3 (20–23) | 22.6 ± 0.9 (20.7–24.3) |
| b | 2.6 ± 0.1 (2.4–2.8) | 3.2 ± 0.3 (2.9–3.7) | 2 ± 0.2 (2.0–2.4) | 3.6 ± 0.1 (3.4–3.7) |
| c | 12.4 ± 1.5 (10.8–15.2) | 11.9 ± 1.4 (10.1–13.5) | 12.1 ± 0.6 (12.1–13.1) | 10.6 ± 1.0 (8.9–11.8) |
| c | 2.8 ± 0.3 (2.4–3.1) | 3.0 ± 0.2 (2.8–3.3) | 4.0 ± 0.6 (3.1–4.1) | 3.4 ± 0.3 (3.0–3.9) |
| Maximum body width | 13.6 ± 1.3 (11.6–15.5) | 17.0 ± 3.3 (13.0–27.0) | 14.0 ± 1.4 (13.0–16.1) | 15.9 ± 0.7 (14.6–16.7) |
| Stylet length | 56.0 ± 3.3 (52.4–61.2) | 52.1 ± 2.8 (47.0–58.6) | 89.0 ± 3.5 (84.1–93.0) | 55.7 ± 1.7 (53.5–58.6) |
| Cone length | 49.1 ± 3.6 (43.0–54.9) | 43.0 ± 2.7 (48.2–48.5) | 78.0 ± 2.9 (74.0–83.1) | 44.7 ± 1.7 (42.2–47.4) |
| Cone%stylet | 87.5 ± 3.8 (80.1–91.0) | 82.4 ± 2.5 (78.0–89.6) | 88.0 ± 2.2 (83–89) | 80 ± 1.8 (76–83) |
| Knob width | 3.2 ± 0.5 (2.3–4.0) | 4.1 ± 0.6 (3.3–5.2) | 4.0 ± 0.2 (4.0–4.3) | 3.9 ± 0.4 (3.1–4.6) |
| Pharynx length | 101 ± 8.3 (87.0–113) | 109 ± 11.7 (92.7–133) | 130 ± 12.8 (114–147) | 100 ± 3.8 (92.1–105) |
| Anterior end to SE pore | 66.7 ± 5.2 (54.3–74.4) | 80.7 ± 8.6 (68.5–99.0) | 93.0 ± 11.7 (82.0–115) | 82.5 ± 2.9 (79.4–87.6) |
| SE pore%L | 25.2 ± 0.9 (23.3–26.4) | 22.9 ± 1.4 (21.0–25.8) | 31 ± 4.2 (28–40) | 23 ± 0.7 (22–24) |
| Anterior end to vulva | 193 ± 16.1 (165–218) | 279 ± 31.9 (216–356) | 233 ± 17.9 (214–260) | 270 ± 18.8 (249–330) |
| V% | 72.5 ± 1.5 (70.8–75.7) | 80.1 ± 1.4 (77.8–82.3) | 78.0 ± 0.9 (77–79) | 81 ± 1.9 (80–84) |
| Body width at anus | 7.6 ± 0.5 (7.0–8.3) | 9.7 ± 0.5 (9.7–10.0) | 7.0 ± 0.1 (7.0–7.1) | 10.1 ± 0.4 (9.1–10.5) |
| Tail length | 20.9 ± 2.3 (18.1–25.1) | 29.0 ± 2.4 (25.8–32.1) | 25.0 ± 2.2 (22.0–28.1) | 34.4 ± 3.7 (31–40.8) |

### 2.1.2. *Paratylenchus elachistus*

*Females* (Sample BE15; Figure 2, Table 3): Heat relaxed specimens open C- to J-shape. Lateral field with four lateral lines. Deirids visible under SEM. Cephalic region conical-rounded to sometimes truncated. *En face* square-shaped, showing poorly developed submedian lobes, two pronounced lateral ridges and small indistinct dorso-ventral ridges around oral opening, two slit-like amphidial openings laterally. Stylet 20–22 μm long, cone 61–68% of stylet length and knobs 3–4 μm across. Pharynx well developed, about one-fourth of body length. Hemizonid commonly above secretory-excretory pore about two body annuli long. Secretory-excretory pore between mid-isthmus and end bulb level. Spermatheca rounded to oval and filled with sperm cells. Vulval flaps rounded, prominent. Vulva located at 80–83% of body length from anterior end. Vagina oblique, reaching to half of body width. Tail 21–29 μm long, conical, thin and terminus from spicate to pointed or minutely rounded.

*Molecular characterisation:* Two D2-D3 of 28S, four ITS, four 18S rRNA and four *COI* gene sequences were generated for the first time from this species without intraspecific sequence variations.

*Remarks:* Males were not found. This species is reported for the first time in Belgium and has only been recorded in Poland and Slovakia before in Europe [19,54,55]. Female morphology and morphometrics agree well with the original description [56] and also with descriptions of other populations [19,31]. *Paratylenchus elachistus* can be separated from its closest species, *Paratylenchus minutus*, Lindford in Lindford, Oliveira & Ishii, 1949, by a longer body length (0.23–0.34 mm vs. 0.19–0.31 mm), a more robust and longer stylet (19–25 μm vs. 15–21 μm) and a more slender tail, commonly with spicate to finely rounded tail termini.

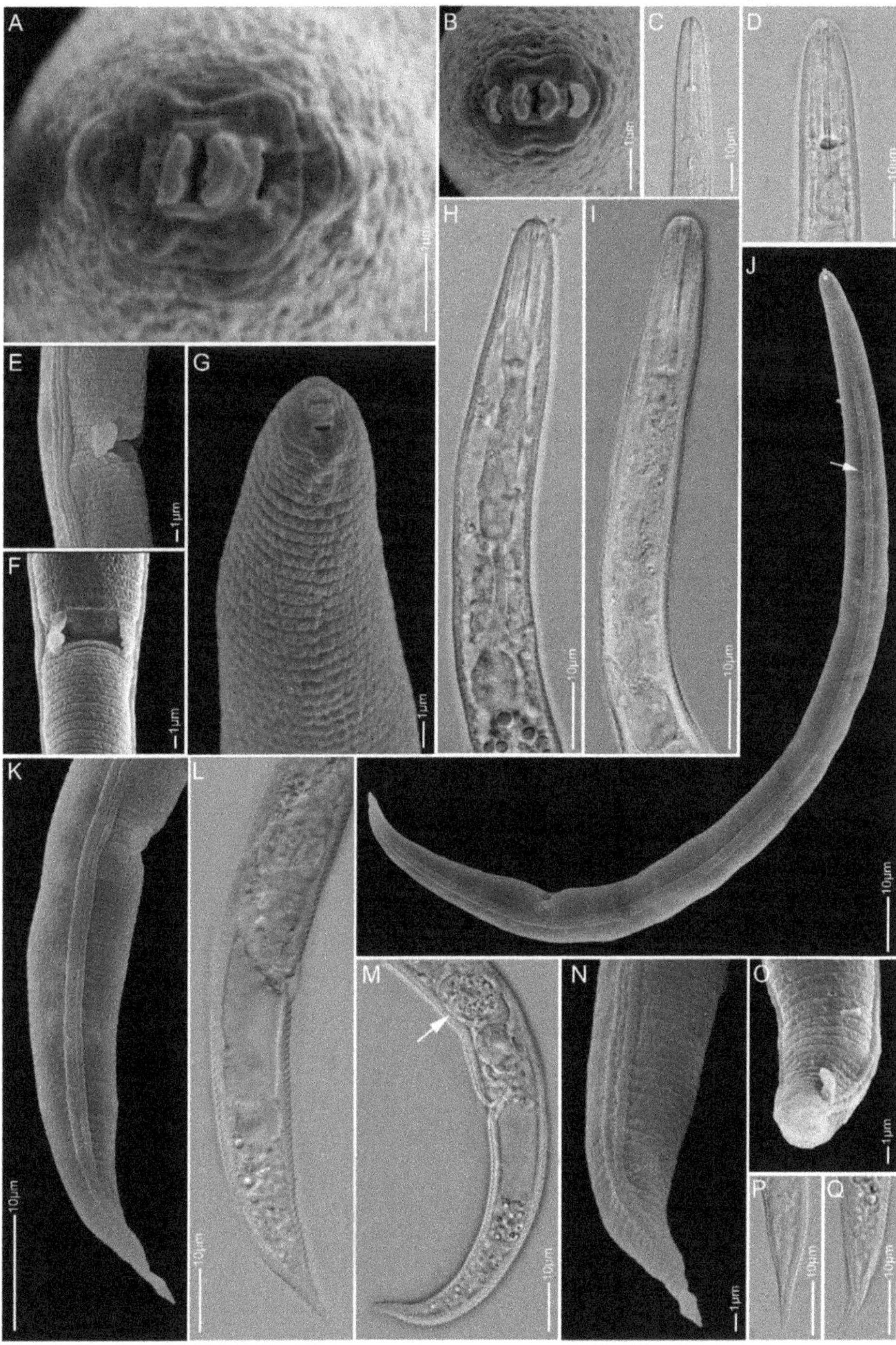

**Figure 2.** Light and scanning electron microscopy images of *Paratylenchus elachistus* females: (**A**,**B**) face view; (**C**,**D**,**G**–**I**) anterior region; (**E**,**F**) vulva region; (**J**) total body; (**K**–**Q**) tail region; arrows pointed to deirid in (**J**) and spermatheca in (**M**).

**Table 3.** Female morphometrics of *Paratylenchus elachistus*, *Paratylenchus holdemani*, *Paratylenchus microdorus* and *Paratylenchus veruculatus* from fixed specimens mounted in glycerine. All measurements except for ratios and percentages are given in µm and in the form mean ± stdev (range).

| Population | *P. elachistus* (BE15) | *P. holdemani* (AR3) | *P. microdorus* (BE9) | *P. veruculatus* (BE20) |
|---|---|---|---|---|
| n | 24 | 31 | 10 | 15 |
| L | 301 ± 12.5 (283–329) | 359 ± 47 (285–475) | 330.7 ± 20 (297–355) | 286 ± 24.7 (251–331) |
| a | 20.4 ± 1.1 (17.7–22.6) | 20.9 ± 1.9 (16.4–25.2) | 21.4 ± 1.4 (18.7–23.1) | 19.8 ± 1.9 (17.2–23.3) |
| b | 4.1 ± 0.2 (3.8–4.3) | 4.1 ± 0.7 (2.2–5.1) | 5.0 ± 0.5 (4.5–6.2) | 3.8 ± 0.3 (3.3–4.2) |
| c | 12.1 ± 0.9 (10.9–14.3) | 14.8 ± 1.4 (12.4–17.7) | 10.6 ± 0.9 (9.2–12) | 17.8 ± 1.8 (14.6–20.6) |
| c′ | 2.8 ± 0.2 (2.4–3.3) | 2.5 ± 0.3 (2.1–3.2) | 3.8 ± 0.5 (2.8–4.5) | 2.2 ± 0.2 (1.8–2.6) |
| Maximum body width | 14.8 ± 1.0 (12.7–16.6) | 17.3 ± 3.0 (11.3–23.8) | 15.5 ± 1.4 (13.3–17.1) | 14.5 ± 1.4 (12.5–16.5) |
| Stylet length | 20.9 ± 0.7 (19.7–22.2) | 22.5 ± 2.0 (19.0–26.1) | 12.4 ± 1.3 (10.6–14.7) | 14.2 ± 0.5 (13.1–14.8) |
| Cone length | 13.3 ± 0.4 (12.4–13.9) | 15.1 ± 1.1 (13.2–18.5) | 6.7 ± 1.3 (4.8–8.1) | 8.9 ± 0.3 (8.3–9.3) |
| Cone%stylet | 63.7 ± 1.5 (60.5–67.5) | 67.3 ± 3.5 (60.9–77.4) | 53.4 ± 5.7 (45.3–60.4) | 62.8 ± 1.1 (60.3–64.8) |
| Knob width | 3.5 ± 0.2 (3.1–4.1) | 3.3 ± 0.4 (2.9–4.2) | - | 3.1 ± 0.3 (2.7–3.5) |
| Pharynx length | 74.5 ± 2.6 (70.4–80.6) | 89.7 ± 21.5 (66.1–161) | 66.6 ± 6.4 (56.2–76.1) | 75.8 ± 7.0 (60.8–88.4) |
| Anterior end to SE pore | 60.7 ± 3.9 (54.0–68.5) | 74.8 ± 9.1 (60.1–99.0) | 63.6 ± 4.7 (57.5–71.3) | 62.3 ± 6.5 (51.2–74.4) |
| SE pore%L | 20.1 ± 0.9 (18.6–21.9) | 21.2 ± 1.8 (16.4–23.7) | 19.2 ± 1.1 (17.3–20.6) | 21.8 ± 2.0 (17.5–25.6) |
| Anterior end to vulva | 245 ± 10.2 (226–269) | 303 ± 40.9 (238–391) | - | 245 ± 22.0 (215–284) |
| V% | 81.3 ± 0.9 (79.7–83.2) | 84.3 ± 1.8 (81.3–90.5) | 81 ± 1.6 (79.1–82.8) | 85.7 ± 1.4 (83.8–89.7) |
| Body width at anus | 8.9 ± 0.7 (7.5–9.9) | 10.0 ± 1.3 (7.2–12.3) | 8.5 ± 1.3 (6.8–10.8) | 7.3 ± 0.6 (6.3–8.8) |
| Tail length | 24.8 ± 2.2 (20.9–29.1) | 25.2 ± 2.8 (20.0–29.5) | 31.8 ± 3.1 (28.1–35.9) | 16.1 ± 1.7 (13.6–19.1) |

### 2.1.3. *Paratylenchus goodeyi*

*Females* (Sample BE22; Figure 3, Table 2): Heat relaxed specimens C- to J-shape. Lateral field with four lateral lines. Deirids observed under LM. Cephalic region conical-rounded, submedian lobes not protruding except in two freshly killed specimens where small protrusions were seen under LM. Stylet 47–59 µm long, cone 78–90% of stylet length, stylet guide faintly seen, knobs 3–5 µm across. Pharynx well developed, about one-third of body length. Secretory-excretory pore around median bulb level. Spermatheca oval to elongate, filled with sperm cells. Vulval flaps present. Vulva located at 78–82% of body length from anterior end, in one female a short post-vulva sac observed. Vagina oblique and reaching to two-third of body width. Tail 26–32 µm long, conoid with variable terminus from finely rounded to bluntly rounded and rarely pointed.

*Molecular characterisation:* Three D2-D3 of 28S, one ITS, two 18S rRNA and three *COI* gene sequences were generated for the first time for this species without intraspecific sequence variations.

*Remarks:* Males were not found. Females morphology and morphometrics agree well with former *P. goodeyi* descriptions [18,19,57]. This species was originally described from the Netherlands and has been reported from many European countries, including Belgium. *Paratylenchus goodeyi* is one of the 22 species of the Group 10 of *Paratylenchus* after Ghaderi et al. [23] with stylet length more than 40 µm, four lateral lines and presence of vulval flaps. This species is comparable to other members of the group namely *Paratylenchus ivorensis* Luc & de Guiran, 1962, *Paratylenchus pandatus* (Raski, 1976) Siddiqi, 1989 and *P. straeleni* with females having more or less conical-rounded heads, stylet lengths in the range 40–61 µm (except for *P. pandatus* for which, a stylet length up to 68 µm was rarely reported). However, the vulvae of *P. goodeyi* and *P. straeleni* are located more posteriorly (77–88%) than that of the other two species (70–78%); *P. ivorensis* and *P. goodeyi* have been reported with variable tail termini, while *P. pandatus* and *P. straeleni* have been found usually with finely rounded to sub-acute female tail termini.

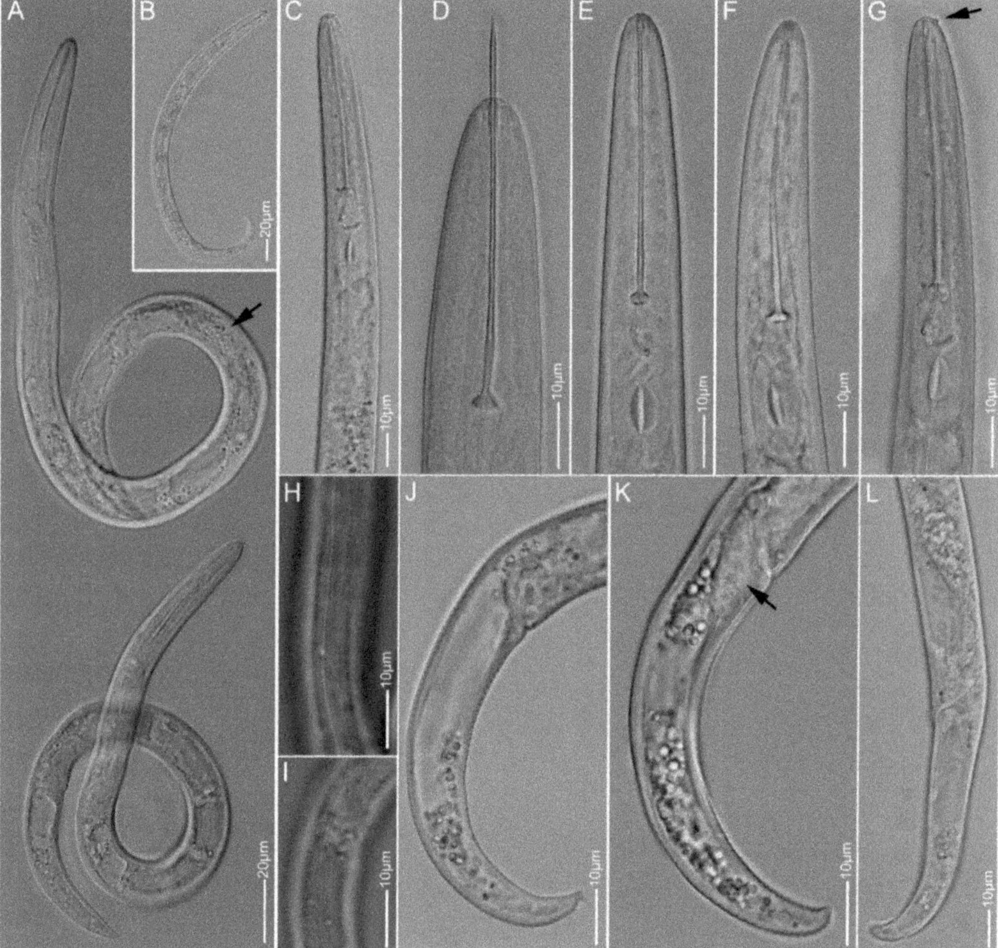

**Figure 3.** Light microscopy images of *Paratylenchus goodeyi* females: (**A,B**) total body; (**C–G**) anterior region; (**H,I**) lateral field; (**J–L**) tail region; arrows pointed to spermatheca in A, protruding submedian lobe in G and post-vulva sac in K.

### 2.1.4. *Paratylenchus holdemani*

*Females* (Sample AR3; Figure 4, Table 3): Heat relaxed specimens C- to J-shape. Lateral field with four lateral lines. Deirids not observed. Cephalic region slightly truncated, submedian lobes not protruded. *En face* showing four small submedian lobes, four irregular ridges around oral opening, slit-like lateral amphidial openings. Stylet 19–26 μm long, cone 61–77% of stylet length, knobs 3–4 μm across. Pharynx well developed, about one-fourth of body length. Secretory-excretory pore commonly between mid-isthmus and end bulb level. Spermatheca rounded, filled with sperm cells. Vulval flaps prominent. Vulva located at 81–90% of body length from anterior end. Vagina oblique and reaching to two-third of body width. Tail 20–30 μm long, conoid with regularly finely rounded to sometimes bluntly rounded or digitate terminus.

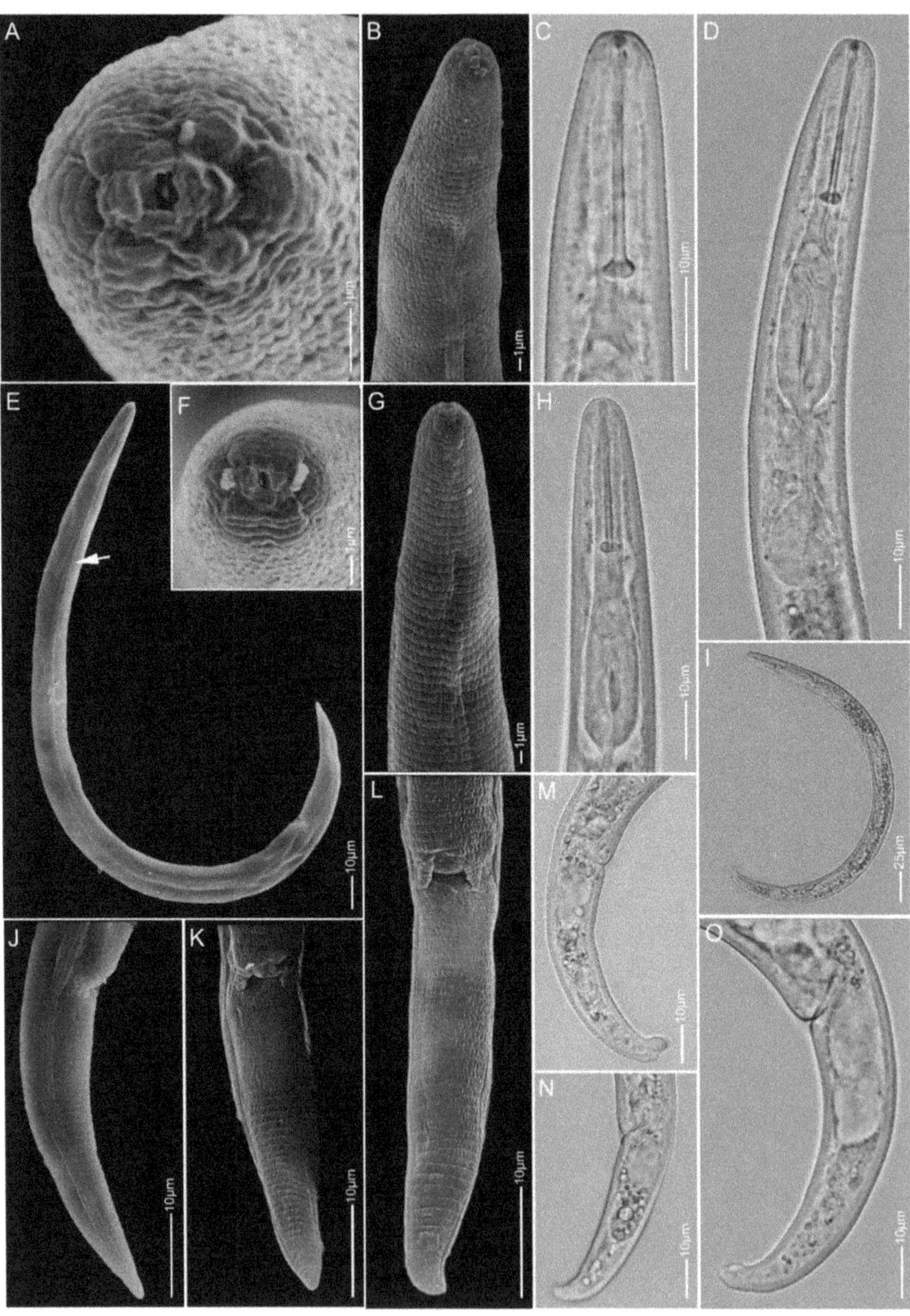

**Figure 4.** Light and scanning electron microscopy images of *Paratylenchus holdemani* females: (**A,F**) face view; (**B–D,G,H**) anterior region; (**E,I**) total body; (**J–O**) tail region; arrow pointed to deirid in (**E**).

*Males*: Two males were obtained from Sample AR3 and one from Sample BE20. Their conspecificity with the females was confirmed by identical D2-D3 of 28S rRNA and *COI* gene sequences identified from the AR3 and BE20 males, respectively. The males had an average stylet length of approximately 12 μm and spicule length of 21 μm.

*Molecular characterisation*: Three D2-D3 of 28S, one 18S rRNA and three *COI* gene sequences were generated from the AR3 females, whereas two D2-D3 of 28S, one ITS, one 18S rRNA and four *COI* gene sequences were generated from the BE20 females. No sequence from either population showed any intraspecific variations. The D2-D3 sequences were found to be identical to *P. bukowinensis* sequences that originated from Italy [37] and Belgium [47]; however, morphological data for these populations are not available for comparison and both are considered here as representatives of *P. holdemani*.

*Remarks:* This species has been reported for the first time in Belgium and has only been reported in the Czech Republic in Europe [19]. The morphology and morphometrics of the AR3 population agree well with the original description [58] and with the population from the Czech Republic [19]. Although our D2-D3 sequences were identical to a *P. bukowinensis* sequence (AY780943), the female morphology of this Belgian population is different from *P. bukowinensis* descriptions.

Most importantly, the average stylet length (22.5 μm) of our population is shorter than for many previously reported *P. bukowinensis* populations. In addition, the cephalic region of *P. bukowinensis* is more rounded than that of *P. holdemani*. *Paratylenchus holdemani* is comparable to *P. hamatus* and *Paratylenchus baldaccii* Raski, 1975, but is distinguishable from both species by a shorter stylet length of 22.5 ± 2.0 (19–26) μm vs. always above 26 μm. In this study, greater variation in the tail termini was observed in our *P. holdemani* population compared to the other two species.

### 2.1.5. *Paratylenchus idalimus*

*Females* (Samples BE19 and BE20, two nearby localities; Figures 5 and 6, Table 2): Heat relaxed specimens J- or open C-shape. Lateral field with three lateral lines. Deirids not observed. Cephalic region conical-truncate, submedian lobes well developed and protruding. Stylet 84–93 μm long of which 83–89% is cone, prominent stylet guide, knobs about 4 μm across. Pharynx well developed, occasionally reaching up to half of body length. Secretory-excretory pore around level of stylet knobs which is above median bulb level. Spermatheca small, rounded, usually filled with sperm cells. Vulval lips slightly protruding. Vulval flaps reduced and small, sometimes easily overlooked. Vulva located at 77–79% of body length from anterior end. Vagina oblique, often reaching to two-third of body width. Anus obscure. Tail 22–28 μm long, conoid with subacute to finely rounded terminus.

*Molecular characterisation:* Two identical D2-D3 of 28S and two identical 18S rRNA gene sequences were generated from the BE20 population, whereas one D2-D3 sequence, identical to that of the BE20 sequence, was generated from the BE19 population. These sequences were generated for the first time for this species.

*Remarks:* Males and swollen females were not found. Only one juvenile was recorded from the BE20 population with a stylet length of 42 μm. This is the first time the species has been reported in Europe. Female morphology and morphometrics based on seven females from both populations (three from BE19 and four from BE20) agree well with the description of the slender female by Raski [7] in the USA. This species and *P. colinus* are the only two members of Group 8 of *Paratylenchus* [23]. It differs from *P. colinus* in having a longer stylet (84–93 μm vs. 56–72 μm), more pronounced protrusion of submedian lobes, slightly posterior position of vulva (77–79% vs. 69–78%) and absence vs. presence of cuticular ornamentation in anterior body.

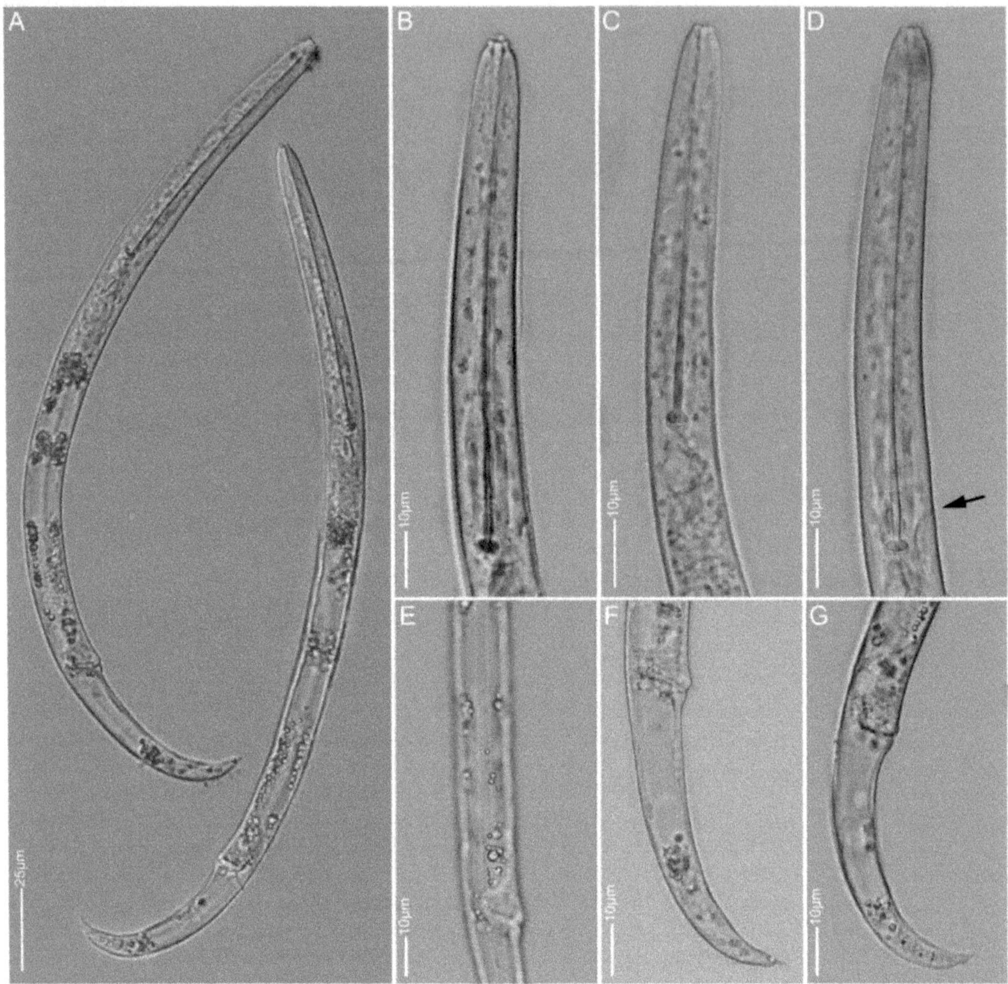

**Figure 5.** Light microscopy images of *Paratylenchus idalimus* females from sample BE19: (**A**) total body; (**B–D**) anterior region; (**E–G**) lateral field and tail region; arrow pointed to secretory–excretory pore in D.

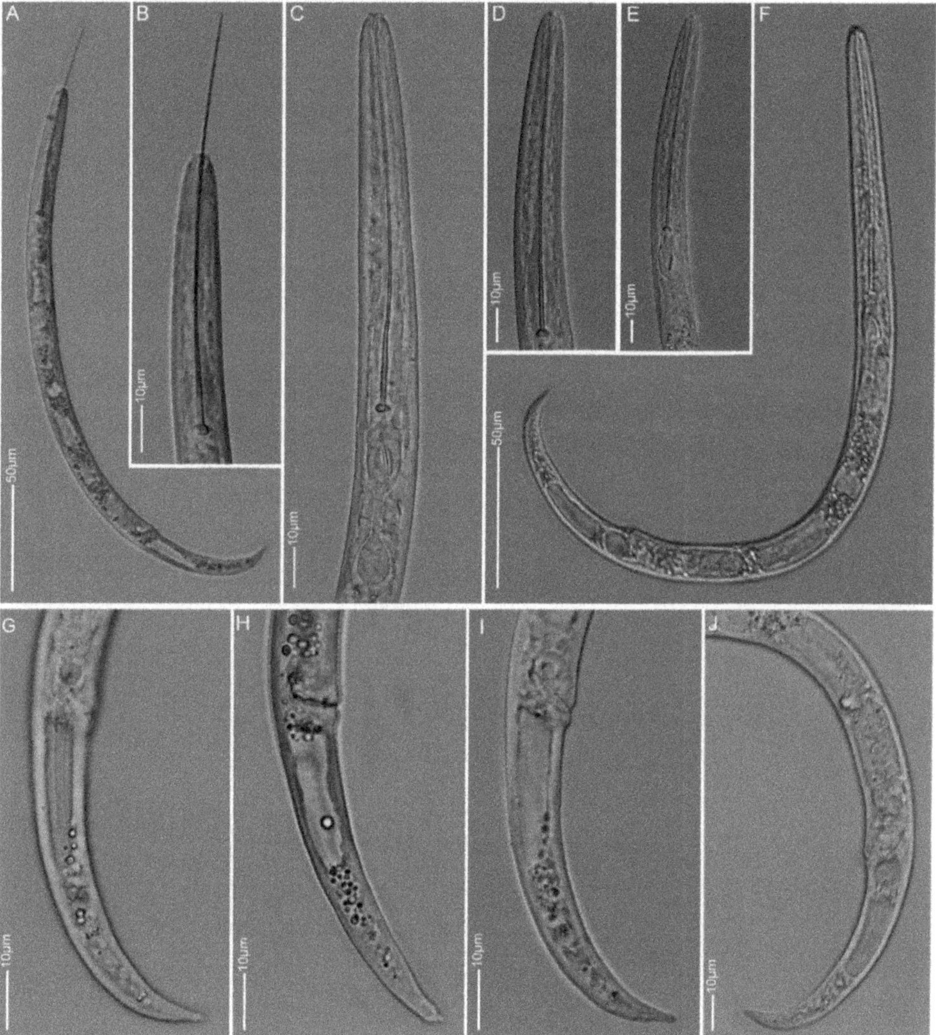

**Figure 6.** Light microscopy images of *Paratylenchus idalimus* females from sample BE20: (**A,F**) total body; (**B–E**) anterior region; (**G–J**) tail region.

### 2.1.6. *Paratylenchus microdorus*

*Females* (Sample BE9; Figure 7, Table 3): Body small, heat relaxed specimens open C- to 6-shape. Lateral field with four lateral lines. Deirids not observed. Cephalic region conical-truncate, submedian lobes sometimes slightly protruding. *En face* showing four submedian lobes and slit-like lateral amphidial openings. Stylet 11–15 μm long, cone 45–60% of stylet length. Pharynx about one-fifth of body length. Secretory-excretory pore between mid-isthmus and end bulb level. Spermatheca rounded, empty or filled with sperm cells. Vulval flaps prominent. Vulva located at 79–83% of body length from anterior end. Vagina oblique, reaching to half of the body width. Post-vulval uterine sac not seen. Tail 28–36 μm long, conoid and terminus pointed to subacute to sometimes finely rounded.

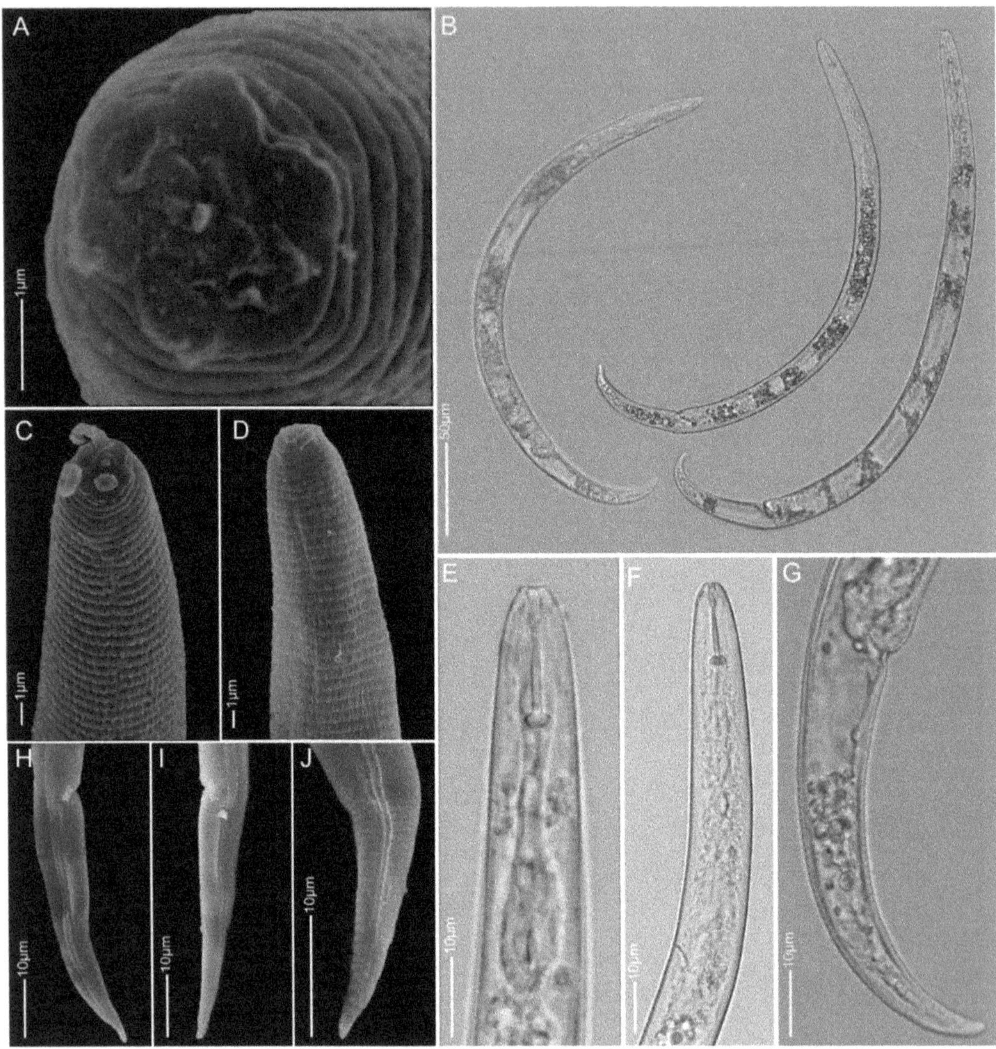

**Figure 7.** Light and scanning electron microscopy images of *Paratylenchus microdorus* females: (**A**) face view; (**B**) total body; (**C–F**) anterior region; (**G–J**) tail region.

*Molecular characterisation:* Three D2-D3 of 28S, three ITS, four 18S rRNA and three *COI* gene sequences were generated without intraspecific variability; the ITS and the *COI* sequences are new for this species. Only 300 bp of the D2-D3 sequences were found to be homologous with four *P. microdorus* sequences from Germany (MF325254–MF325257; 98% similarity; 5 bp difference). The 18S rRNA sequences are 98–99% similar with *P. microdorus* from the Netherlands (AY284632 and AY284633; 8–15 out of 880 bp difference).

*Remarks:* Males were not found. Female morphologies and morphometrics agree well with the original description [59] and other populations [16,31], except for a slightly shorter stylet length (11–15 μm vs. 13–18 μm). Wide variations in the tail termini have been reported for this species [31]. However, for the BE9 population, finely rounded to subacute female tail termini were commonly observed. This species is comparable to *Paratylenchus recisus* Siddiqi, 1996, *Paratylenchus variabilis* Raski, 1975 and *P. veruculatus*, with a female stylet length within 11–17 µm, four lateral lines, presence of vulval flaps, secretory–excretory pore at the posterior part of pharynx and vulva located at 78–87% of body length. However, they differ from each another in having conical-truncate heads with sometimes slightly protruded submedian lobes in *P. microdorus*, broadly rounded to truncated head with central swallow depression in *P. recisus*, rounded to almost hemispherical head in *P. variabilis* and low and broadly rounded head in *P. veruculatus*. Only *P. microdorus* and *P. veruculatus* males have been reported to have weak stylets, while stylets in the males of the other two species are degenerated.

### 2.1.7. Paratylenchus nanus

*Females* (Sample BE11; Figure 8, Table 4): Heat relaxed specimens open C- to J-shape. Lateral field with four lateral lines. Deirids not observed. Cephalic region conical-rounded, in some specimens with sloping sides to rounded end, submedian lobes not protruding under LM. *En face* square shaped, revealing four submedian lobes, four distinct ridges around oral opening, lateral ridges slightly larger than dorso-ventral ridges, and two slit-like lateral amphidial openings. Stylet 27–31 µm long, cone 67–78% of stylet length and knobs 3–5 µm across. Pharynx well developed, about one-fourth of body length. Hemizonid just above secretory-excretory pore, about two body annuli long. Secretory-excretory pore between isthmus and end bulb level. Spermatheca rounded and filled with sperm cells. Vulval flaps present. Vulva located at 82–86% of body length from anterior end. Vagina oblique, reaching up to half of body width. Tail 19–26 µm long, conoid, often more pronounced curvature on dorsal side ending with sub-acute to finely rounded terminus.

*Molecular characterisation:* Seven D2-D3 of 28S, four ITS, four 18S rRNA and seven *COI* gene sequences were generated without any intraspecific sequence variations among four *P. nanus* populations—AR3, BE1, BE11 and BE18. The D2-D3 and the ITS sequences were, respectively, identical to KF242194, KF242197 and KF242267, KF242268 of *P. nanus* from Van den Berg et al. [22]

*Remarks:* Only in the BE11 population was a sufficient number of females recovered to allow morphological and morphometrical data comparisons, which agreed well with the original description [60] and subsequent descriptions of *P. nanus* [19,22,58]. Van den Berg et al. [22] reported two sibling species of *P. nanus* with different genotypes—type A and type B (the latter of which was recently transferred to *P. projectus*) [61]. This correction suggests that the available 28S (MN720102–MN720103) and *COI* (MN734387 and MN734388) sequences of *P. nanus* from South Korea [48] were misidentified as they were found to be identical to the *P. projectus* sequences. *Paratylenchus nanus* is very similar to *P. projectus* and *P. neoamblycephalus*. It differs from *P. projectus* in having a conical-rounded vs. more trapezoid head shape and sperm-filled vs. empty spermathecae. It is differentiated from *P. neoamblycephalus* by more rounded vs. oval spermathecae and a conoid tailwith pronounced curvature on the dorsal side ending with a subacute or finely rounded terminus vs. a conoid tail with subacute terminus or almost acute tip. Furthermore, in our study we also observed that the ridges around the oral opening of the freshly killed specimens protruded more in *P. neoamblycephalus* compared to *P. nanus* when observed under LM.

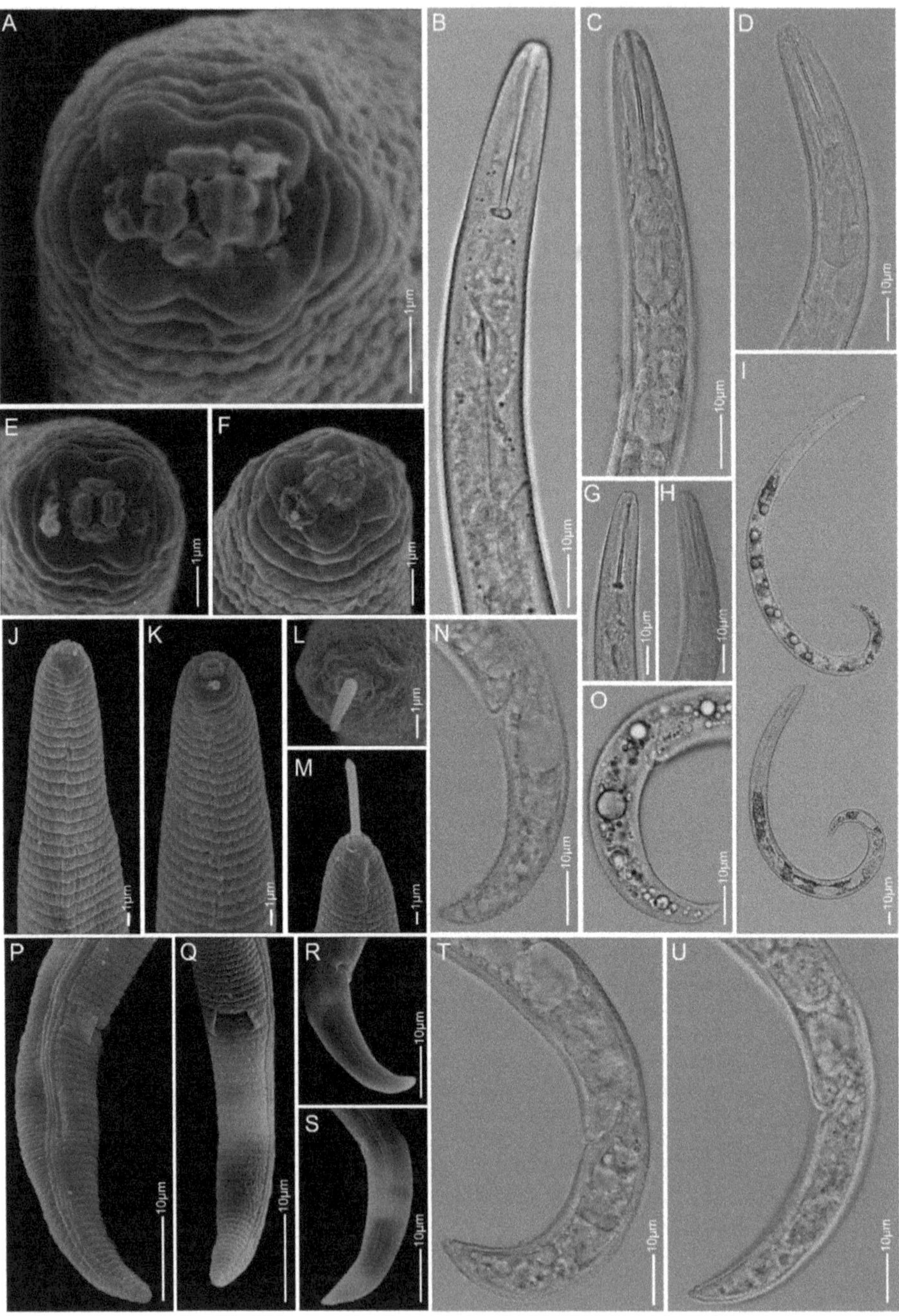

**Figure 8.** Light and scanning electron microscopy images of *Paratylenchus nanus* females: (**A,E,F,L**) face view; (**B–D,G,H,J,K,M**) anterior region; (**I**) total body; (**N,O,P–U**) tail region.

**Table 4.** Female morphometrics of *Paratylenchus nanus*, *Paratylenchus neoamblycephalus*, *Paratylenchus* sp.2, *Paratylenchus* sp.D and *Paratylenchus* sp.F from fixed specimens mounted in glycerine. All measurements except for ratios and percentages are given in µm and in the form mean ± stdev (range).

| Population | *P. nanus* (BE11) | *P. neoamblycephalus* (BE10) | *Paratylenchus* sp.2 (BE15) | *Paratylenchus* sp.D (BE20) | *Paratylenchus* sp.F (BE22) |
|---|---|---|---|---|---|
| n | 30 | 15 | 16 | 11 | 17 |
| L | 318 ± 15.8 (287–352) | 337 ± 20.2 (301–367) | 347 ± 20.7 (308–389) | 328 ± 36.1 (285–387) | 300 ± 21.1 (264–339) |
| a | 18.1 ± 1.3 (15.6–20.4) | 18.4 ± 1.0 (16.8–19.9) | 23.2 ± 1.8 (20.1–28.7) | 21.5 ± 1.6 (19.1–24.4) | 20.3 ± 1.3 (18.4–23.0) |
| b | 3.9 ± 0.3 (3.5–4.8) | 4.3 ± 0.5 (3.4–4.9) | 3.9 ± 0.3 (3.5–4.4) | 3.6 ± 0.3 (3.3–4.1) | 4.0 ± 0.3 (3.7–4.6) |
| c | 14.8 ± 1.3 (12.8–16.8) | 14.6 ± 1.5 (12.9–16.6) | 13.2 ± 0.7 (12.3–14.4) | 14.5 ± 1.4 (12.0–15.8) | 12.9 ± 0.8 (11.8–14.0) |
| c′ | 2.0 ± 0.2 (1.7–2.6) | 2.2 ± 0.2 (1.9–2.4) | 3.1 ± 0.2 (2.9–3.5) | 2.7 ± 0.3 (2.5–3.2) | 2.6 ± 0.2 (2.3–2.9) |
| Max. body width | 17.6 ± 1.2 (15.4–20.5) | 18.3 ± 1.4 (16.8–20.7) | 15.1 ± 1.3 (13.2–16.8) | 15.3 ± 2.2 (13.2–19.9) | 14.8 ± 1.1 (13.3–16.6) |
| Stylet length | 28.8 ± 1.2 (26.7–31.2) | 33.4 ± 0.9 (32.0–34.3) | 28.4 ± 1.5 (26.5–31.4) | 27.5 ± 1.0 (25.7–28.9) | 27.6 ± 1.2 (25.3–29.6) |
| Cone length | 20.2 ± 1.4 (17.7–23.4) | 22.6 ± 0.9 (21.4–24.5) | 19.2 ± 1.0 (17.5–20.8) | 17.7 ± 0.8 (17.0–19.2) | 18.5 ± 0.9 (17.1–20.3) |
| Cone%stylet | 70.1 ± 3.6 (64.9–78.5) | 67.8 ± 2.8 (63.0–72.5) | 67.5 ± 2.0 (64.4–71.2) | 64.3 ± 2.0 (60.7–67.3) | 67.1 ± 1.3 (65.2–69.5) |
| Knob width | 3.9 ± 0.4 (3.1–4.6) | 4.8 ± 0.2 (4.4–5.1) | 4.1 ± 0.3 (3.5–4.7) | 4.0 ± 0.3 (3.5–4.6) | 3.6 ± 0.3 (3.3–4.1) |
| Pharynx length | 81.6 ± 5.4 (65.4–91.1) | 79.7 ± 9.1 (65.7–93.8) | 88.2 ± 4.1 (78.0–96.7) | 89.6 ± 7.3 (76.0–104) | 74.8 ± 4.9 (67.7–83.1) |
| Ant. end to SE pore | 65.7 ± 6.0 (54.7–75.0) | 63.9 ± 4.7 (52.2–70.0) | 72.3 ± 4.5 (64.2–90.0) | 74.6 ± 6.8 (66.7–90.2) | 63.0 ± 5.4 (51.5–70.6) |
| SE pore%L | 20.6 ± 1.4 (17.0–22.5) | 19.0 ± 1.5 (15.8–21.7) | 20.8 ± 1.3 (19.4–23.5) | 23.6 ± 0.9 (22.1–24.8) | 21.0 ± 1.6 (18.6–24.2) |
| Ant. end to vulva | 270 ± 18.8 (249–330) | 276 ± 15.0 (247–296) | 286 ± 17.1 (252–313) | 270 ± 27.9 (239–320) | 217 ± 17.2 (217–278) |
| V% | 83.8 ± 1.1 (81.7–85.7) | 81.9 ± 0.9 (80.7–83.8) | 82.2 ± 0.8 (81.2–83.5) | 83.7 ± 1.0 (81.8–85.2) | 82 ± 0.7 (80.9–83.5) |
| Body width at anus | 10.9 ± 0.9 (9.3–12.6) | 10.6 ± 0.8 (9.4–12.0) | 8.6 ± 0.7 (7.6–9.8) | 8.0 ± 0.6 (7.3–8.7) | 9.0 ± 0.6 (8.0–9.9) |
| Tail length | 21.7 ± 1.9 (18.9–25.5) | 23.2 ± 2.6 (20.2–27.8) | 26.1 ± 2.0 (23.0–28.7) | 22.0 ± 3.0 (18.0–25.6) | 23.1 ± 2.4 (20.0–26.5) |

### 2.1.8. *Paratylenchus neoamblycephalus*

*Females* (Sample BE10; Figure 9, Table 4): Heat relaxed specimens open C-shape. Lateral field with four lateral lines. Deirids clearly visible on SEM images. Cephalic region truncated-rounded, submedian lobes sometimes very slightly protruding under LM. *En face* showing four rounded to oval submedian lobes, four ridges around oral opening, lateral ridges thicker than dorso-ventral ridges and seen as a protruding bi-lobed structure under LM. Stylet 32–34 µm long, cone 63–73% of stylet length, knobs 4–5 µm across. Pharynx about one-fourth of body length. Secretory-excretory pore between mid-isthmus and end bulb level, with swellings sometimes seen at the duct near the opening in freshly killed specimens. Spermatheca oval and filled with sperm cells. Vulval flaps present. Vulva located at 81–84% of body length from anterior end. Vagina oblique and reaching up to half of body width. Tail 20–28 µm long, conoid and terminating to sub-acute to almost acute tip.

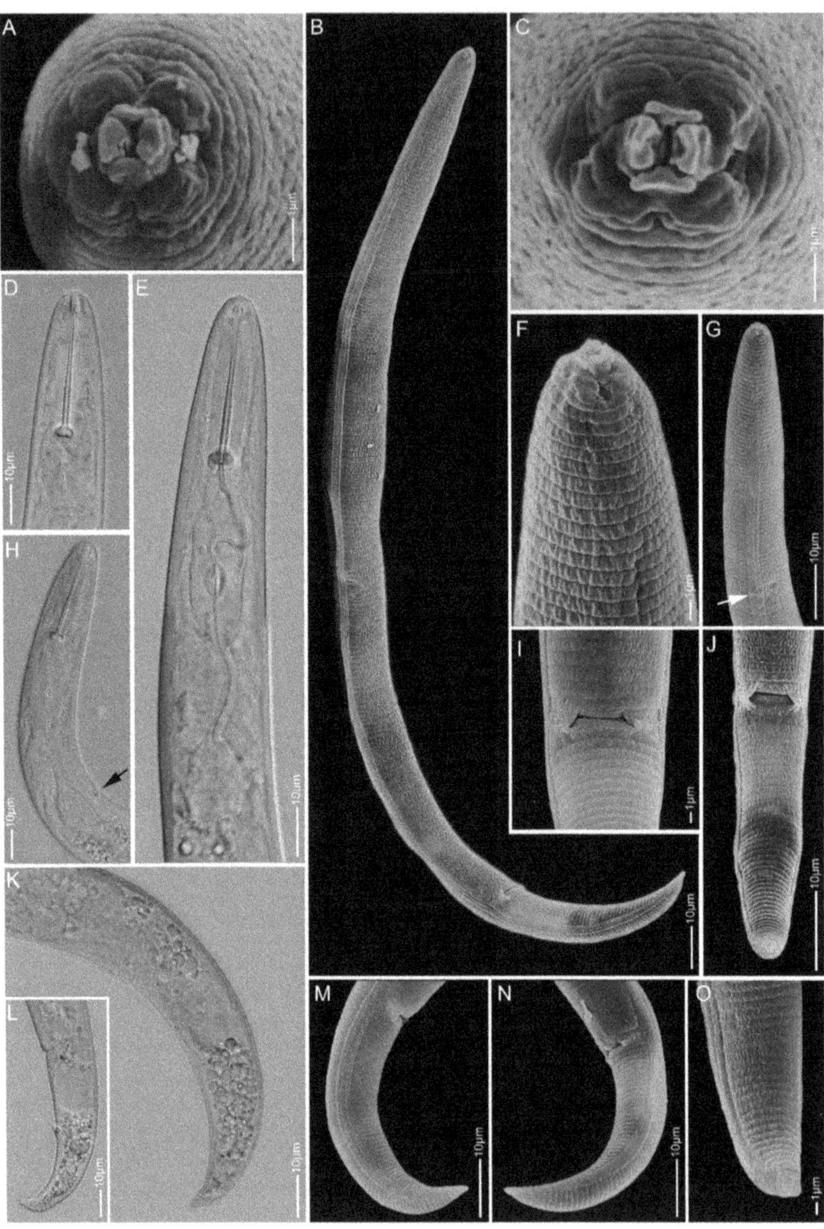

**Figure 9.** Light and scanning electron microscopy images of *Paratylenchus neoamblycephalus* females: (**A,C**) face view; (**B**) total body; (**D–H**) anterior region; (**I–O**) tail region; arrows pointed to secretory–excretory pore in (**H**) and deirid in (**G**).

*Males:* Two males were obtained with very thin stylets in freshly killed specimens, which were not visible after fixation, and spicules of 24 μm long. Their conspecificity with the females was confirmed by identical *COI* and D2-D3 sequences.

*Molecular characterisation:* Four D2-D3 of 28S, five ITS, six 18S rRNA and eight *COI* gene sequences were generated without intraspecific sequence variations. The 18S and the *COI* sequences are new for this species. The D2-D3 sequences were found to be identical to

KF242189 and KF242190 of an unidentified *Paratylenchus* sp.6 from the USA [22], which is considered here as *P. neoamblycephalus*. However, the D2-D3 sequences were only 89% similar (79 out of 710 bp difference) with MG925221 and 92% similar (43 out of 546 bp difference) with MK506807 of *P. neoamblycephalus* from the USA and Iran, and named here as type A and type B, respectively. Interestingly, we observed 17 ambiguous nucleotide sites in the American *P. neoamblycephalus* type B sequence, which was found to be similar to *P. projectus* (previously *P. nanus* type B; KF242198–KF242201; 98% similarity; 16–20 out of 690 bp difference) after Van den Berg et al. [22,61]. On the other hand, the Iranian *P. neoamblycephalus* sequence [24] was similar to *P. nanus* (KF242194 and KF242197; 95% similarity; 27 out of 575 bp difference) [22]. Furthermore, our ITS sequences were only 74% similar (222 out of 865 bp difference) to MK506794 of *P. neoamblycephalus* type A generated from the same Iranian population.

*Remarks:* This species is reported for the first time in Belgium. Female morphology and morphometrics agree well with the original description from Germany [18] and to subsequent descriptions from Poland [19]. *Paratylenchus neoamblycephalus* is very similar to *P. nanus* and a comparison is provided above.

### 2.1.9. Paratylenchus straeleni

*Females* (Sample BE15; Figure 10, Table 2): Heat relaxed specimens J- to C-shape. Lateral field with four lateral lines. Deirids clearly visible under SEM. Cephalic region conical-rounded to sometimes slightly truncated, submedian lobes not protruded. Stylet straight to slightly curved, 54–59 μm long, cone 76–83% of stylet length, knobs 3–5 μm across. Pharynx roughly one-fourth of body length. Secretory-excretory pore between isthmus and end bulb level. Spermatheca rounded to sometimes slightly ovoid and filled with sperm cells. Vulval flaps distinct. Vulva located at 80–84% of body length from anterior end. Vagina oblique, occasionally reaching to two-third of body width. Tail conical, 31–41 μm long, and terminus sharply pointed to minutely rounded.

*Males:* Two males were recovered without stylets and with spicule lengths of 20 and 22 μm, respectively. Their conspecificity with the females was confirmed by identical *COI* sequences.

*Molecular characterisation:* One D2-D3 of 28S, two ITS (99% similarity; 4 out of 830 bp difference), three identical 18S rRNA and five identical *COI* gene sequences were generated from the BE15 population. From another population (BE11), single D2-D3, ITS and 18S sequences and three identical *COI* sequences were also generated. All the sequences from both populations showed no intraspecific variation, except for the ITS sequences. The 18S sequences were 99% similar (3–5 out of 930 bp difference) with *P. straeleni* from the Netherlands (AY284630 and AY284631). The D2-D3 sequences were 97–99% similar (11–18 out of 700 bp difference) with four *P. straeleni* sequences—i.e., MK506804 from Iran [24], KM875547 from Turkey [42], and KF242235 and KF242236 from the USA [22]. The *COI* sequences were generated for the first time for this species. Remarkably, the Belgian ITS sequences were only 62% similar (295 bp difference) to the Iranian *P. straeleni* sequence (MK506791) of Hesar et al. [24].

*Remarks:* Female morphology and morphometrics agree well with the original description, also from Belgium [62], and subsequent descriptions of globally distributed *P. straeleni* populations [31]. This species is comparable to *P. goodeyi* as described above.

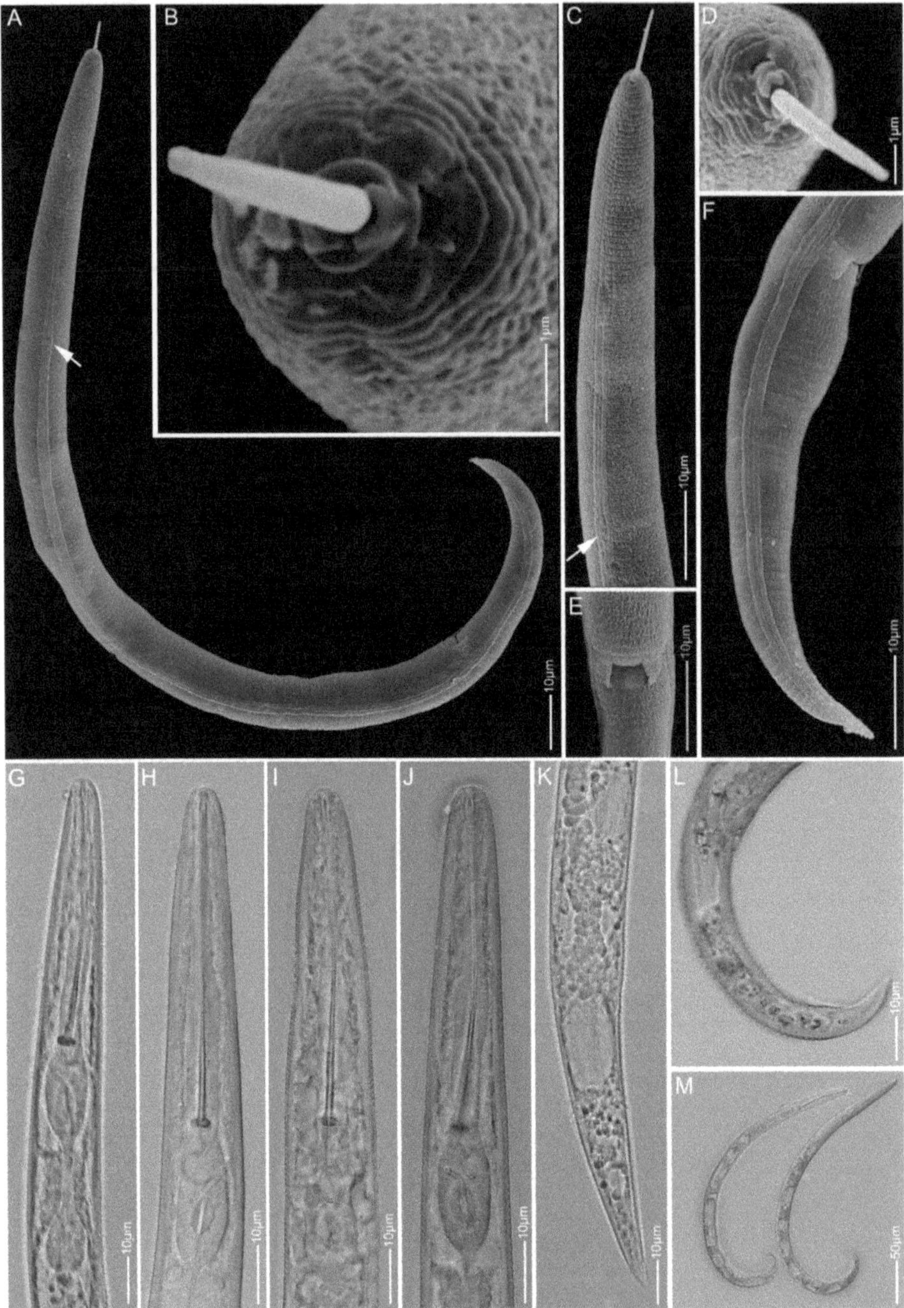

**Figure 10.** Light and scanning electron microscopy images of *Paratylenchus straeleni* females: (**A**) whole body; (**B,D**) face view; (**C,G–J**) anterior region; (**E**) vulva region; (**F,K,L**) tail region; (**M**) total body; arrows pointed to deirids in (**A,C**).

### 2.1.10. *Paratylenchus veruculatus*

*Females* (Sample BE20; Figure 11, Table 3): Heat relaxed specimens open C-shape to slightly ventrally curved. Lateral field with four lateral lines. Deirids not observed. Cephalic region broadly rounded, submedian lobes not protruding. *En face* rectangular with indistinct submedian lobes, four irregular ridges around oral opening and lateral amphidial openings. Stylet 13–15 μm long, cone 60–65% of stylet length, knobs about 3 μm across. Pharynx roughly one-fourth of total body length. Hemizonid two body annuli long, usually visible just above secretory-excretory pore. Secretory-excretory pore between mid-isthmus and end bulb level. Spermatheca rounded and filled with sperm cells, young females with empty spermatheca also seen. Vulval flaps prominent. Vulva located at 84–90% of body length from anterior end. Vagina oblique and long, reaching up to three-fourth of body width. Tail 14–19 μm long, conoid with often broadly rounded to sometimes finely rounded terminus.

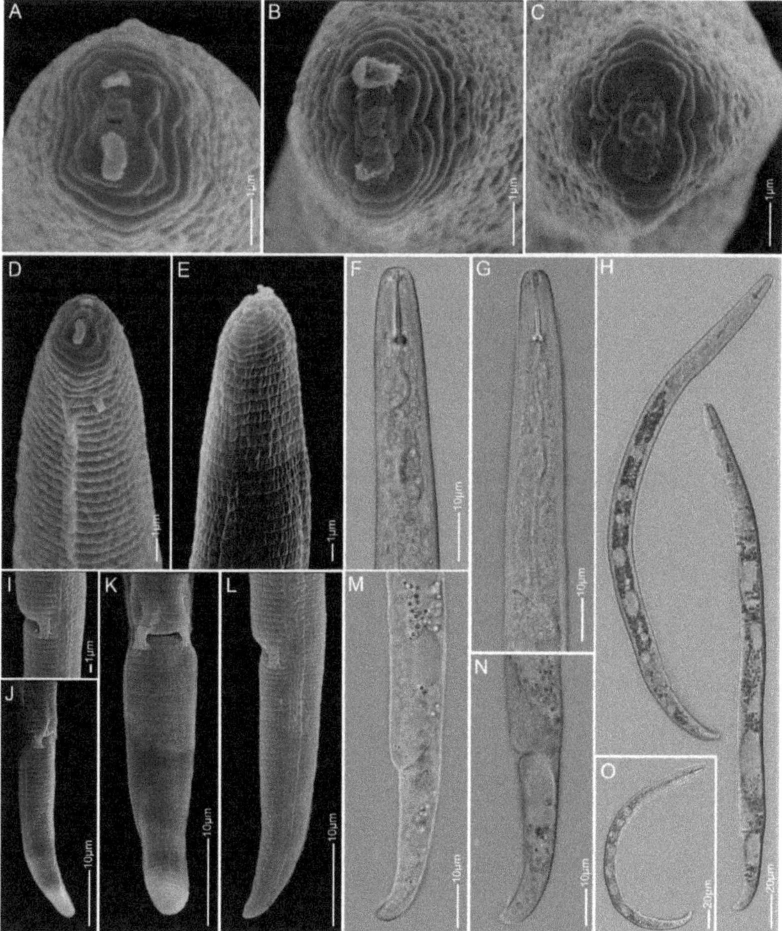

**Figure 11.** Light and scanning electron microscopy images of *Paratylenchus veruculatus* females: (**A**–**C**) face view; (**D**–**G**) anterior region; (**H**,**O**) total body; (**I**–**N**) lateral field and tail region.

*Molecular characterisation:* Five D2-D3 of 28S (99% similarity; 1–2 out of 720 bp difference) and two identical 18S rRNA and seven *COI* gene (97–100% similarity; 12–13 out of 410 bp difference) sequences were generated for the first time for this species.

*Remarks:* Female morphology and morphometrics agree well with the original description [63] and with other populations [19,31]. This species is comparable to other species with a short stylet such as *P. microdorus*, *P. recisus* and *P. variabilis* (see also above).

### 2.1.11. *Paratylenchus* sp.2

*Females* (Sample BE15; Figure 12, Table 4): Heat relaxed specimens open C-shape. Lateral field with four lateral lines. Deirids observed under SEM. Cephalic region conical-rounded, sometimes slightly trapezoid, submedian lobes not protruding under LM. *En face* square-shaped, showing four rounded, poorly separated submedian lobes, four ridges around oral opening, lateral ridges more prominent and larger than dorso-ventral ridges. Stylet 27–31 μm long, cone 64–71% of stylet length, knobs about 4 μm across. Pharynx well developed, about one-fourth of body length. Hemizonid just above secretory-excretory pore, about two body annuli long. Secretory-excretory pore between mid-isthmus and end bulb level. Spermatheca rounded to occasionally slightly ovoid, filled with sperm cells. Vulval flaps prominent. Vulva located at 81–84% of body length from anterior end. Vagina oblique, reaching up to two-third of body width. Tail 23–29 μm long, conoid, slender and terminating with finely rounded tip.

*Molecular characterisation:* Two identical sequences each of D2-D3 of 28S, ITS, 18S rRNA as well as the *COI* gene were generated. The D2-D3 and ITS sequences were found to be, respectively, identical to KF242220 and KF242221 and 99% similar (five out of 750 bp difference) to KF242243 of *Paratylenchus* sp.2, which was identified as a member of the *P. hamatus* species complex [22]. The 18S and *COI* sequences were generated for the first time.

*Remarks:* Males were not found. The female morphology and morphometrics are in agreement with the description of *P. hamatus* [64]. Based on morphology and D2-D3 and ITS sequences, Van den Berg et al. [22] considered *P. hamatus* as a species complex containing several species, including *P. hamatus sensu stricto* collected from the type locality, and *Paratylenchus* sp.1 and *Paratylenchus* sp.2, collected from other places in California. *Paratylenchus* sp.1 is identified as representative of *P. tenuicaudatus*. *Paratylenchus* sp.2 is morphologically similar with *P. hamatus sensu stricto* but differs based on D2-D3 and ITS sequences [22], and this species appears to be not only present in the USA (California) but also in Belgium and Kyrgyzstan.

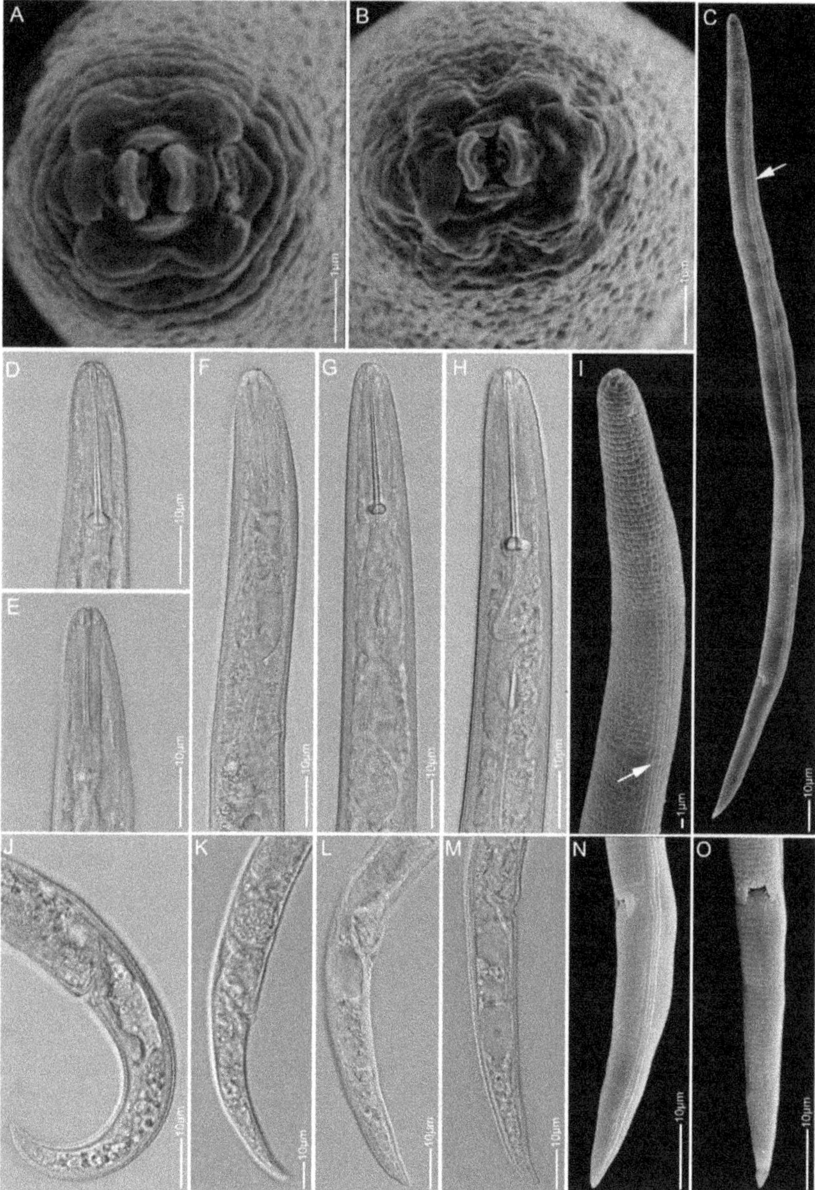

**Figure 12.** Light and scanning electron microscopy images of *Paratylenchus* sp.2 females: (**A,B**) face view; (**C**) total body; (**D–I**) anterior region; (**J–O**) tail region; arrows pointed at deirids in (**C,I**).

### 2.1.12. *Paratylenchus* sp.BE11

*Females* (Sample BE11; *n* = 3; Figure 13): Body about 0.3 mm long with maximum body width of about 15 μm, heat relaxed specimens open C- to 6-shape. Lateral field with four lateral lines. Deirids not observed. Head broadly rounded, submedian lobes not protruded, cephalic sclerotization strong. Stylet about 15 μm long, cone 60% of stylet length, knobs 3 μm across. Pharynx about one-fourth of body length. Secretory-excretory pore at the level of pharyngeal end bulb or about 70 μm from anterior end. Spermatheca rounded and filled with sperm cells. Vulval flaps small and rounded. Vulva located at 80–82% of body length from anterior end. Tail 25–32 μm, conoid with bluntly rounded tip.

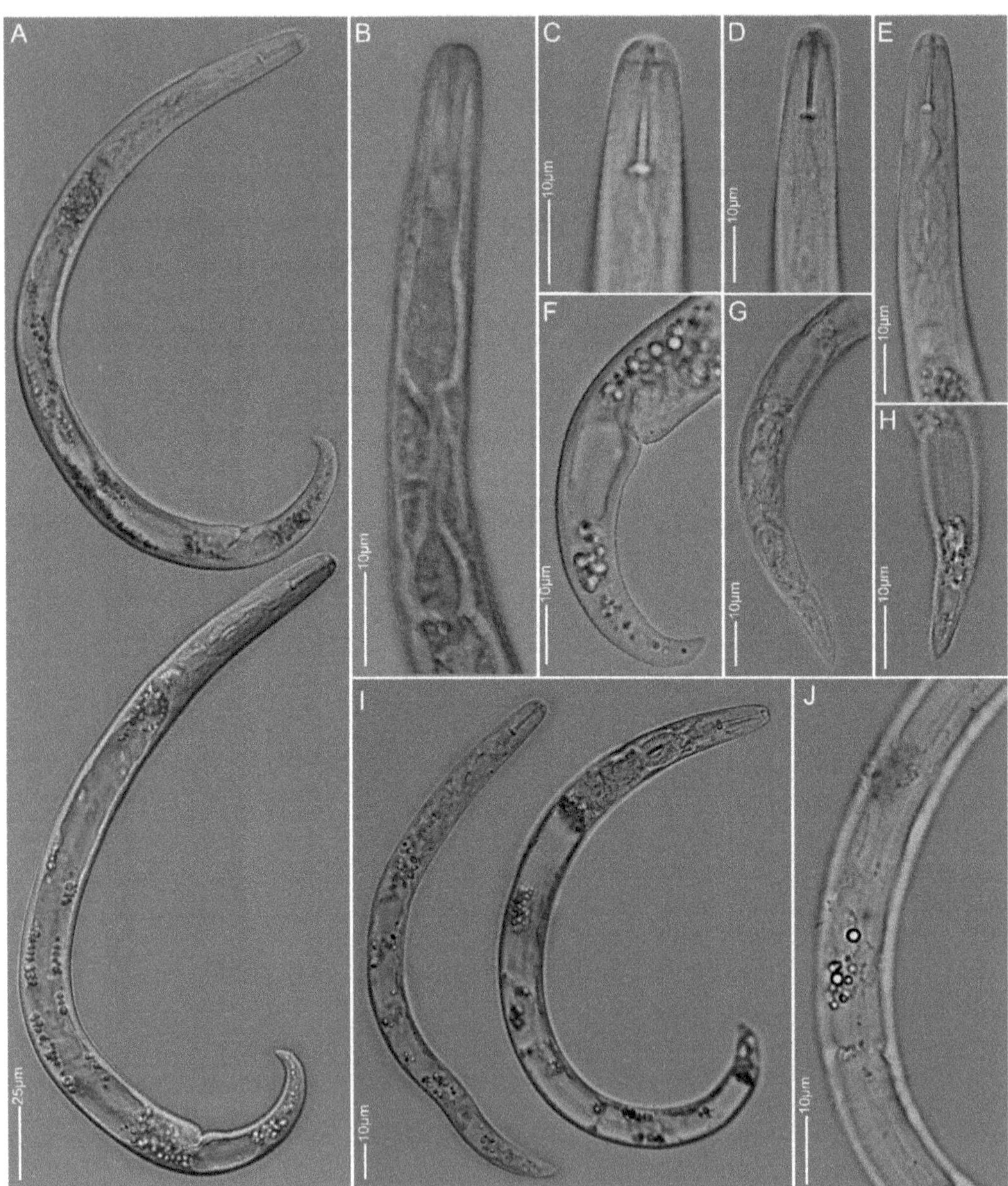

**Figure 13.** Light microscopy images of *Paratylenchus* sp.BE11 females: (**A**) total body; (**B–E**) anterior region; (**F–H**) tail region; (**I**) total body; (**J**) lateral field.

*Molecular characterisation:* Three identical D2-D3 of 28S, one ITS rRNA and two identical *COI* gene sequences were generated.

*Remarks:* No males were found. Female description is based on only three freshly killed specimens, while sufficient specimens are needed for a comprehensive species characterisation. The female morphology is close to *P. variabilis*, *P. veruculatus* and *Paratylenchus vexans* Thorne and Malek, 1986. These four species have more or less broadly rounded heads with non-protruding submedian lobes, stylet lengths in the range of 12–18 µm, four lateral lines, sperm-filled spermathecae, vulval flaps and conoid tails with more or less rounded termini. However, our population appears to have a stronger cephalic sclerotisation and slightly more anteriorly located vulvae (80–82% vs. 80–87%) compared to the other three species. This species is a sister to *P. microdorus* in the D2-D3 tree (96% similarity; 27 out of 740 bp difference), ITS tree (93% similarity; 37 out of 530 bp difference) as well as the COI tree (91% similarity; 36 out of 420 bp difference). It can, however, be readily morphologically distinguished from *P. microdorus* (see above).

### 2.1.13. *Paratylenchus* sp.D

*Females* (Sample BE20; Figure 14, Table 4): Heat relaxed specimens open C-shape. Lateral field with four lateral lines. Cephalic region conical-rounded to sometimes slightly trapezoid, submedian lobes not protruding under LM. Deirids visible under SEM. *En face* showing four well-separated rounded submedian lobes and four ridges around oral opening. Stylet 26–29 µm long, cone 61–67% of stylet length, knobs about 4 µm across. Pharynx about one-fourth of body length. Hemizonid just above secretory-excretory pore, about two body annuli long. Secretory-excretory pore between mid-isthmus and end bulb level. Spermatheca empty. Vulval flaps prominent, commonly rounded. Vulva located at 82–85% of body length from anterior end. Vagina oblique reaching up to half of body width. Tail 18–26 µm long, conoid with finely rounded to bluntly rounded terminus, sometimes dorsally sinuate.

*Molecular characterisation:* Seven D2-D3 of 28S (99% similarity; one out of 730 bp difference), four ITS, five 18S rRNA and eleven *COI* gene sequences were generated without intraspecific sequence variation.

*Remarks:* Males were not found. The female morphology and morphometrics is close to *P. projectus* and *Paratylenchus neoprojectus* Wu and Hawn, 1975. The cephalic region of females were seen with both rounded to trapezoid shape, secretory-excretory pore located between mid-isthmus to end bulb level, empty spermatheca and tail termini which fit both the above two species. However, the molecular data appears to be different from any available sequences including that of *P. projectus*. A comparative study of this species with type specimens of *P. neoprojectus* and its molecular information should further confirm whether or not this species is *P. neoprojectus*.

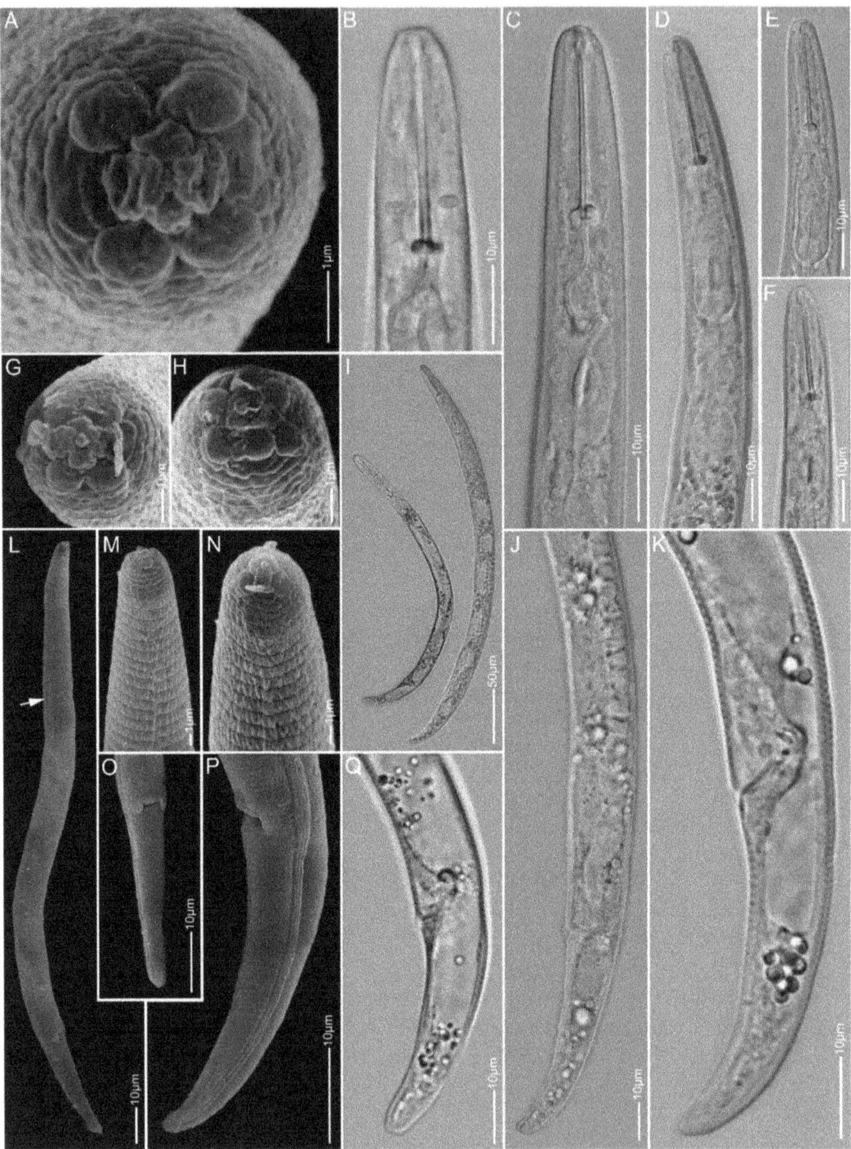

**Figure 14.** Light and scanning electron microscopy images of *Paratylenchus* sp.D females; (**A,G,H**) face view; (**B–F,M,N**) anterior region; (**I,L**) total body; (**J,K,O–Q**) tail region; arrow pointed to deirid in L.

### 2.1.14. *Paratylenchus* sp.F

*Females* (Sample BE22; Figure 15, Table 4): Heat relaxed specimen open C- to 6-shape. Lateral field with four lateral lines. Deirids present. Cephalic region conical-truncate, slightly rounded in few specimens, submedian lobes not protruding under LM. *En face* square-shaped, showing four rounded submedian lobes, four ridges around oral opening, dorso-ventral ridges much larger than the lateral ridges, slit like amphidial apertures laterally. Stylet 25–30 μm long, cone 65–70% of stylet length, knobs about 4 μm across. Pharynx about one-fourth of body length. Hemizonid just above secretory-excretory pore, about two body annuli long. Secretory-excretory pore between mid-isthmus and end bulb level. Spermatheca oval to elongated and filled with sperm cells. Vulval flaps rounded to oval and very prominent. Vulva located at 81–84% of body length from anterior end. Vagina oblique, reaching up to two-third of body width. Tail 20–26 µm long, conoid with regularly bluntly rounded terminus.

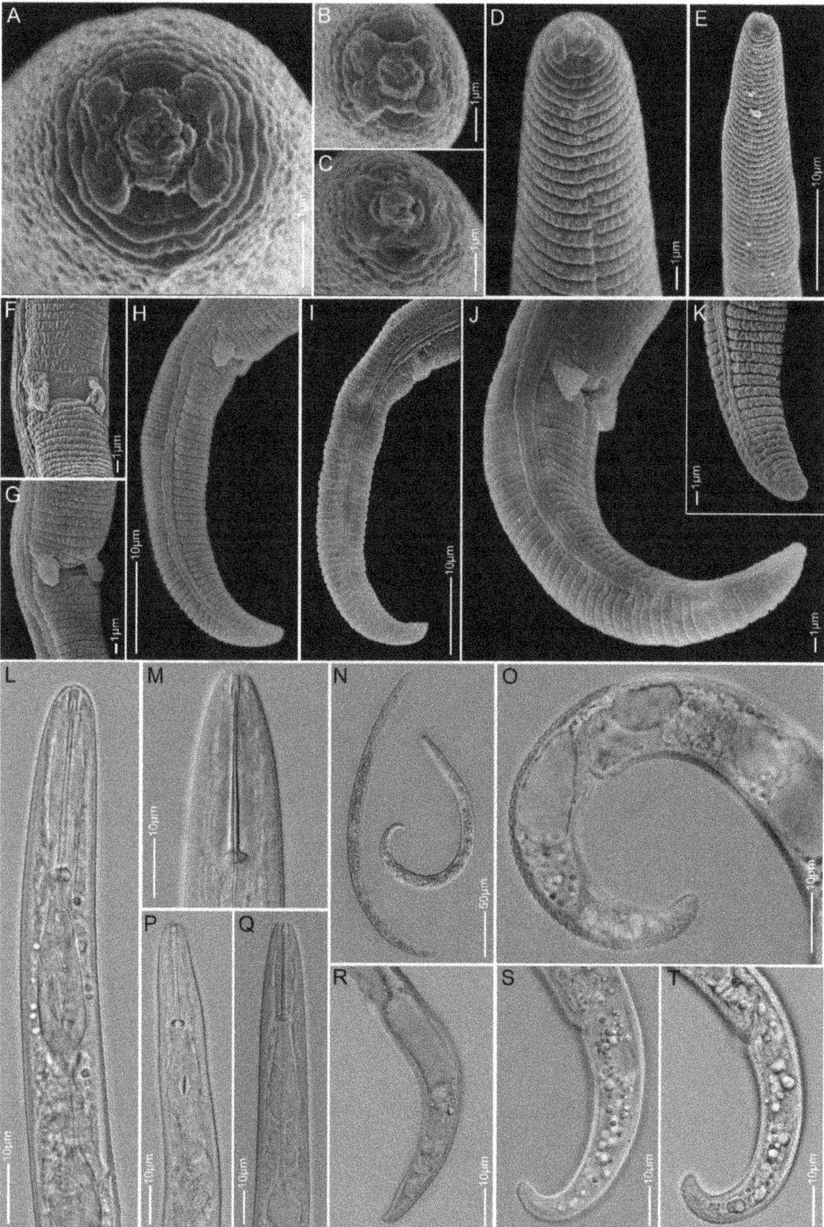

**Figure 15.** Light and scanning electron microscopy images of *Paratylenchus* sp.F females: (**A–C**) face view; (**D,E,L,M,P,Q**) anterior region; (**F,G**) vulva region; (**H–K,O,R–T**) tail region; (**N**) total body.

*Males:* Heat relaxed specimen curved slightly ventrally, about the same body length as females but slightly slender. Cephalic region conoid and rounded. Stylet and pharynx degenerated. Secretory-excretory pore at about one-fifth of body length from anterior end. Spicule arcuate ventrally, about 21.5 μm in length. Gubernaculum 3–5 μm long. Tail conical with finely rounded tip. Conspecificity of males with females was confirmed by identical D2-D3, 18S and ITS sequences.

*Molecular characterisation:* Five D2-D3 of 28S, three ITS, four 18S rRNA and three *COI* gene sequences were generated without intraspecific sequence variations. The D2-D3, 18S and *COI* sequences were found to be identical, respectively, to MN783707, MN783708, MN783668–MN783670 and MN782407–MN782413 of *Paratylenchus* sp.F [47], while the ITS sequences were generated for the first time.

*Remarks:* Specimens belong to the same population as *Paratylenchus* sp.F in Etongwe et al.'s work [47]. Detailed morphological reanalysis revealed very close similarity to *P. nanus*. Nevertheless, the submedian lobes of this species appear to be somewhat more rounded than that of *P. nanus* based on SEM images and the vulval flaps also appear to be more pronounced and rounded compared to that of the latter. However, these characteristics need careful additional observations based on more specimens from both species. All four gene sequences of *Paratylenchus* sp.F were closest to the sequences of *P. elachistus* and phylogenetic analysis revealed their highly supported (PP > 90%) sister relationship. However, this species is morphologically different from *P. elachistus en face*, with rounded vs. poorly differentiated submedian lobes, stylet lengths of 25–30 μm vs. 20–22 μm and bluntly rounded vs. spicate to pointed tail termini.

### 2.2. Phylogenetic and Species Delimitation Analysis

The D2-D3 domains of the 28S rRNA gene alignment (744 bp long) included 128 sequences of 31 *Paratylenchus* species and three outgroup species. Forty-nine new sequences were included in this analysis. The Bayesian 50% majority rule consensus tree inferred from the analysis of the D2-D3 alignment contained three highly supported major clades and a weakly supported one (Figure 16, PP < 70%). The molecular species delimitation based on the generalized mixed-yule coalescent (GMYC) and Poisson tree process (bPTP) methods revealed 66 and 63 putative species, respectively, a result that is largely congruent with former species delineations. However, *P. projectus*, *P. straeleni*, *P. minor* and *P. shenzhenensis* were further divided into 6, 5 (four according to bPTP), 2 and 2 separate lineages, respectively.

The ITS rRNA gene alignment (995 bp long) included 99 sequences of 37 *Paratylenchus* species and three outgroup species. Thirty-six new sequences were included in this analysis. The Bayesian 50% majority rule consensus tree inferred from the analysis of the ITS alignment contained four highly supported major clades (Figure 17). Results of molecular species delimitation showed a high discrepancy between the models used—i.e., 48 putative species based on GMYC vs. 56 species based on bPTP. Additionally, molecular species delimitation based on the GMYC and bPTP methods did not correspond to species demarcation based on morphology and clade support; for example, virtually all individual sequences of *P. chongqingensis* and *P. shenzhenensis* were delineated as separate species.

**Figure 16.** Phylogenetic relationships within populations and species of *Paratylenchus*, as inferred from Bayesian analysis using the D2-D3 of 28S rRNA gene sequence dataset with the GTR + I + G model. Posterior probability of more than 70% is given for the appropriate clades. Newly obtained sequences are indicated in bold. [1] = originally identified as *P. nanus*, [2] = originally identified as *P. bukowinensis*, [3] = originally identified as *Paratylenchus* sp., [4] = originally identified as *Paratylenchus* sp.8, [5] = originally identified as *Paratylenchus* sp.E, [6] = originally identified as *Gracilacus* sp. [7] = originally identified as *Paratylenchus* sp.5 and [8] = originally identified as *Paratylenchus* sp.6. Black and grey bars represent species boundaries estimated by generalized mixed-yule coalescent (GMYC) and Poisson tree process (bPTP) methods, respectively (only differences with GMYC provided).

**Figure 17.** Phylogenetic relationships within populations and species of *Paratylenchus* as inferred from Bayesian analysis using the ITS rRNA gene sequence dataset with the GTR + I + G model. Posterior probability more than 70% is given for appropriate clades. Newly obtained sequences are indicated in bold. [1] = originally identified as *P. nanus* and [2] = originally identified as *Paratylenchus* sp. Black and grey bars represent species boundaries estimated by GMYC and bPTP methods, respectively.

The 18S rRNA gene alignment (899 bp long) included 88 sequences of 31 *Paratylenchus* species and two outgroup species. Fifty-four new sequences were obtained for this study. The Bayesian 50% majority rule consensus tree inferred from the analysis of the partial 18S sequence alignment contained four highly supported major clades (Figure 18). Molecular species delimitation failed to delimit well established species—for example *P. goodeyi*, *P. veruculatus*, *P. nanus* and *P. neoamblycephalus* were identified as belonging to the same species. Furthermore, both models provided highly varied results (14 putative species according to GMYC vs. 26 according to bPTP), reducing the confidence in said results.

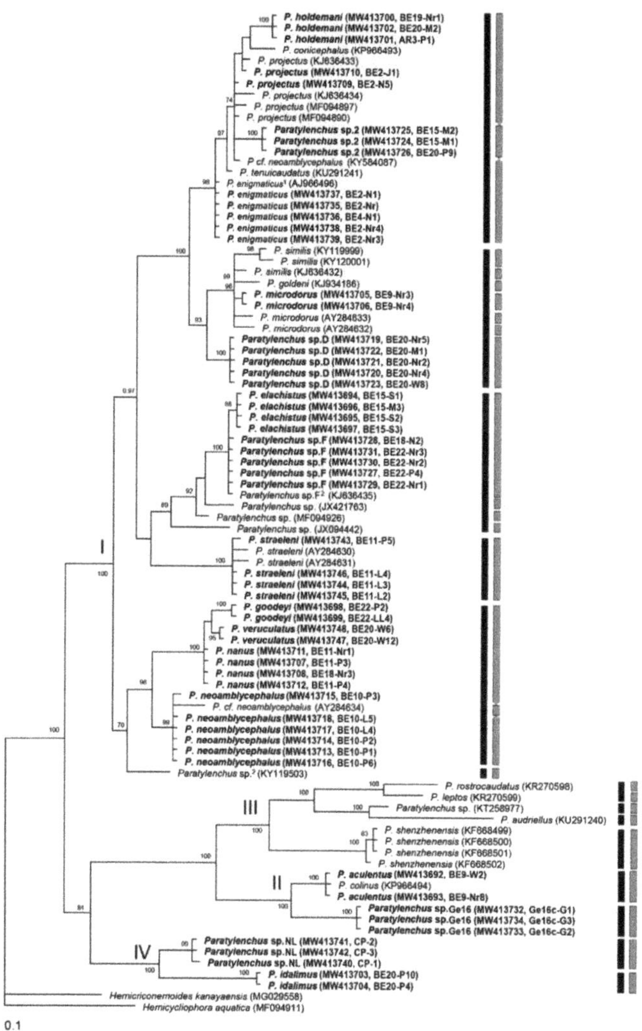

**Figure 18.** Phylogenetic relationships within populations and species of *Paratylenchus*, as inferred from Bayesian analysis using the 18S rRNA gene sequence dataset with the GTR + I + G model. Posterior probability more than 70% is given for appropriate clades. Newly obtained sequences are indicated in bold. [1] = originally identified as *P. dianthus* and [2] = originally identified as *P. nanus*. Black and grey bars represent species boundaries estimated by GMYC and bPTP methods, respectively.

The *COI* gene alignment (745 bp long) included 130 sequences of 31 *Paratylenchus* species and three outgroup species. Seventy-one new sequences were included in this analysis. The Bayesian 50% majority rule consensus tree inferred from the analysis of the *COI* sequence alignment contained four moderate (Figure 19, PP = 70–90%) or highly supported major clades. Both employed species delineation methods, GMYC and bPTP, provided exactly the same 54 putative species delineations. These results were largely consistent with those obtained using other methods. However, *P. enigmaticus*, *P. microdorus* and *P. veruculatus* were subdivided into different species despite these sequences originating from the same population and their corresponding D2-D3 sequences being similar.

*Paratylenchus straeleni* was appointed as a species complex with nine putative species. Statistical parsimony networks showing the phylogenetic relationships between different isolates of *P. straeleni* and *P. enigmaticus* based on *COI* sequences are given in Figure 19B,C. The maximum variation of sequences for *P. straeleni* was found to be 9.1%.

Taking both morphological and molecular evidence together, we have been able to reassign a total of 49 *Paratylenchus* sequences, including 18 D2-D3 of 28S, 3 ITS, 3 18S rRNA and 25 *COI* gene sequences, to their appropriate species (Table 5). However, we cannot exclude that in future, the identification of *Paratylenchus* species made in this study may be improved in light of new datasets.

**Table 5.** List of some existing unidentified or incorrectly classified *Paratylenchus* sequences on the GenBank reassigned to corrected species. In total, 18 D2-D3 of 28S, 3 ITS, 3 18S rRNA and 25 *COI* gene sequences have been reassigned.

| Gene | GenBank Accession No. | Linked Species | Country of Origin | Reference | Reassigned Species Name |
|---|---|---|---|---|---|
| D2-D3 | MN437514 | *Gracilacus* sp. | Myanmar | Du, Y. (Unpublished) | *P. sinensis* |
| D2-D3 | AY780943 | *P. bukowinensis* | Italy | Subbotin et al. [37] | *P. holdemani* |
| D2-D3 | MN088372 | *P. bukowinensis* | Iran | Mirbabaei et al. [46] | *P. holdemani* |
| D2-D3 | MN783703 | *P. bukowinensis* | Belgium | Etongwe et al. [47] | *P. holdemani* |
| D2-D3 | AY780944 | *Paratylenchus* sp. | Italy | Subbotin et al. [37] | *P. tenuicaudatus* |
| D2-D3 | MH156807 | *Paratylenchus* sp. | China | Fan et al. (Unpublished) | *P. lepidus* |
| D2-D3 | KF242223 | *Paratylenchus* sp.1 | USA | Van den Berg et al. [22] | *P. tenuicaudatus* |
| D2-D3 | KF242224 | *Paratylenchus* sp.1 | USA | Van den Berg et al. [22] | *P. tenuicaudatus* |
| D2-D3 | KF242225 | *Paratylenchus* sp.1 | USA | Van den Berg et al. [22] | *P. tenuicaudatus* |
| D2-D3 | KF242237 | *Paratylenchus* sp.5 | USA | Van den Berg et al. [22] | *P. idalimus* |
| D2-D3 | KF242238 | *Paratylenchus* sp.5 | USA | Van den Berg et al. [22] | *P. idalimus* |
| D2-D3 | KT258978 | *Paratylenchus* sp. | China | Liu et al. (Unpublished) | *P. minor* |
| D2-D3 | KF242189 | *Paratylenchus* sp.6 | USA | Van den Berg et al. [22] | *P. neoamblycephalus* |
| D2-D3 | KF242190 | *Paratylenchus* sp.6 | USA | Van den Berg et al. [22] | *P. neoamblycephalus* |
| D2-D3 | KF242233 | *Paratylenchus* sp.8 | USA | Van den Berg et al. [22] | *P. straeleni* |
| D2-D3 | KF242234 | *Paratylenchus* sp.8 | USA | Van den Berg et al. [22] | *P. straeleni* |
| D2-D3 | MN783711 | *Paratylenchus* sp.8 | Belgium | Etongwe et al. [47] | *P. straeleni* |
| D2-D3 | MN783712 | *Paratylenchus* sp.E | Belgium | Etongwe et al. [47] | *P. microdorus* |
| ITS | KT258979 | *Paratylenchus* sp. | China | Liu et al. (Unpublished) | *P. minor* |
| ITS | KF242260 | *Paratylenchus* sp.1 | USA | Van den Berg et al. [22] | *P. tenuicaudatus* |
| ITS | KF242259 | *Paratylenchus* sp.1 | USA | Van den Berg et al. [22] | *P. tenuicaudatus* |
| 18S | AJ966496 | *P. dianthus* | Belgium | Meldal et al. [65] | *P. enigmaticus* |
| 18S | KJ636435 | *P. nanus* | The Netherlands | Van Megen et al. (Unpublished) | *Paratylenchus* sp.F |
| 18S | KY119503 | *P. nanus* | Ireland | Ortiz et al. [66] | *Paratylenchus* sp. |
| *COI* | MF770960 | *Gracilacus* sp. | USA | Munawar et al. (Unpublished) | *P. straeleni* |
| *COI* | MN710983 | *Gracilacus* sp. | USA | Powers et al. [49] | *P. straeleni* |
| *COI* | MN710984 | *Gracilacus* sp. | USA | Powers et al. [49] | *P. straeleni* |
| *COI* | MN711354 | *Paratylenchus* sp. | USA | Powers et al. [49] | *P. straeleni* |
| *COI* | MN711355 | *Paratylenchus* sp. | USA | Powers et al. [49] | *P. hamatus* |
| *COI* | MN711356 | *Paratylenchus* sp. | USA | Powers et al. [49] | *P. hamatus* |
| *COI* | MN711357 | *Paratylenchus* sp. | USA | Powers et al. [49] | *P. hamatus* |
| *COI* | MN711358 | *Paratylenchus* sp. | Canada | Powers et al. [49] | *P. straeleni* |
| *COI* | MN711359 | *Paratylenchus* sp. | Canada | Powers et al. [49] | *P. straeleni* |
| *COI* | MN711360 | *Paratylenchus* sp. | Canada | Powers et al. [49] | *P. straeleni* |
| *COI* | MN711363 | *Paratylenchus* sp. | USA | Powers et al. [49] | *P. straeleni* |
| *COI* | MN711367 | *Paratylenchus* sp. | Ireland | Powers et al. [49] | *P. straeleni* |
| *COI* | MN711368 | *Paratylenchus* sp. | Ireland | Powers et al. [49] | *P. straeleni* |
| *COI* | MN711369 | *Paratylenchus* sp. | Ireland | Powers et al. [49] | *P. straeleni* |
| *COI* | MN711374 | *Paratylenchus* sp. | USA | Powers et al. [49] | *P. straeleni* |
| *COI* | MN711375 | *Paratylenchus* sp. | Canada | Powers et al. [49] | *P. straeleni* |
| *COI* | MN711376 | *Paratylenchus* sp. | Canada | Powers et al. [49] | *P. straeleni* |
| *COI* | MN711378 | *Paratylenchus* sp. | Poland | Powers et al. [49] | *P. holdemani* |
| *COI* | MN711380 | *Paratylenchus* sp. | Canada | Powers et al. [49] | *P. enigmaticus* |
| *COI* | MN711364 | *Paratylenchus* sp. | Ireland | Powers et al. [49] | *P. nanus* |
| *COI* | MN711365 | *Paratylenchus* sp. | Ireland | Powers et al. [49] | *P. nanus* |
| *COI* | MN782401 | *Paratylenchus* sp.8 | Belgium | Etongwe et al. [47] | *P. straeleni* |
| *COI* | MN782404 | *Paratylenchus* sp.B | Belgium | Etongwe et al. [47] | *P. holdemani* |
| *COI* | MN782405 | *Paratylenchus* sp.B | Belgium | Etongwe et al. [47] | *Paratylenchus* sp.D |
| *COI* | MN782406 | *Paratylenchus* sp.E | Belgium | Etongwe et al. [47] | *P. microdorus* |

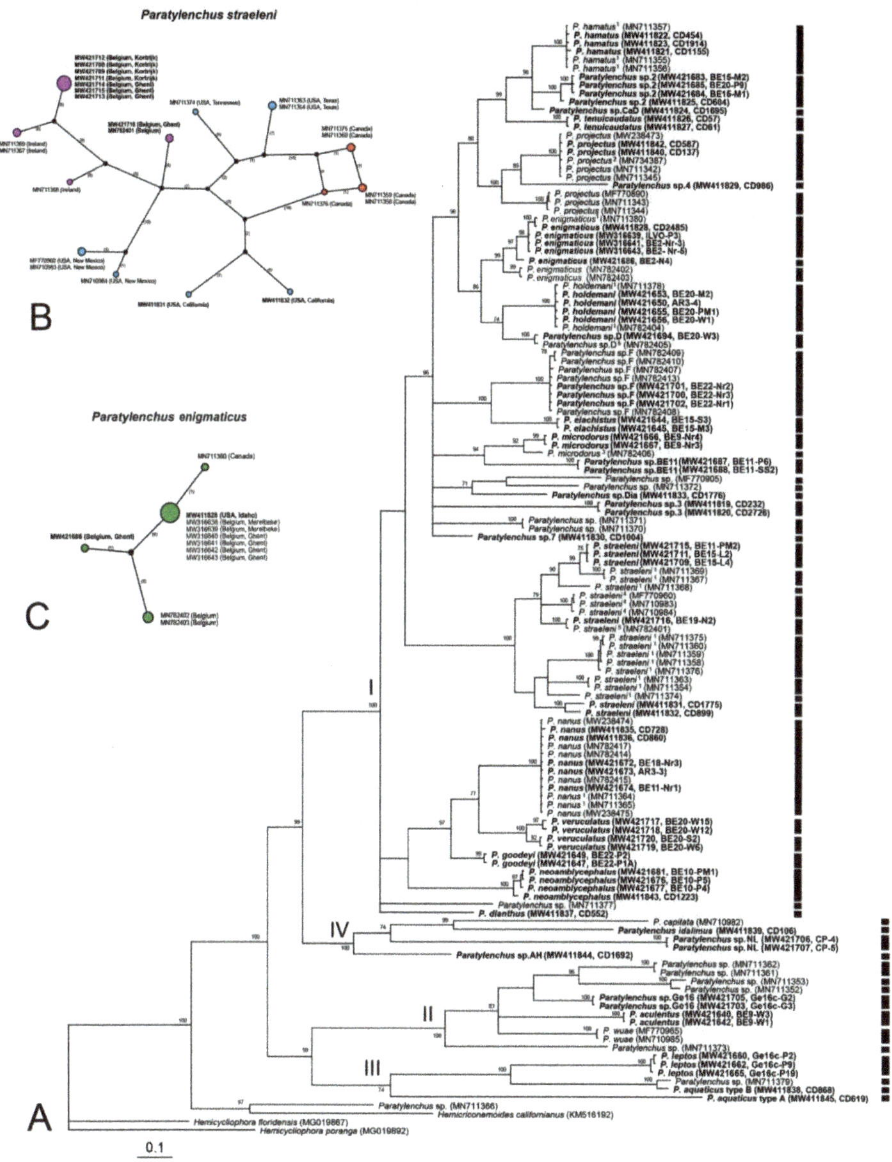

**Figure 19.** (**A**). Phylogenetic relationships within populations and species of *Paratylenchus*, as inferred from Bayesian analysis using the *COI* gene sequence dataset with the GTR + I + G model. Posterior probability more than 70% is given for appropriate clades. Newly obtained sequences are indicated in bold. [1] = originally identified as *Paratylenchus* sp., [2] = originally identified as *P. nanus*, [3] = identified as *Paratylenchus* sp.E, [4] = originally identified as *Gracilacus* sp., [5] = originally identified as *Paratylenchus* sp.8, [6] = originally identified as *Paratylenchus* sp.B; (**B**). Statistical parsimony network showing the phylogenetic relationships between *COI* haplotypes for *P. straeleni*; (**C**). Statistical parsimony network showing the phylogenetic relationships between *COI* haplotypes for *P. enigmaticus*. Pies (circles) represent the sequences with the same haplotype and their size is proportional to the number of these sequences in the samples. Numbers of nucleotide differences between the sequences are indicated on lines connecting the pies. Small black circles represent missing haplotypes. Bars represent species boundaries estimated by both GMYC and bPTP methods (identical results).

## 3. Discussion

The genus *Paratylenchus sensu lato*, with 124 valid species, is an important plant-parasitic group consisting of several commonly occurring and economically important species such as *P. bukowinensis*, *P. dianthus*, *P. hamatus*, *P. nanus*, *P. neoamblycephalus* and *P. projectus*, which are difficult to separate solely based on morphology [14,23,25,31–36]. Female morphological traits are the most commonly used features for the identification of *Paratylenchus* populations, with the relative lengths of stylet cones and the positions of the secretory-excretory pores and vulvae as the most informative traits [31,67], while several ratios such as a, c and c′ show high intraspecific variation. Given the limited species-specific female traits, some characteristics of males and juveniles—such as the presence or absence of stylet and male spicule length—may also be used to supplement the available data. However, care must be taken—for example, the occasional observance of a thin stylet in freshly killed juveniles or males that was invisible once the specimens were fixed highlights the importance of reporting this characteristic from both freshly killed and fixed specimens. Further complicating *Paratylenchus* taxonomy is the presence of mixture of species within one locality and sample [31]—an observation which calls for precaution concerning the conspecificity of several life stages. Indeed, the presence of multiple species in a soil sample was amply illustrated in our study. Seventy five percent of our investigated soil samples contained multiple species, with up to five different *Paratylenchus* species present in the same sample. This is, to the best of our knowledge, one of the highest numbers of species of one plant-parasitic nematode genus present in a single soil sample. More suitable morphological characters such as ridges around the oral opening or distinct to fused submedian lobes in face view also appear to be usefully informative but were only clearly revealed in our study with supporting evidence from SEM; additionally, the small vulval flaps in *P. aculentus* confirmed in this study have often been overlooked in previous studies under LM. Scanning electron microscopy is known to be important in nematode taxonomy [68–70], and this is especially true for the genus *Paratylenchus* as demonstrated in this study.

Nevertheless, even if all existing morphological tools are carefully employed, it remains impossible for all *Paratylenchus* species to be morphologically delineated, owing to the existence of cryptic species such as *P. aquaticus* [22]. The extensive use of new molecular data in the current study has demonstrated a remarkable molecular diversity in *Paratylenchus*, with several additional cryptic species being potentially present. The most obvious example is *P. straeleni*, which comprises 9, 5 and 4 putative species according to *COI*- (both GMYC and bPTP), D2-D3 (GMYC) and D2-D3 (bPTP)-based molecular species delimitation methods, respectively. It is noteworthy that the *P. straeleni COI* sequences have clearly clustered according to geographical location, as revealed by the COI haplotype network. The problems of morphologically delineating the *Paratylenchus* species have been further demonstrated in our study by the difficulties experienced in distinguishing between *Paratylenchus* sp.2, *Paratylenchus* sp.D and *Paratylenchus* sp.F, which were found to be very similar to *P. hamatus*, *P. projectus*/*P. neoprojectus* and *P. nanus*, respectively. A formal description with an appropriate diagnosis can only be developed for these putative new species following detailed observations of additional specimens and a thorough comparison with type materials of the known species.

Taken together, it is abundantly clear that molecular data are essential in advancing *Paratylenchus* taxonomy. Unfortunately, the several sequences published for *Paratylenchus* have serious limitations. One such issue is that the majority of the available D2-D3, 18S and ITS rRNA sequences have either not been linked to morphological data or have been associated with poor morphological data, thereby rendering them unreliable for use in identification purposes. For example, sequences of *P. aculentus*, *P. leptos*, *P. microdorus*, *P. neoamblycephalus*, etc., are not currently linked to reliable and clear morphological data and any subsequent identification based on these sequences may, therefore, lead to the deposition of further sequences under incorrect names [51]. An additional problem identified with the currently available 18S sequences, which are often relatively short (700–800 bp), is that several *Paratylenchus* species were detected with almost identical sequences. It is clear that to render these conserved sequences useful, complete or nearly complete lengths of the 18S rRNA gene (1600–1800 bp) will be required to allow species delimitation [71–74].

In the present study, we have also applied DNA-based species delimitation approaches to infer putative species boundaries on a given phylogenetic input, based on two different models [75,76] and four gene fragments (D2-D3 of 28S, ITS, 18S rRNA and *COI*). These coalescence-based species delimitation methods are rapidly gaining popularity in studies on closely related species that are difficult to distinguish based on phenotypic features, and have been applied to various eukaryotic groups [77,78]. However, despite plant-parasitic nematodes being a morphologically minimalistic group par excellence, such methods have been rarely applied to this group; nevertheless, they appear to be largely congruent with traditional methods [79–81]. Conversely, we have observed a remarkable discrepancy among the genes used, showing a poor link between DNA species delimitation and other methods, including a discrepancy between the employed models. The ITS and 18S rRNA genes gave, respectively, a likely overestimation and underestimation of the number of putative species, while for *COI* and D2-D3 of 28S rRNA genes, we observed, to a certain extent, an agreement with traditional methods, albeit with a likely overestimation of the number of species in several cases. This was not unexpected, as it has been exemplified by several studies that methods of species delimitation based on the coalescent model tend to overestimate phylogenetic lineages [52,77,82]. Both approaches (bPTP and GMYC) are similar in the fact that they identify significant changes in the pace of branching events on the tree. However, GMYC uses time to identify branching rate transition points, whereas, in contrast, bPTP directly uses the number of substitutions. Based on real and simulated data, both methods yield, in general, similar results [76,83]. This is the case for our COI-based output (identical results) and the D2-D3-based output (two differences, bPTP being more conserved). If differences have been observed, bPTP usually yields a more conservative delimitation than GMYC [76,80,83]. This is contrary to our unexpected ITS and 18S results and reduces the trust in the latter. Counterintuitively, the mutation rate of a chosen marker does not have a direct influence on its effectiveness to detect species. Mitochondrial markers reveal clearer discontinuities between interspecific divergence and intraspecific variation because of their faster coalescence within species lineages compared with nuclear loci, not necessarily because of their higher mutation rates [84,85]. The discrepancy between ITS and other delimitation methods in this study agrees with previous observations pointing to an unclear transition between species-level and population-level genetic distance for ITS [78]. Furthermore, it has been indicated that species delimitation based on single gene trees has serious limitations due to gene tree-species tree incongruence—confusions caused by processes including incomplete lineage sorting, trans-species polymorphism, hybridisation and introgression [78]. Multilocus approaches provide a posteriori double-check for contamination, sequencing errors or mitochondria-specific pitfalls [86]—for example, the high *COI* gene sequence variations within *P. enigmaticus*, *P. microdorus* and *P. veruculatus* observed in this study, despite these sequences originating from the same population. Although both nuclear and mitochondrial sequences were provided consistently from the same morphologically vouchered individuals, this study was restricted to the use of only single-locus data since only a limited number of other *Paratylenchus* individuals (and plant-parasitic nematodes in general) are linked to the same two genes. A further rigorous acquisition of both D2-D3 of 28S and *COI* gene sequences, which appear most promising for species delimitation in plant-parasitic nematodes (see [74]), will allow for more substantiated coalescence-based, multilocus species delimitation in plant-parasitic nematodes. Nevertheless, based on all obtained evidence, our findings support the proposition of Puillandre et al. [87], Padial et al. [88] and Qing et al. [80], that DNA-based species delimitation methods are important tools for the exploration of species delineation in diverse groups, but that identification of any new putative species will require further corroboration by an integrative taxonomic approach.

## 4. Materials and Methods

### 4.1. Nematode Populations

Nematode samples used in this study were collected from various localities (Table 1). Bulk soil samples of about 500 mL from 15–20 cm depths were collected from twelve locations in Belgium using a shovel. They were subsequently stored at 4 °C until nematode extraction. Nematodes were extracted from soil using a modified Baermann's method [89] or a rapid centrifugal flotation method [90]. Nematode extracts were observed under a stereo microscope. *Paratylenchus* populations were picked out in an embryo glass dish and stored in tap water at 4 °C for further analysis.

### 4.2. Morphological Study

Morphological study of nematodes was carried out using both heat relaxed and fixed specimens mounted on temporary and permanent slides, respectively. For preparation of a temporary mount of a nematode, a Cryo-Pro label (VWR International) was cut into two halves and stuck at the centre of a glass slide creating a small parallel gap between them. A single nematode was then transferred in a drop of distilled water to the glass slide in the centre of the gap. The nematode was then heat relaxed by passing over a flame a few times and covered with a glass coverslip. The specimen was then examined, photographed and measured using an Olympus BX51 DIC Microscope (Olympus Optical, Tokyo, Japan), equipped with an Olympus C5060Wz camera [91]. After recording morphological data, the specimen was recovered from the slide by adding a few drops of water from one end of the gap and collecting the nematode that was flushed out on the other end of the gap. The recovered specimens were subsequently used to extract genomic DNA as described in the next section.

A small nematode suspension of the remaining nematodes was heated in an embryo glass dish with a few drops of Trump's fixative ((2% paraformaldehyde, 2.5% glutaraldehyde in 0.1 M Sorenson buffer (sodium phosphate buffer at pH = 7.5)) in a microwave (700 Watts) for 3–4 sec and leaving it at room temperature for 1 h and at 4 °C for 24 h and followed by gradually transferring to anhydrous glycerine, as described in Singh et al. [92]. The fixed specimens were then mounted in glycerine on glass slides and were studied as above using the camera-equipped microscope. Species identification was carried out both at Nematology Research Unit of Ghent University and National Plant Protection Organization, Wageningen, the Netherlands.

For scanning electron microscopy, specimens fixed in Trump's fixative were washed in 0.1 M phosphate buffer (pH = 7.5) and dehydrated in a graded series of ethanol solutions, critical point-dried with liquid $CO_2$, mounted on stubs with carbon tabs (double conductive tapes), coated with gold of 25 nm, and photographed with a JSM-840 EM (JEOL) at 12 kV [92].

### 4.3. Extraction of DNA, PCR and Sequencing

Genomic DNA was extracted from individual heat relaxed nematode specimen, which had been morphologically vouchered. The cuticle of the specimen was punctured using a fine entomological pin mounted on a thin bamboo stick, which was also used as nematode picking tool and the nematode was subsequently transferred to a PCR tube with 20 µL of worm lysis buffer (50 mM KCl, 10 mM Tris at pH = 8.3, 2.5 mM $MgCl_2$, 0.45% NP 40 (Tergitol Sigma), 0.45% Tween 20) and incubated at −20 °C (at least 10 min). This was followed by adding 1 µL proteinase K (1.2 mg/mL), incubation at 65 °C (1 h) and 95 °C (10 min) and ending by centrifuging the mixture at 14,000 rpm for 1 min [92]. Genomic DNA from a single nematode was used to amplify four DNA fragments—D2-D3 of 28S, partial ITS and partial 18S rRNA gene and partial *COI* gene of mtDNA. PCR and sequencing were completed in two laboratories: Nematology Research Unit, Gent University, Belgium and Nematology lab, Plant Pest Diagnostic Center, CDFA, Sacramento, California, USA. For PCR amplifications of the D2-D3 of 28S, ITS and 18S rRNA gene sequences, the primer pairs D2A: 5′-ACA AGT ACC GTG AGG GAA AGT TG-3′/D3B: 5′-TCC TCG GAA GGA

ACC AGC TAC TA-3′ [93], Vrain2F: 5′-CTT TGT ACA CAC CGC CCG TCG CT-3′/Vrain2R: 5′-TTT CAC TCG CCG TTA CTA AGG GAA TC-3′ [94] or TW81: 5′-GTT TCC GTA GGT GAA CCT GC-3′/AB28: 5′-ATA TGC TTA AGT TCA GCG GGT-3′ [95], and SSU18A: 5′-AAA GAT TAA GCC ATG CAT G-3′/SSU26R: 5′-CAT TCT TGG CAA ATG CTT TCG-3′ [96] were used, respectively, with thermal profiles described by Singh et al. [97] and Tahna Maafi et al. [98]. Partial *COI* gene was amplified using the primer pairs JB3: 5′-TTT TTT GGG CAT CCT GAG GTT TAT-3′/JB4.5: 5′-TTT TTT GGG CAT CCT GAG GTT TAT-3′ according to Bowles et al. [99] or COI-F5: 5′-AAT WTW GGT GTT GGA ACT TCT TGA AC-3′/COI-R9: 5′-CTT AAA ACA TAA TGR AAA TGW GCW ACW ACA TAA TAA GTA TC-3′ according to Powers et al. [100]. The PCR products were purified [101] and sent to Macrogen [102] and Genewiz [103] for sequencing. New sequences were assembled using Geneious Prime 2020.0.5 and deposited to the GenBank under the accession numbers given in Table 1.

*4.4. Phylogenetic and Species Delimitation Analysis*

The new sequences for each gene (D2-D3 of 28S, ITS, 18S rRNA and *COI*) were aligned using Clustal X 1.83 [104] with their corresponding published gene sequences [22,24,32–39,42–50]. Outgroup taxa for each dataset were chosen based on previously published data [105]. Sequence datasets were analysed with Bayesian inference (BI) using MrBayes 3.1.2 [106] under the GTR + I + G model. BI analysis was initiated with a random starting tree and was run with four chains for $1.0 \times 10^6$ generations for 18S and ITS rRNA gene alignments, $5.0 \times 10^6$ generations for D2-D3 of 28S rRNA gene alignment and $9.0 \times 10^6$ generations for *COI* gene alignment. The Markov chains were sampled at intervals of 100 generations. Two runs were performed for each analysis. The log-likelihood values of the sample points stabilised after approximately 1,000 generations. After discarding burn-in samples and evaluating convergence, the remaining samples were retained for further analysis. The topologies were used to generate a 50% majority rule consensus tree. Posterior probabilities (PPs) are given on appropriate clades. Sequence analyses of alignments were performed with PAUP* 4b10 [107]. Pairwise divergences between taxa were computed as absolute distance values and as percentage mean distance values based on whole alignment with adjustments for missing data.

The *COI* gene alignments for *P. straeleni* and *P. enigmaticus* were used to construct phylogenetic network estimation using statistical parsimony, as implemented in POPART software [108].

Species delimitation of *Paratylenchus* in this study was undertaken using an integrated approach that considered morphological and morphometric evaluations combined with molecular-based phylogenetic inference (tree-based methods) and coalescent-based molecular species-delimitation methods. Putative species boundaries on a given phylogenetic input tree were inferred using a Bayesian implementation of the Poisson tree processes (bPTP) method [76] and using the generalized mixed-yule coalescent (GMYC) method [75]; see Qing et al. [80] for more details. Ultrametric trees were constructed using BEAST v1.10.4 [109] based on D2-D3, ITS, 18S and *COI* sequences, respectively. Default prior distributions were used and analyses were run for $1 \times 10^7$ generations, saving trees every $1 \times 10^3$ generations. The final trees were produced after removing 2,000 samples (20%) as burn-ins, and the maximum clade credibility tree was calculated using TreeAnnotator 1.10.4 [109]. Finally, for the bPTP method, an unrooted Bayesian 50% majority-rule consensus tree, containing only ingroups and unique haplotypes, was uploaded on the online server [110] and $1 \times 10^5$ Markov chain Monte Carlo (MCMC) generations were performed. The same tree was also uploaded on the GMYC web server [111] using the single threshold method.

## 5. Conclusions

An integrative approach by linking DNA sequences and morphological characters represents the best way to move nematode taxonomy forward. Creating this link involves the rigorous generation of multiple DNA sequences from individual morphologically vouchered nematode specimens, which, in the current study, resulted in the first molecular characterisations for five species, the first *COI* sequences for eight species and, most importantly, the reassignments of 18 D2-D3 of 28S, 3 ITS, 3 18S rRNA and 25 *COI* gene sequences, which had been unidentified or misidentified.

This study showed that *Paratylenchus* is a case in point, representing an incredibly diverse yet morphologically minimalistic plant-parasitic genus. Our recommendations for future protocol in *Paratylenchus* taxonomy, which are also valid for integrative nematode taxonomy, are: (1) to include SEM in new descriptions or re-descriptions; (2) to use juvenile and male traits after their conspecificity is irrefutably proven using molecular data; (3) to unequivocally link elaborate morphological data with both nuclear D2-D3 of 28S rRNA and mitochondrial *COI* gene sequences; (4) to employ caution when performing molecular identification using partial 18S rRNA gene fragments only; (5) to make use of the promising molecular species delineation methods to establish species boundaries, but base this on multilocus data and merely use it as one of the elements of integrative taxonomy.

**Author Contributions:** Conceptualisation, P.R.S., W.B. and S.A.S.; methodology, P.R.S., M.C., W.B. and S.A.S.; software, P.R.S., W.B. and S.A.S.; validation, P.R.S., G.K., W.B. and S.A.S.; formal analysis, P.R.S., G.K., W.B. and S.A.S.; investigation, P.R.S., G.K., W.B. and S.A.S.; resources, P.R.S., G.K., M.C., W.B. and S.A.S.; data curation, P.R.S., G.K. and S.A.S.; writing—original draft preparation, P.R.S.; writing—review and editing, W.B., S.A.S. and G.K.; visualisation, P.R.S. and M.C.; supervision, W.B. and G.K.; project administration, W.B.; funding acquisition, W.B. and S.A.S. All authors have read and agreed to the published version of the manuscript.

**Funding:** This research was funded by Ghent University, Belgium, UGent BOF01D05918 and the Plant Pest Diagnostic Center, CDFA, USA.

**Institutional Review Board Statement:** Not applicable.

**Informed Consent Statement:** Not applicable.

**Data Availability Statement:** The datasets generated during and/or analyzed during the currentstudy are available from the corresponding author on reasonable request.

**Acknowledgments:** The authors thank E. Van den Berg, A. Westphal, J. Zogfar, C.M. Etongwe and A.W. Aseffa for providing nematode specimens and soil samples.

**Conflicts of Interest:** The authors declare no conflict of interest.

# References

1. Micoletzky, H. *Die freilebenden Erd-Nematoden*; Archiv für Naturgeschichte: Berlin, Germany, 1922.
2. Siddiqi, M.R. *Tylenchida: Parasites of Plants and Insects*, 2nd ed.; CABI Publishing: Wallingford, UK, 2000.
3. Manzanilla-López, R.H.; Marbán-Mendoza, N. *Practical plant nematology*; BBA: Colegio de Postgraduados, Mexico, 2012.
4. Fourie, H.; Spaull, V.W.; Jones, R.K.; Daneel, M.S.; De Waele, D. *Nematology in South Africa: A View from the 21st Century*; Springer: Cham, Switzerland, 2017.
5. Subbotin, S.A.; Chitambar, J.J. *Plant Parasitic Nematodes in Sustainable Agriculture of North America*; Springer: Cham, Switzerland, 2018.
6. Tarjan, A.C. A review of the genus *Paratylenchus* Micoletzky, 1922 (Paratylenchinae: Nematoda) with a description of two new species. *Ann. N. Y. Acad. Sci.* **1960**, *84*, 329–390. [CrossRef]
7. Raski, D.J. Paratylenchidae n. fam. with descriptions of five new species of *Gracilacus* n. g. and an emendation of *Cacopaurus* Thorne, 1943, *Paratylenchus* Micoletzky, 1922 and Criconematidae Thorne, 1943. *Proc. Helm. Soc. Wash.* **1962**, *29*, 189–207.
8. Thorne, G.; Malek, R.B. *Nematodes of the Northern Great Plains part I. Tylenchida [Nemata Secernentra]*; South Dakota State University, Brookings: Sioux Falls, SD, USA, 1968.
9. Raski, D.J.; Luc, M. A reappraisal of Tylenchina (Nemata). 10: The superfamily Criconematoidea Taylor, 1936. *Rev. Nématol.* **1987**, *10*, 409–444.
10. Maggenti, A.R.; Luc, M.; Raski, D.J.; Fortuner, R.; Geraert, E. A reappraisal of Tylenchina (Nemata). 11. List of generic and supra-generic taxa, with their junior synonyms. *Rev. Nématol.* **1988**, *11*, 177–188.
11. Raski, D.J. Tylenchulidae of agricultural soils. In *Manual of Agricultural Nematology*; Nickle, W.R., Ed.; Marcel Dekker Inc.: New York, NY, USA, 1991; pp. 761–794.
12. Esser, R.P. A diagnostic compendium to species included in Paratylenchinae Thorne, 1949 and Tylenchocriconematinae Raski & Siddiqi, 1975 (Nematoda: Criconematoidea). *Nematologica* **1992**, *38*, 146–163.

13. Andrássy, I. *Free-living nematodes of Hungary, II (Nematoda, Errantia)*; Hungarian Natural History Museum: Budapest, Hungary, 2007.
14. Yu, Q.; Ye, W.; Powers, T. Morphological and Molecular Characterization of *Gracilacus wuae* n. sp. (Nematoda: Criconematoidea) Associated with Cow Parsnip (*Heracleum maximum*) in Ontario, Canada. *J. Nematol.* **2016**, *48*, 203–213. [CrossRef]
15. Siddiqi, M.R. *Tylenchida: Parasites of Plants and Insects*; Commonwealth Agricultural Bureaux: Slough, UK, 1986.
16. Brzeski, M.W.; Szczygiel, A. Studies On the Nematodes of the Genus *Paratylenchus* Micoletzky (Nematoda: Paratylenchinae) in Poland. *Nematologica* **1963**, *9*, 613–625. [CrossRef]
17. Siddiqi, M.R.; Goodey, J.B. The status of the genera and subfamilies of the Criconematidae (Nematoda); with a comment on the position of *Fergusobia*. *Nematologica* **1964**, *9*, 363–377. [CrossRef]
18. Geraert, E. The genus *Paratylenchus*. *Nematologica* **1965**, *11*, 301–334. [CrossRef]
19. Brzeski, M.W. *Nematodes of Tylenchina in Poland and temperate Europe*; Muzeum i Instytut Zoologii, Polska Akademia Nauk: Warszawa, Poland, 1998.
20. Nguyen, C.N.; Baldwin, J.G.; Choi, Y.E. New records of *Paratylenchus* Micoletzky, 1922 (Nematoda: Paratylenchinae) from Viet Nam with description of *Paratylenchus laocaiensis* sp. n. *J. Nematode Morphol. Syst.* **2004**, *7*, 51–75.
21. Decraemer, W.; Hunt, D.J. Structure and classification. In *Plant nematology*; Perry, R.N., Moens, M., Eds.; CABI Publishing: Wallingford, UK, 2006; pp. 3–32.
22. Van den Berg, E.; Tiedt, L.R.; Subbotin, S.A. Morphological and molecular characterisation of several *Paratylenchus* Micoletzky, 1922 (Tylenchida: Paratylenchidae) species from South Africa and USA, together with some taxonomic notes. *Nematology* **2014**, *16*, 323–358. [CrossRef]
23. Ghaderi, R.; Kashi, L.; Karegar, A. Contribution to the study of the genus *Paratylenchus* Micoletzky, 1922 *sensu lato* (Nematoda: Tylenchulidae). *Zootaxa* **2014**, *3841*, 151–187. [CrossRef]
24. Hesar, A.M.; Karegar, A.; Ghaderi, R. Phylogenetic relationships of *Cacopaurus pestis* Thorne, 1943 within representatives of the Tylenchulidae Skarbilovich, 1947 as inferred from ITS and D2-D3 expansion segments of 28S-rRNA sequences. *Nematology* **2019**, *21*, 971–994. [CrossRef]
25. Maria, M.; Miao, W.; Castillo, P.; Zheng, J. A new pin nematode, *Paratylenchus sinensis* n. sp. (Nematoda: Paratylenchinae) in the rhizosphere of white mulberry from Zhejiang Province, China. *Eur. J. Plant Pathol.* **2020**, *156*, 1023–1039. [CrossRef]
26. Raski, D.J. *Paratylenchoides* gen. n. and two new species (Nematoda: Paratylenchidae). *Proc. Helm. Soc. Wash.* **1973**, *40*, 230–233.
27. Ganguly, S.; Khan, E. *Gracilpaurus corbetti* gen. n., sp. n. (Nematoda: Tylenchida: Paratylenchidae) from India with a key to the species of this genus. *Pak. J. Nematol.* **1990**, *8*, 65–71.
28. Thorne, G. *Cacopaurus pestis*, n. g., n. spec. (Nematoda: Criconematinae), a destructive parasite of the walnut, *Juglans regia* Linn. *Proc. Helm. Soc. Wash.* **1943**, *10*, 78–83.
29. Goodey, T. *Soil and Freshwater Nematodes*, 2nd ed.; Goodey, J.B., Ed.; Methuen: London, UK, 1963; p. 544.
30. Ebsary, B.A. *Catalog of the Order Tylenchida (Nematoda)*; Agriculture Canada, Research Branch: Ottawa, ON, Canada, 1991.
31. Ghaderi, R.; Geraert, E.; Karegar, A. *The Tylenchulidae of the world: Identification of the family Tylenchulidae (Nematoda: Tylenchida)*; Academia Press: Ghent, Belgium, 2016.
32. Wang, K.; Li, Y.; Xie, H.; Xu, C.L.; Wu, W.J. Morphology and molecular analysis of *Paratylenchus guangzhouensis* n. sp. (Nematoda: Paratylenchinae) from the soil associated with *Bambusa multiplex* in China. *Eur. J. Plant Pathol.* **2016**, *145*, 255–264. [CrossRef]
33. Wang, K.; Xie, H.; Li, Y.; Wu, W.J.; Xu, C.L. Morphology and molecular analysis of *Paratylenchus nanjingensis* n. sp. (Nematoda: Paratylenchinae) from the rhizosphere soil of *Pinus massoniana* in China. *J. Helminthol.* **2016**, *90*, 166–173. [CrossRef]
34. Maria, M.; Cai, R.; Ye, W.; Powers, T.O.; Zheng, J. Description of *Gracilacus paralatescens* n. sp. (Nematoda: Paratylenchinae) found from the rhizosphere of Bamboo in Zhejiang, China. *J. Nematol.* **2018**, *50*, 611–622. [CrossRef]
35. Li, Y.; Wang, K.; Xie, H.; Xu, C.L. Morphology and molecular analysis of *Paratylenchus chongqingensis* n. sp. (Nematoda: Paratylenchinae) from soil associated with *Ophiopogon japonicus* in China. *Eur. J. Plant Pathol.* **2019**, *154*, 597–605. [CrossRef]
36. Phani, V.; Somvanshi, V.S.; Rao, U.; Khan, M.R. *Paratylenchus jasmineae* sp. n. (Nematoda: Paratylenchinae) from rhizosphere of *Jasminum sambac* in India. *Nematology* **2019**, *21*, 469–478. [CrossRef]
37. Subbotin, S.A.; Vovlas, N.; Crozzoli, R.; Sturhan, D.; Lamberti, F.; Moens, M.; Baldwin, J.G. Phylogeny of Criconematina Siddiqi, 1980 (Nematoda: Tylenchida) based on morphology and D2-D3 expansion segments of the 28SrRNA gene sequences with application of a secondary structure model. *Nematology* **2005**, *7*, 927–944.
38. Chen, D.Y.; Ni, H.F.; Tsay, T.T.; Yen, J.H. Identification of *Gracilacus bilineata* and *G. aculenta* (Nematoda: Criconematoidea, Tylenchulidae) among bamboo plantations in Taiwan. *Plant Pathol. Bull.* **2008**, *17*, 209–219.
39. Chen, D.Y.; Ni, H.F.; Yen, J.H.; Tsay, T.T. Identification of a new recorded pin nematode *Paratylenchus minutus* (Nematoda: Criconematoidea, Tylenchulidae) in Taiwan. *Plant Pathol. Bull.* **2009**, *18*, 167–174.
40. van Megen, H.; van den Elsen, S.; Holterman, M.; Karssen, G.; Mooyman, P.; Bongers, T.; Holovachov, O.; Bakker, L.; Helder, J. A phylogenetic tree of nematodes based on about 1200 full-length small subunit ribosomal DNA sequences. *Nematology* **2009**, *11*, 927–950. [CrossRef]
41. López, M.A.C.; Robbins, R.T.; Szalanski, A.L. Taxonomic and molecular identification of *Hemicaloosia, Hemicycliophora, Gracilacus* and *Paratylenchus* species (Nematoda: Criconematidae). *J. Nematol.* **2013**, *45*, 145–171.
42. Akyazi, F.; Felek, A.F.; Čermák, V.; Čudejková, M.; Foit, J.; Yildiz, S.; Háněl, L. Description of *Paratylenchus* (*Gracilacus*) *straeleni* (De Coninck, 1931) Oostenbrink, 1960 (Nematoda: Criconematoidea, Tylenchulidae) from hazelnut in Turkey and its comparison with other world populations. *Helminthologia* **2015**, *52*, 270–279. [CrossRef]

43. Esmaeili, M.; Heydari, R.; Castillo, P.; Bidhendi, M.Z.; Palomares-Rius, J.E. Molecular characterisation of two known species of *Paratylenchus* Micoletzky, 1922 from Iran with notes on the validity of *Paratylenchus audriellus* Brown, 1959. *Nematology* **2016**, *18*, 591–604. [CrossRef]

44. Zhuo, K.; Liu, X.; Tao, Y.; Wang, H.; Lin, B.; Liao, J. Morphological and molecular characterisation of three species of *Paratylenchus* Micoletzky, 1922 (Tylenchida: Paratylenchidae) from China, with a first description of the male *P. rostrocaudatus*. *Nematology* **2018**, *20*, 837–850. [CrossRef]

45. Maria, M.; Miao, W.; Ye, W.; Zheng, J. Updated description of *Paratylenchus lepidus* Raski 1975 and *P. minor* Sharma, Sharma and Khan, 1986 by integrating molecular and ultra-structural observations. *J. Nematol.* **2019**, *51*, 1–13. [CrossRef]

46. Mirbabaei, H.; Eskandari, A.; Ghaderi, R. On the synonymy of *Trophotylenchulus asoensis* and *T. okamotoi* with *T. arenarius*, and intra-generic structure of *Paratylenchus* (Nematoda: Tylenchulidae). *J. Nematol.* **2019**, *51*, 1–22. [CrossRef]

47. Etongwe, C.M.; Singh, P.R.; Bert, W.; Subbotin, S.A. Molecular characterisation of some plant-parasitic nematodes (Nematoda: Tylenchida) from Belgium. *Russ. J. Nematol.* **2020**, *28*, 1–28.

48. Mwamula, A.O.; Kabir, M.F.; Lee, G.; Choi, I.H.; Kim, Y.H.; Bae, E.J.; Lee, D.W. Morphological characterisation and molecular phylogeny of several species of Criconematina (Nematoda: Tylenchida) associated with turfgrass in Korea, as inferred from ribosomal and mitochondrial DNA. *Nematology* **2020**, *22*, 939–956. [CrossRef]

49. Powers, T.; Harris, T.S.; Higgins, R.S.; Mullin, P.G.; Powers, K.S. Nematode biodiversity assessments need vouchered databases: A BOLD reference library for plant-parasitic nematodes in the superfamily Criconematoidea. *Genome* **2020**, *5*, 1–10. [CrossRef] [PubMed]

50. Wang, W.H.; Chen, P.J. First Report of a Pin Nematode (*Paratylenchus dianthus*) on Chrysanthemum in Taiwan. *Plant Dis.* **2020**, *104*, 995. [CrossRef]

51. Janssen, T.; Karssen, G.; Couvreur, M.; Waeyenberge, L.; Bert, W. The pitfalls of molecular species identification: A case study within the genus *Pratylenchus* (Nematoda: Pratylenchidae). *Nematology* **2017**, *19*, 1179–1199. [CrossRef]

52. Palomares-Rius, J.E.; Cantalapiedra-Navarrete, C.; Castillo, P. Cryptic species in plant-parasitic nematodes. *Nematology* **2014**, *16*, 1105–1118. [CrossRef]

53. Bert, W.; Geraert, E. Nematode species of the order Tylenchida, new to the Belgian nematofauna with additional morphological data. *Belg. J. Zool.* **2000**, *130*, 47–57.

54. Brzeski, M.W. Paratylenchinae: Morphology of some known species and descriptions of *Gracilacus bilineata* sp. n. and *G. vera* sp. n. (Nematoda: Tylenchulidae). *Nematologica* **1995**, *41*, 535–565. [CrossRef]

55. Lišková, M.; Sasanelli, N.; D'Addabbo, T. Some notes on the occurrence of plant parasitic nematodes on fruit trees in Slovakia. *Plant Prot. Sci.* **2007**, *43*, 26–32. [CrossRef]

56. Steiner, G. Plant nematodes the grower should know. *Proc. Soil Sci. Soc.* **1949**, *2*, 37–39.

57. Oostenbrink, M. A note on *Paratylenchus* in the Netherlands with the description of *P. goodeyi* n. sp. (Nematoda, Criconematidae). *Tijdschr. Plantenziekten* **1953**, *59*, 207–216. [CrossRef]

58. Raski, D.J. Revision of the genus *Paratylenchus* Micoletzky, 1922 and descriptions of new species. Part II of three parts. *J. Nematol.* **1975**, *7*, 274–295. [PubMed]

59. Andrássy, I. Neue und wenig bekannte nematoden aus Jugoslawien. *Ann. Hist. Nat. Musei Natl. Hung.* **1959**, *51*, 259–275.

60. Cobb, N.A. Notes on *Paratylenchus*, a genus of nemas. *J. Wash. Acad. Sci.* **1923**, *13*, 254–257.

61. Subbotin, S.A.; Yan, G.; Kantor, M.; Handoo, Z. On the molecular identity of *Paratylenchus nanus* Cobb, 1923 (Nematoda: Tylenchida). *J. Nematol.* **2020**, accepted.

62. De Coninck, L. Sur trois espèces nouvelles de nématodes libres trouvés en Belgique. *Bul. Mus. Roy. D'hist. Nat. Belg.* **1931**, *7*, 1–5.

63. Wu, L.Y. *Paratylenchus veruculatus* n. sp. (Criconematidae: Nematoda) from Scotland. *Can. J. Zool.* **1962**, *40*, 773–775. [CrossRef]

64. Thorne, G.; Allen, M.W. *Paratylenchus hamatus* n. sp, and *Xiphinema index* n. sp., two nematodes associated with fig roots, with a note on *Paratylenchus* anceps Cobb. *Proc. Helm. Soc. Wash.* **1950**, *17*, 27–35.

65. Meldal, B.H.; Debenham, N.J.; De Ley, P.; De Ley, I.T.; Vanfleteren, J.R.; Vierstraete, A.R.; Bert, W.; Borgonie, G.; Moens, T.; Tyler, P.A.; et al. An improved molecular phylogeny of the Nematoda with special emphasis on marine taxa. *Mol. Phylogenetics Evol.* **2007**, *42*, 622–636. [CrossRef]

66. Ortiz, V.; Phelan, S.; Mullins, E. A temporal assessment of nematode community structure and diversity in the rhizosphere of cisgenic *Phytophthora infestans*-resistant potatoes. *BMC Ecol.* **2016**, *16*, 1–23. [CrossRef] [PubMed]

67. Háněl, L.; Brzeski, M. Paratylenchinae: Evaluation of diagnostic morpho-biometrical characters of females in the genus *Paratylenchus* Micoletzky, 1922 (Nematoda: Tylenchulidae). *Nematology* **2000**, *2*, 253–261. [CrossRef]

68. Hirschmann, H. Scanning electron microscopy as a tool in nematode taxonomy. In *Concepts in Nematode Systematics*; Stone, A.R., Platt, H.M., Kahlil, L.F., Eds.; Academic Press: New York, NY, USA, 1983; pp. 95–111.

69. Wergin, W.P. Scanning electron microscopic techniques and applications for use in nematology. In *Plant Parasitic Nematodes*; Zuckerman, B.M., Rhode, R.A., Eds.; Academic Press: New York, NY, USA, 1981; Volume 3, pp. 175–204.

70. Eisenback, J.D. Techniques for preparing nematodes for scanning electron microscopy. In *An Advanced Treatise on Meloidogyne*; Barker, K.R., Carter, C.C., Sasser, N.S., Eds.; Raleigh, State University Graphics: Raleigh, NC, USA, 1985; Volume 2, pp. 79–105.

71. Nassonova, E.; Smirnov, A.; Fahrni, J.; Pawlowski, J. Barcoding amoebae: Comparison of SSU, ITS and *COI* genes as tools for molecular identification of naked lobose amoebae. *Protist* **2010**, *161*, 102–115. [CrossRef]

72. Morise, H.; Miyazaki, E.; Yoshimitsu, S.; Eki, T. Profiling nematode communities in unmanaged flowerbed and agricultural field soils in Japan by DNA barcode sequencing. *PLoS ONE* **2012**, *7*, 1–10. [CrossRef] [PubMed]

73. Armenteros, M.; Rojas-Corzo, A.; Ruiz-Abierno, A.; Derycke, S.; Backeljau, T.; Decraemer, W. Systematics and DNA barcoding of free-living marine nematodes with emphasis on tropical desmodorids using nuclear SSU rDNA and mitochondrial COI sequences. *Nematology* **2014**, *16*, 979–989. [CrossRef]

74. Kiewnick, S.; Holterman, M.; van den Elsen, S.; van Megen, H.; Frey, J.E.; Helder, J. Comparison of two short DNA barcoding loci (COI and COII) and two longer ribosomal DNA genes (SSU & LSU rRNA) for specimen identification among quarantine root-knot nematodes (*Meloidogyne* spp.) and their close relatives. *Eur. J. Plant Pathol.* **2014**, *140*, 97–110.

75. Pons, J.; Barraclough, T.G.; Gomez-Zurita, J.; Cardoso, A.; Duran, D.P.; Hazell, S.; Kamoun, S.; Sumlin, W.D.; Vogler, A.P. Sequence-based species delimitation for the DNA taxonomy of undescribed insects. *Syst. Biol.* **2006**, *55*, 595–609. [CrossRef]

76. Zhang, J.; Kapli, P.; Pavlidis, P.; Stamatakis, A. A general species delimitation method with applications to phylogenetic placements. *Bioinformatics* **2013**, *29*, 2869–2876. [CrossRef] [PubMed]

77. Carstens, B.C.; Pelletier, T.A.; Reid, N.M.; Satler, J.D. How to fail at species delimitation. *Mol. Ecol.* **2013**, *22*, 4369–4383. [CrossRef]

78. Leliaert, F.; Verbruggen, H.; Vanormelingen, P.; Steen, F.; López-Bautista, J.M.; Zuccarello, G.C.; De Clerck, O. DNA-based species delimitation in algae. *Eur. J. Phycol.* **2014**, *49*, 179–196. [CrossRef]

79. Janssen, T.; Karssen, G.; Orlando, V.; Subbotin, S.; Bert, W. Molecular characterization and species delimiting of plant parasitic nematode of the genus *Pratylenchus* from the *Penetrans* group (Nematoda: Pratylenchidae). *Mol. Phylogenet. Evol.* **2017**, *117*, 30–48. [CrossRef]

80. Qing, X.; Bert, W.; Gamliel, A.; Bucki, P.; Duvrinin, S.; Alon, T.; Miyara, S.B. Phylogeography and molecular species delimitation of *Pratylenchus capsici* n. sp., a new root-lesion nematode in Israel on pepper (*Capsicum annuum*). *Phytopathology* **2018**, *109*, 847–858. [CrossRef]

81. Xu, X.; Qing, X.; Xie, J.L.; Yang, F.; Peng, Y.L.; Ji, H.L. Population structure and species delimitation of rice white tip nematode, *Aphelenchoides besseyi* (Nematoda: Aphelenchoididae), in China. *Plant Pathol.* **2020**, *69*, 159–167. [CrossRef]

82. Padial, J.M.; De la Riva, I. A paradigm shift in our view of species drives current trends in biological classification. *Biol. Rev.* **2020**, *23*, 1–21.

83. Prévot, V.; Jordaens, K.; Sonet, G.; Backeljau, T. Exploring Species Level Taxonomy and Species Delimitation Methods in the Facultatively Self-Fertilizing Land Snail Genus Rumina (Gastropoda: Pulmonata). *PLoS ONE* **2013**, *8*, e60736.

84. Zink, R.M.; Barrowclough, G.F. Mitochondrial DNA under siege in avian phylogeography. *Mol. Ecol.* **2008**, *17*, 2107–2121. [CrossRef]

85. Palumbi, S.R.; Cipriano, F.; Hare, M.P. Predicting nuclear gene coalescence from mitochondrial data: The three-times rule. *Evolution* **2001**, *55*, 859–868. [CrossRef]

86. Jörger, K.M.; Norenburg, J.L.; Wilson, N.G.; Schrödl, M. Barcoding against a paradox? Combined molecular species delineations reveal multiple cryptic lineages in elusive meiofaunal sea slugs. *BMC Evol. Biol.* **2012**, *12*, 1–18. [CrossRef] [PubMed]

87. Puillandre, N.; Modica, M.V.; Zhang, Y.; Sirovich, L.; Boisselier, M.C.; Cruaud, C.; Holford, M.; Samadi, S. Large-scale species delimitation method for hyperdiverse groups. *Mol. Ecol.* **2012**, *21*, 2671–2691. [CrossRef]

88. Padial, J.M.; Miralles, A.; De la Riva, I.; Vences, M. The integrative future of taxonomy. *Front. Zool.* **2010**, *7*, 1–16. [CrossRef]

89. Whitehead, A.G.; Hemming, J.R. A comparison of some quantitative methods of extracting small vermiform nematodes from soil. *Ann. Appl. Biol.* **1965**, *55*, 25–38. [CrossRef]

90. Jenkins, W.R.B. A rapid centrifugal-flotation technique for separating nematodes from soil. *Plant Dis. Rep.* **1964**, *48*, 692.

91. Available online: https://www.youtube.com/watch?v=qWITH1yB1U0&t=2s (accessed on 15 February 2021).

92. Singh, P.R.; Nyiragatare, A.; Janssen, T.; Couvreur, M.; Decraemer, W.; Bert, W. Morphological and molecular characterisation of *Pratylenchus rwandae* n. sp. (Tylenchida: Pratylenchidae) associated with maize in Rwanda. *Nematology* **2018**, *20*, 781–794. [CrossRef]

93. Nunn, G.B. Nematode Molecular Evolution. An Investigation of Evolutionary Patterns among Nematodes Based Upon DNA Sequences. Ph.D. Thesis, University of Nottingham, Nottingham, UK, 1992.

94. Vrain, T.C.; Wakarchuk, D.A.; Levesque, A.C.; Hamilton, R.I. Intraspecific rDNA restriction fragment length polymorphism in the *Xiphinema americanum* group. *Fundam. Appl. Nematol.* **1992**, *15*, 563–573.

95. Curran, J.; Driver, F.; Ballard, J.W.O.; Milner, R.J. Phylogeny of *Metarhizium*: Analysis of ribosomal DNA sequence data. *Mycol. Res.* **1994**, *98*, 547–552. [CrossRef]

96. Mayer, W.E.; Herrmann, M.; Sommer, R.J. Phylogeny of the nematode genus *Pristionchus* and implications for biodiversity, biogeography and the evolution of hermaphroditism. *BMC Evol. Biol.* **2007**, *7*, 1–13. [CrossRef] [PubMed]

97. Singh, P.R.; Couvreur, M.; Decraemer, W.; Bert, W. Survey of slug-parasitic nematodes in East and West Flanders, Belgium and description of *Angiostoma gandavensis* n. sp. (Nematoda: Angiostomidae) from arionid slugs. *J. Helminthol.* **2019**, *94*, 1–11. [CrossRef]

98. Tanha Maafi, Z.; Subbotin, S.A.; Moens, M. Molecular identification of cyst-forming nematodes (Heteroderidae) from Iran and a phylogeny based on the ITS sequences of rDNA. *Nematology* **2003**, *5*, 99–111. [CrossRef]

99. Bowles, J.; Blair, D.; McManus, D.P. Genetic variants within the genus *Echinococcus* identified by mitochondrial DNA sequencing. *Mol. Biochem. Parasitol.* **1992**, *54*, 165–173. [CrossRef]

100. Powers, T.O.; Bernard, E.C.; Harris, T.; Higgins, R.; Olson, M.; Lodema, M.; Mullin, P.; Sutton, L.; Powers, K.S. COI haplotype groups in *Mesocriconema* (Nematoda: Criconematidae) and their morphospecies associations. *Zootaxa* **2014**, *3827*, 101–146. [CrossRef]

101. Singh, P.R.; Karssen, G.; Couvreur, M.; Bert, W. Morphological and molecular characterization of *Heterodera dunensis* n. sp.(Nematoda: Heteroderidae) from Gran Canaria, Canary Islands. *J. Nematol.* **2020**, *52*, 1–14.

102. Available online: https://dna.macrogen.com (accessed on 30 December 2020).

103. Available online: https://www.genewiz.com (accessed on 15 December 2020).

104. Thompson, J.D.; Gibson, T.J.; Plewniak, F.; Jeanmougin, F.; Higgins, D.G. The CLUSTAL_X windows interface: Flexible strategies for multiple sequence alignment aided by quality analysis tools. *Nucleic Acids Res.* **1997**, *25*, 4876–4882. [CrossRef]

105. Subbotin, S.A.; Sturhan, D.; Adams, B.J.; Powers, T.O.; Mullin, P.G.; Chizhov, V.N.; Vovlas, N.; Baldwin, J.G. Molecular phylogeny of the order tylenchida: Analysis of nuclear ribosomal RNA genes. *J. Nematol.* **2006**, *38*, 296.

106. Huelsenbeck, J.P.; Ronquist, F. MRBAYES: Bayesian inference of phylogenetic trees. *Bioinformatics* **2001**, *17*, 754–755. [CrossRef] [PubMed]

107. Swofford, D.L.; Sullivan, J. Phylogeny inference based on parsimony and other methods using PAUP* Practice. In *The Phylogenetic Handbook, a Practical Approach to DNA and Protein Phylogeny*; Salemi, M., Vandamme, A.M., Eds.; Cambridge University Press: Cambridge, UK, 2003; pp. 182–206.

108. Bandelt, H.J.; Forster, P.; Röhl, A. Median-joining networks for inferring intraspecific phylogenies. *Mol. Biol. Evol.* **1999**, *16*, 37–48. [CrossRef]

109. Drummond, A.J.; Suchard, M.A.; Xie, D.; Rambaut, A. Bayesian phylogenetics with BEAUti and the BEAST 1.7. *Mol. Biol. Evol.* **2012**, *29*, 1969–1973. [CrossRef] [PubMed]

110. Available online: https://species.h-its.org (accessed on 10 December 2020).

111. Available online: https://species.h-its.org/gmyc (accessed on 10 December 2020).

*Article*

# Sensitive, Accurate and Rapid Detection of the Northern Root-Knot Nematode, *Meloidogyne hapla*, Using Recombinase Polymerase Amplification Assays

Sergei A. Subbotin * and Julie Burbridge

Plant Pest Diagnostic Center, California Department of Food and Agriculture, 3294 Meadowview Road, Sacramento, CA 95832, USA; burbridge.julie@gmail.com
* Correspondence: sergei.subbotin@cdfa.ca.gov

**Abstract:** Rapid and reliable diagnostics of root-knot nematodes are critical for selections of effective control against these agricultural pests. In this study, recombinase polymerase amplification (RPA) assays were developed targeting the IGS rRNA gene of the northern root-knot nematode, *Meloidogyne hapla*. The RPA assays using TwistAmp® Basic, TwistAmp® exo and TwistAmp® nfo kits (TwistDx, Cambridge, UK) allowed for the detection of *M. hapla* from crude extracts of females, eggs and juveniles without a DNA extraction step. The results of the RPA assays using real-time fluorescence detection (real-time RPA) in series of crude nematode extracts showed reliable detection after 13 min with a sensitivity of 1/100 of a second-stage juvenile and up to 1/1000 of a female in reaction tubes. The results of the RPA assays using lateral flow dipsticks (LF-RPA) showed reliable detection within 30 min with a sensitivity of 1/10 of a second-stage juvenile and 1/1000 of a female in reaction tubes. The RPA assay developed here is a successful tool for quick, accurate and sensitive diagnostics of *M. hapla*. The application of the LF-RPA assay has great potential for diagnosing infestation of this species in the lab, field or in areas with a minimal laboratory infrastructure.

**Keywords:** diagnostics; root-knot nematode; recombinase polymerase amplification

**Citation:** Subbotin, S.A.; Burbridge, J. Sensitive, Accurate and Rapid Detection of the Northern Root-Knot Nematode, *Meloidogyne hapla*, Using Recombinase Polymerase Amplification Assays. *Plants* **2021**, *10*, 336. https://doi.org/10.3390/plants10020336

Academic Editor: Zafar Handoo
Received: 15 January 2021
Accepted: 8 February 2021
Published: 10 February 2021

**Publisher's Note:** MDPI stays neutral with regard to jurisdictional claims in published maps and institutional affiliations.



## 1. Introduction

The northern root-knot nematode, *Meloidogyne hapla* is one of the four most common root-knot nematode species worldwide. This nematode is extremely polyphagous, attacking a wide variety of crops and weeds. *Meloidogyne hapla* causes important economic losses for several horticultural, vegetable and pasture crops, including carrots, lettuce, lucerne, onion, potato, rose, sugarbeet, strawberry, white clover and others [1,2].

Accurate and rapid identification of nematodes is essential for their control. It has been shown that sequences of nuclear ribosomal genes: 18S rRNA, ITS rRNA, the D2–D3 of 28S rRNA, IGS rRNA and mitochondrial genes: *COII*-16S rRNA fragment, *COI* and *COII* clearly differentiate *M. hapla* from all other root-knot nematodes [3]. Several specific primers have been designed for the diagnostics of this species using conventional PCR [4–7]. Several authors also developed a TaqMan real-time PCR assay with species-specific primers for the detection of *M. hapla* from root galls and soil samples [8–12]. Recently, Peng et al. [13] developed loop-mediated isothermal amplification methods (LAMP) combined with a Flinders Technology Associates card for the identification of *M. hapla*.

Recombinase polymerase amplification (RPA), an isothermal in vitro nucleic acid amplification technique, has recently appeared as a novel molecular technology for simple, robust, rapid, reliable, and low-resource diagnostics. RPA represents a hugely versatile alternative to PCR [14–16]. RPA uses a highly efficient displacement polymerase that amplifies a few copies of target nucleic acid in 20 min at a constant temperature (37–42 °C). It does so by utilizing three core enzymes: recombinase, single-stranded binding protein (SSB), and strand-displacing polymerase. The recombinase enzyme forms a complex with a primer to facilitate their binding to the targeted DNA template. Then, the SSB binds to the displaced strands of DNA and prevents the displacement of the recombinase–primer complex by branch migration. The strand-displacing polymerase then recognises the bound recombinase–primer complex and initiates DNA synthesis. Like PCR, RPA produces an amplicon constrained in size to the binding sites of the primers. The advantages of RPA include highly efficient and rapid amplification and a low constant operating temperature. RPA products can be detected by agarose gel electrophoresis or carried out by using fluorescent probes in real time (real-time RPA) or lateral flow strips (LF-RPA). RPA assays show high sensitivity and specificity for detecting various plant viruses, bacteria, fungi, vertebrate parasitic trematodes, nematodes and other organisms [17–22]. Real-time RPA detection assay of plant parasitic nematodes was first designed and published by Subbotin et al. [23] for *Meloidogyne enterolobii*. RPA assays were also developed for *Meloidogyne javanica*, *M. arenaria* and *M. incognita* [24,25] and *Bursaphelenchus xylophilus* [26,27]. Recently, Song et al. [28] described diagnostics of *Meloidogyne hapla* using RPA combined with a lateral flow dipstick assay, where species-specific primers and a probe were designed based on the effector gene *16D10* sequence. This LF-RPA assay allows detecting *M. hapla* from infested plant roots and soil samples and the entire detection process can be completed within 1.5 h.

In our study, we developed real-time RPA and LF-RPA assays for the detection of *Meloidogyne hapla* using crude nematode and infected plant root extracts, with results within 13–30 min. Species-specific primers and probes were designed based on the IGS ribosomal RNA gene sequence.

## 2. Results

### 2.1. RPA Primers and Probe Design

All available sequences of the IGS rRNA for *M. hapla* and other *Meloidogyne* were downloaded from the Genbank and aligned with ClustalX. Several regions with high sequence dissimilarity between *M. hapla* and other *Meloidogyne* were assessed and several species-specific *M. hapla* candidate primers sets and probes were manually designed. The Blastn search of these species-specific candidate primer sequences and probe sequences showed high similarity (100%) only with the IGS rRNA fragments of *M. hapla* deposited in the GenBank.

### 2.2. RPA Detection

Nine primer combination candidate sets were screened for the best performance under the same RPA conditions. The species-specific forward F3-IGS-Hapl and the species-specific reverse R3-IGS-Hapl primers were found to be optimal with clearly visible bands and had no cross-reactions with other root-knot nematodes (Table 1). The final sequences of primers and probes used for the assays are listed in Table 2 and are indicated in the IGS rRNA gene alignment in Figure 1. This primer set reliably and specifically amplified the target gene fragment, approximately 164 bp in length from the IGS region (Figure 2) and was also confirmed by a direct sequencing of the product. Additional non-specific weak additional bands having other sizes were observed sometimes in experiments with *M. hapla* as well as samples with other root-knot nematode species (data not shown).

**Table 1.** Samples of *Meloidogyne hapla* and other root-knot nematodes tested in the present study.

| Species | Location | Plant | Sample Code | Source |
|---|---|---|---|---|
| *M. hapla* | USA, California | Tomato | VW9 | V. Williamson |
| *M. hapla* | USA, California | Tomato | C44 | V. Williamson |
| *M. hapla* | Moldova, Tiraspol | Sweet pepper | CD3384 | V. N. Chizhov |
| *M. hapla* | USA, Balm, Florida | Strawberry | CD2461 | R.N. Inserra |
| *M. hapla* | USA, Michigan, Van Buren County | Grapevine | CD3385e | S. Álvarez-Ortega |
| *M. arenaria* | USA, Florida | Unknown | CD3093 | J.A. Brito |
| *M. arenaria* | USA, Florida | Unknown | CD3100 | J.A. Brito |
| *M. baetica* | Spain | Olive | CD3382 | P. Castillo |
| *M. christiei* | USA Florida, | Turkey oak | CD1471 | J.A. Brito |
| *M. enterelobii* | USA, UCR collection | Tomato | CD3386 | P. Roberts |
| *M. floridensis* | USA, California, Kern county | Grapevine | CD3324 | S.A. Subbotin |
| *M. incognita* | USA, Florida | Tomato | CD3038 | J.A. Brito |
| *M. javanica* | USA, Florida | Tomato | CD3050 | J.A. Brito |
| *M. javanica* | USA, UCR collection | Tomato | Isolate 40 | P. Roberts |
| *M. naasi* | Germany | Grasses | CD3381 | D. Sturhan |
| *M. naasi* | USA, California | Grasses | CD2158 | S.A. Subbotin |
| *M. nataliei* | USA, Michigan, Van Buren County | Grapevine | CD3385a, b, c | S. Álvarez-Ortega |
| *Meloidogyne* sp.1 | Germany | Grasses | CD3380 | D. Sturhan |
| *Meloidogyne* sp.2 | Russia | Unknown | CD3383 | V. N. Chizhov |

**Table 2.** RPA primers and probe for amplification of *Meloidogyne hapla* DNA.

| Primer or Probe | Sequence (5′–3′) |
|---|---|
| F3-IGS-Hapl | TGC CAG TAC TCT GTT AGA AGT TGG TGA AGT GAT |
| R3-IGS-Hapl | GAA AAA TCC CCT CGA AAA ATC CAC CAT TTT AAT CCC T |
| R3-IGS-Hapl-biotin | [Biotin] GAA AAA TCC CCT CGA AAA ATC CAC CAT TTT AAT CCC T |
| Probe-hapla-exo1 | T GTC TTG TGC AAA GGA GAT TAT AAT TTG CTG GCT [FAM-dT] GT [THF] AT [BHQ1-dT] TTA ATC TTT AAT CAT ATT[C3-spacer] * |
| Probe-hapla-nfo1 | [FAM] T GTC TTG TGC AAA GGA GAT TAT AAT TTG CTG GCT TGT [THF] ATT TTA ATC TTT AAT CAT ATT[C3-spacer] * |

* FAM—fluorophore, THF—tetrahydrofuran, BHQ—quencher, C3—spacer block.

*2.3. Real-Time RPA Detection Assays*

Using the results of nine experimental runs, which included positive and negative controls with water and non-target DNA, the threshold level for reliable *M. hapla* detection was established as equal to 8 cycles (~3 min) with a baseline of 250,000 ($\Delta$Rn) fluorescence using the TwistAmp® exo kit on the Applied Biosystems™ QuantStudio™ 7 Flex Real-Time PCR System (Figures 3 and 4A). Samples that produced an exponential amplification curve above the threshold were considered as positive for *M. hapla* and below the threshold were considered as negative. Detection of *M. hapla* was confirmed with all samples.

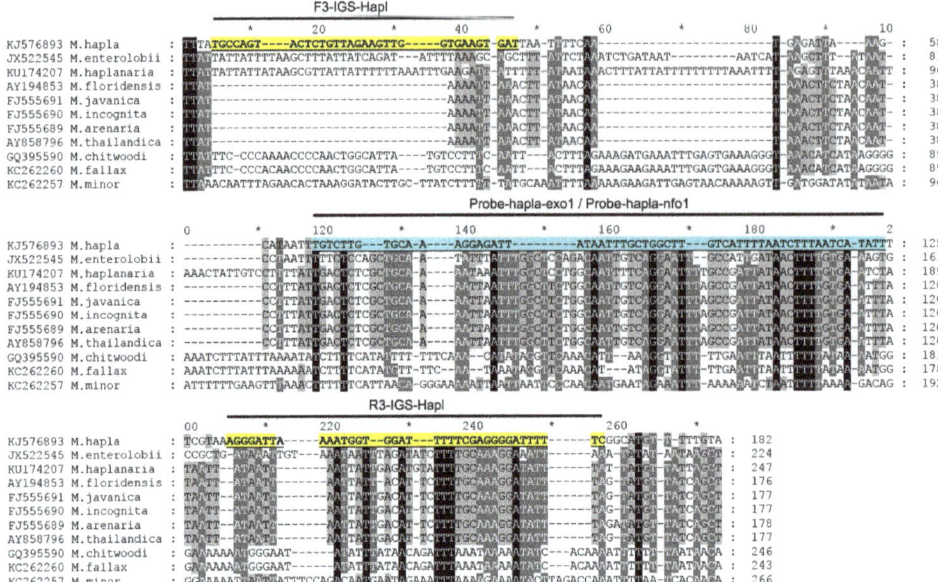

**Figure 1.** The fragment of alignment of the IGS rRNA gene sequences for several root-knot nematodes, *Meloidogyne*, with the positions of recombinase polymerase amplification (RPA) primers and probes used in the present study.

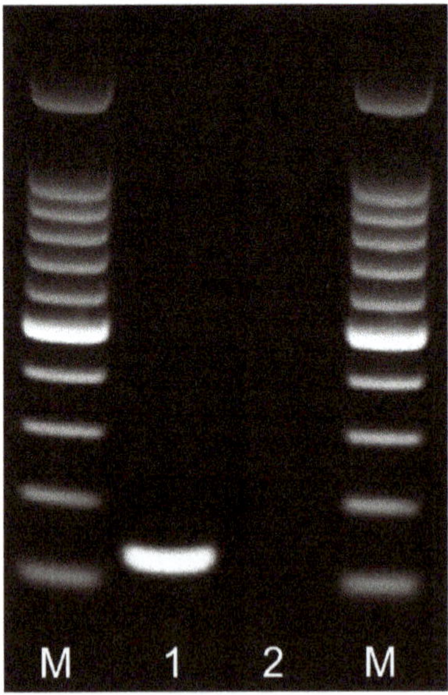


**Figure 2.** RPA amplicon of the partial IGS rRNA gene on agarose gel. Lanes: M: 100 bp DNA marker (Promega, Madison, WI, USA); 1: RPA amplicon obtained after 24 min at 39 °C with F3-IGS-Hapl and R3-IGS-Hapl primers using TwistAmp® Basic kit; 2: negative control.

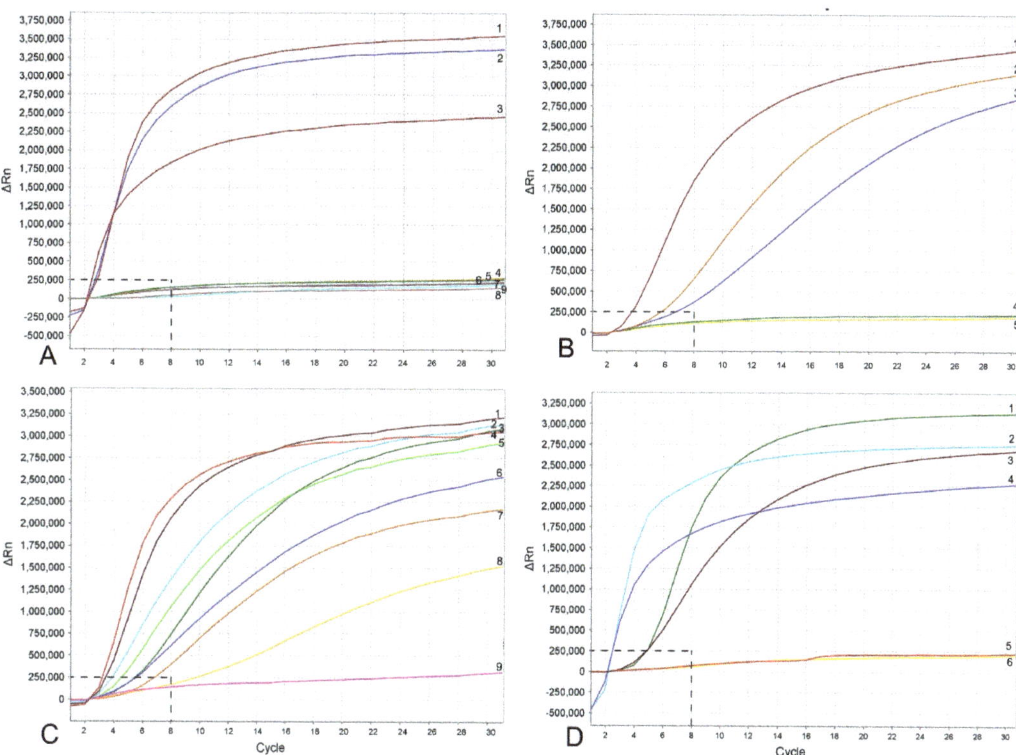

**Figure 3.** RPA assays using real-time fluorescent detection with examples of amplification plots. (**A**) Specificity assay with DNA samples of *Meloidogyne* spp. and crude second-stage juvenile (J2) extracts of *M. hapla*. Line: 1: *M. hapla* (CD2461); 2: *M. hapla* (VW9); 3: *M. hapla* (C44); 4: *M. incognita* (CD3038); 5: *M. arenaria* (CD3100); 6: *Meloidogyne naasi* (CD3381); 7: *M. javanica* (isolate 40); 8 and 9: negative control; (**B**) sensitivity assay with a dilution series of a crude J2 extract of *M. hapla*, line: 1: 1 J2 per tube; 2: 1/10 J2 per tube; 3: 1/100 J2 per tube; 4: 1/1000 J2 per tube; 5: negative control; (**C**) crude extract of *M. hapla* with or without crude extracts of non-target nematodes. Line: 1 and 4: 1 J2 per tube; 2, 5L 1 J2 with 20 non-target nematodes per tube; 3 and 6: 1 J2 with 10 non-target nematodes per tube; 7 and 8: 1 J2 per tube containing half of a reaction mixture; 9: negative control; (**D**) testing of crude extracts of *M. hapla*. Line: 1: 1 J2 per tube; 2 and 4: extracts from infected plant roots containing females with egg-masses; 3 and 6: extracts from infected plant roots containing old females without egg-masses; 5: negative control. The vertical line on a graph: fluorescence ΔRn. ΔRn is calculated at each cycle as DRn (cycle) = Rn (cycle)—Rn (baseline), where Rn = normalized reporter. The horizontal line on a graph: cycles, each cycle = 20 s.

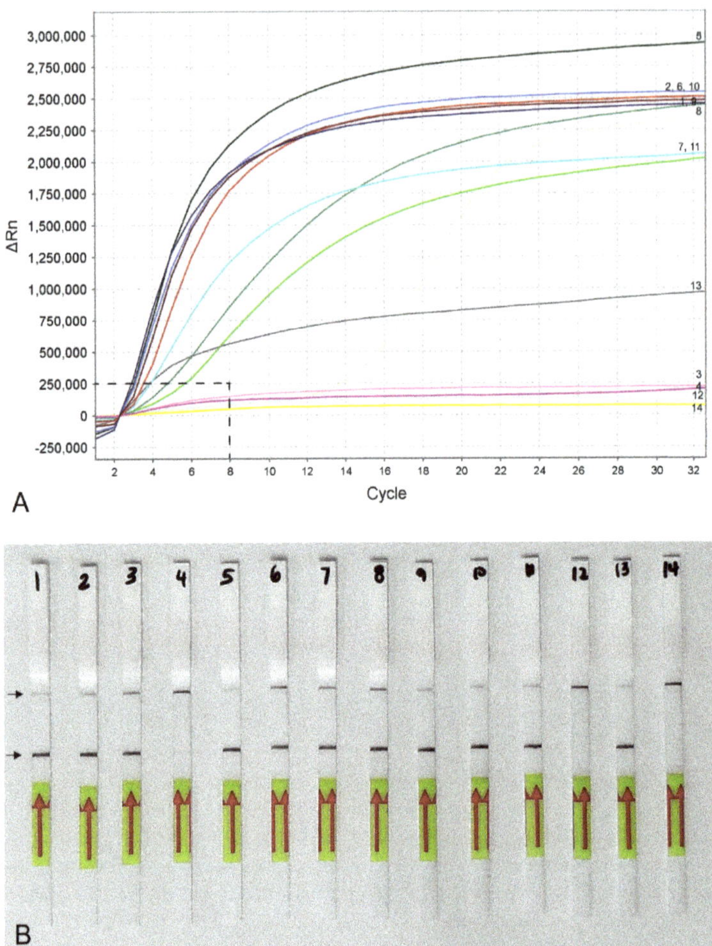

**Figure 4.** RPA sensitivity assays using (**A**) real-time fluorescent detection and (**B**) lateral flow strips. A dilution series of three crude young females (without egg-masses) extracts of *M. hapla*. Line (Strip): 1, 5, 9: 1/10 female per tube; 2, 6, 10: 1/100 female per tube; 3, 7, 11: 1/1000 female per tube; 4, 8, 12: 1/10,000 female per tube; 13: positive control; 14: negative control. Control (upper) and test (lower) lines are indicated by arrows.

The RPA assay was tested for specificity using DNA extracted from several root-knot nematodes. These nematodes include: *Meloidogyne arenaria*, *M. baetica*, *M. christiei*, *M. enterelobii*, *M. floridensis*, *M. incognita*, *M. javanica*, *M. naasi* and *M. nataliei*. The RPA results using real-time fluorescent detection showed high specificity to *M. hapla* only and no cross-reactions were observed against other root-knot nematode species (Figure 3A).

The sensitivity assay was designed for evaluation of the detection limit. Variants with serial dilutions (1, 1/10, 1/100, 1/1000 and 1/10,000 per reaction tube) of crude nematode extractions were obtained from second-stage juveniles (J2s) or females without egg-masses. The reliable detection level of *M. hapla* was estimated at 1/100 of one J2 per a RPA reaction tube (Figure 3B). The detection level of *M. hapla* females varied among replicates and reached 1/100, 1/1000 and 1/10,000 of a female for three, two and one replicates, respectively (Figure 4A).

The detection of J2 for *M. hapla* was confirmed in the presence of background crude extracts from at least 20 non-target nematodes. No decrease in fluorescent signals was observed between the variant of 1 J2 without other nematodes and the variants with 1 J2 with 10 and 20 non-target nematodes (Figure 3C). Lowering in half, a single reaction assay volume showed a decrease in fluorescence signal and reaction rate (Figure 3C). These samples could be considered as positive with threshold level of 12 cycles (~6 min).

*Meloidogyne hapla* detection was also confirmed using extracts obtained from infected tomato and pepper plant roots containing females with egg-masses. Although most replicates from extracts obtained from infected plant roots containing old females without egg-masses gave strong signals, one replicate showed no fluorescence signal (Figure 3D).

### 2.4. LF-RPA Assay

Lateral flow detection of RPA products also showed specific and sensitive results. Positive test lines on the LF strips were observed for all *M. hapla* samples, whereas samples with other nematode species showed only a control line (Figure 5A). The detection of J2 for *M. hapla* was confirmed from extracts of infected pepper roots with *M. hapla* (Figure 5B) as well as in the presence of background crude extract from 10 to 20 non-target nematodes (Figure 5C). Lowering in half, a single reaction assay volume still detected *M. hapla* samples (Figure 5C). The results of RPA assays showed reliable detection with a sensitivity of 1/10 of a J2 (Figure 5D) and 1/1000 of a female (Figure 4B) in reaction tubes.

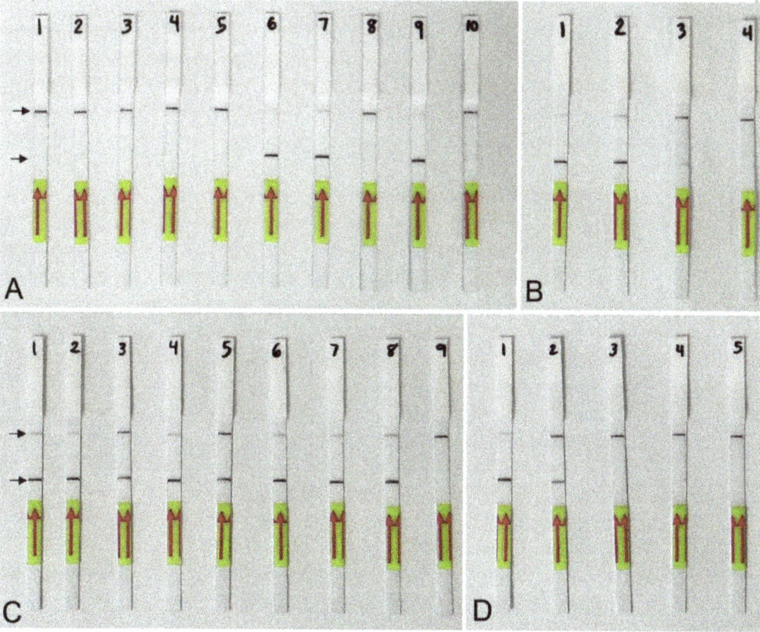

**Figure 5.** Lateral flow recombinase polymerase amplification (LF-RPA) assay with examples of lateral flow strips. (**A**) Specificity assay with DNA samples of *Meloidogyne* spp. and crude J2 extracts of *M. hapla*. Strip: 1 and 2: *Meloidogyne* sp.1 (CD3380); 3: *M. naasi* (CD3381); 4: *M. baetica* (CD3382); 5: *Meloidogyne* sp.2 (CD3383); 6 and 7: *M. hapla*, (CD3384); 8: *M. arenaria* (CD33093); 9: *M. hapla* (C44); 10: negative control; (**B**) crude extracts of *M. hapla*. Strip: 1 and 2: extracts from infected tomato roots containing plant materials and females with egg-masses; 3: 1 j2 per tube; 4: negative control; (**C**) testing of crude extract of *M. hapla* with or without crude extracts of non-target nematodes. Strip: 1 and 2: 1 J2 per tube; 3 and 4: 1 J2 with 10 non-target nematodes per tube; 5 and 6: 1 J2 with 20 non-target nematodes per tube; 7 and 8: 1 J2 per tube containing half of a reaction mixture; 9: negative control; (**D**) sensitivity assay with a dilution series of a crude j2 extract of *M. hapla*, Strip: 1: 1 J2 per tube; 2: 1/10 J2 per tube; 3: 1/100 J2 per tube; 4: 1/1000 J2 per tube; 5: negative control (upper) and test (lower) lines are indicated by arrows.

### 3. Discussion

Polymerase chain reaction is considered the gold standard of molecular detection, however, this method is only available in a laboratory with thermal cycling equipment. In this work, we have developed an affordable, simple, fast and sensitive real-time RPA and LF-RPA assays to detect *M. hapla* from nematode specimens extracted from plant and soil samples. An LF-RPA assay can be performed in a field condition without any special equipment or in areas with a minimal laboratory infrastructure.

Song et al. [28] described the LF-RPA diagnostic assay of *M. hapla* using species-specific primers and a probe designed using the effector gene *16D10* sequence. Authors stated that the entire detection process can be completed within 1.5 h, including 30–60 min for DNA extraction, 20 min for the RPA reaction, and 3–5 min for visual detection on the LF strips [28]. In our assay, the species-specific primers and probes were designed using the IGS rRNA gene sequence. The entire detection process for the LF-RPA assay can be completed within approximately 30 min, including 4 min for crude nematode extract preparation, 20 min (4 + 16) for the RPA reaction, 1 min for mixing and centrifugation of tubes, and 5 min for visual detection on the LF strips. The entire detection process for real-time RPA assay can be completed within approximately 13 min, including 4 min for crude nematode extract preparation, 8 min (5 + 3) for the RPA reaction with 1 min for mixing and centrifugation of tubes. This calculation does not include the time for preparation of the RPA reaction mixture.

In our study the IGS rRNA gene was selected for the RPA assays because of its high copy number and because previously published PCR studies have demonstrated its usefulness to distinguish *Meloidogyne* species [7,29,30]. Zhang et al. [16] noticed that different DNA targets are likely to have extremely different amplification efficiencies, even sharing a series of common characteristics including GC content, primer melting temperature and RPA product length. These authors also concluded that primers are the most important determinant for RPA performance including sensitivity, specificity and reaction rate. Although amplicons obtained from fragments of effector gene *16D10* and the IGS rRNA gene are comparable (148 vs. 164 bp) in a length, it seems that the RPA reaction rate is higher for the IGS rRNA gene than the effector gene fragment. The RPA assays developed based on the IGS rRNA gene are more sensitive for detection than assays based on the effector gene fragment.

Song et al. [28] reported about 1/1000 female (after DNA extraction with proteinase K) as the detection limit of the LF-RPA assay, whereas in our LF-RPA assay, the detection limit can reach up to 1/10,000 (without a special DNA extraction step). Our RPA assays for *M. hapla* also showed higher amplification rates compared with similar assays developed for *M. enterolobii*, in which species-specific primers were also designed based on the IGS rRNA gene [23]. The threshold level for the reliable *M. enterolobii* detection was established as equal to 30 cycles (=10 min) and a baseline of 500,000 ($\Delta$Rn) fluorescence level with the TwistAmp® exo kit and Applied Biosystems™ QuantStudio™ Flex Real-Time PCR System, whereas in our present study for *M. hapla*, the threshold level was estimated equal to 8 cycles (~3 min) and a baseline of 250,000.

The real-time RPA and LF-RPA assays developed in our study allowed the detection of J2, young females without and with eggs-masses. The old and dead females without body contents might not be always detectable using this method. The results of our study estimated that the reliable detection of RPA assays using real-time fluorescence were 1/100 of J2 or female and using lateral flow dipsticks were 1/10 of a J2 and 1/1000 of a female. However, in some replicates the detection limit can reach up to 1/10,000 of a female in reaction tube. Reproducibility of the assays in low concentrations of nematode extracts, extracts from old females or directly from soil samples should be carefully tested further to understand factors, which might have an influence on the performing stability of RPA reactions. RPA diagnostics of root-knot nematodes has several other important advantages over PCR methods. The first advantage is that crude nematode extracts or crude extracts from nematode-infected tomato and pepper plant tissues can be directly used for RPA assays, whereas PCR assays require a DNA extraction step with special treatment of these extracts before use. The second advantage of RPA assays is that results are available in 8–20 min, whereas the results of PCR assays can be observed in 1.5–3 h. The third advantage is higher sensitivity levels of RPA detection over PCR methods; the RPA assay is 10 or 100 times more sensitive than PCR.

However, the application of RPA assays for nematode diagnostics may still face several problems, with cost being a major consideration. Factors affecting the expense of assays depends on the pest, reagent costs, requirement for equipment, infrastructure capacity, employee wages and numbers of samples for testing, among others. The RPA reagents and kits are presently manufactured by only one company, TwistDx, Inc., making the cost of the RPA assay relatively higher than other PCR assays. Reagent costs for RPA assays are currently in range of USD 4.3–5.5 per sample [15] which is higher than for conventional and real-time PCR. Lillis et al. [31] showed that lowering the assay volume from 50 μL, which is the recommended in the manufacturer's protocol, to 5 μL showed similar sensitivity. Our experiments also revealed acceptable diagnostic performance when reducing the reaction volume by half. This approach could be used in cases of resource limitations. It has been noticed that special attention should be paid to the potential of cross-contamination due to the high sensitivity of this reaction. The requirement for use of clean gloves, tubes, and pipets should be carefully considered during the use of RPA assays in a field condition. Thus, RPA has the potential to be a promising alternative to PCR and other methods for rapid detection of nematodes. This assay requires minimal sample preparation, making it ideal for use in the lab, the field, or minimal laboratory infrastructure.

## 4. Materials and Methods

### 4.1. Nematode Samples

Five isolates of *Meloidogyne hapla* were obtained for RPA assay development. Second-stage juveniles (J2s) and females were extracted from the root or soil samples. The D2–D3 expansion segments of the 28S rRNA gene were sequenced from each isolate to confirm its identity. DNA of several root-knot nematodes, *M. arenaria*, *M. baetica*, *M. christiei*, *M. enterelobii*, *M. floridensis*, *M. incognita*, *M. javanica*, *M. naasi* and *M. nataliei* were also used in specificity experiments (Table 1). These species were also identified by molecular methods. Free-living and plant parasitic nematodes from several field samples collected in California were extracted using the centrifugal flotation method and their extracts were used as background non-target DNA.

### 4.2. Nematode Extracts

Second-stage juveniles or females of *M. hapla* were placed in distilled water on a microscope slide. The nematodes were cut using a dental needle under a stereo microscope and put into a 0.2 mL PCR tube with a total volume of 10 μL. This stock crude extract was used to make a series of dilutions sequentially: 1:2, 1:4, 1:8, 1:10, 1:16, 1:100, 1:1000 and 1:10,000 in water. Several extracts were prepared: (i) J2s; (ii) J2s with other non-target nematodes; (iii) female and (iv) plant gall tissue with one or more females and egg-masses. Crude extract of plant gall tissue with nematodes and crude extract of several hundred non-target nematodes soil free-living and plant parasitic nematodes were also obtained by crushing the samples on a microscope slide using a plastic pipe tip or dental needle.

### 4.3. RPA Primer Design and Testing

A total of three forward and three reverse RPA primers specific to *M. hapla* were manually designed based on species sequence polymorphisms in the IGS rRNA gene. Primers were synthesized by Integrated DNA Technologies, Inc. (Redwood City, CA, USA). Nine primer sets were screened in different combinations using the TwistAmp® Basic kit (TwistDx, Cambridge, UK). Reactions were prepared according to the manufacturer's instructions. The lyophilized reaction pellets were suspended in 29.5 μL of the rehydration buffer, 2.4 μL of each forward and reverse primers (10 μM) (Table 2), 1 μL of the DNA template or nematode extract and 12.2 μL of distilled water. For each sample, 2.5 μL of 280 mM magnesium acetate was added to the lid of the tube and the lids were closed carefully. The tubes were inverted 10–15 times and briefly centrifuged to initiate reactions simultaneously. Tubes were incubated at 39 °C (4 min) in a MyBlock Mini Dry Bath (Benchmark Scientific, Edison, NJ, USA) and then they were inverted 10–15 times, briefly centrifuged and returned to the incubator block (39 °C) for 20 min. Sample tubes were then placed in a freezer to stop the reaction. Amplification products were purified with a QIAquick PCR Purification Kit (Qiagen, Valencia, CA, USA). Five μL of purified product were run in a 1% TAE (Tris-acetate-EDTA)-buffered agarose gel (100 V, 60 min) and visualized with Gel Green stain. Amplification products were directly sequenced by Genewiz (San Francisco, CA, USA) using amplification primers.

### 4.4. Real-Time RPA Assay

Two TwistAmp® exo probes were designed according to the manufacturer's instructions and were synthesized by Biosearch Technologies, Inc. (Petaluma, CA, USA). Two probes were tested and only one probe (Probe-hapla-exo1) was selected for the assay (Table 2) based on best amplification performance. The real time detection of RPA assay products was accomplished using the TwistAmp® exo kit (TwistDx, Cambridge, UK). The lyophilized reaction pellets were suspended in 29.5 μL of the rehydration buffer, 2.1 μL of each forward and reverse primers (10 μM) (Table 2), 0.6 μL of the probe (10 μM), 1 μL of the DNA template or nematode extract and 12.2 μL of distilled water. Magnesium acetate in a volume of 2.5 μL was added to the lid of each tube, the lids were carefully closed, tubes were inverted 10–15 times and briefly centrifuged. The reaction tubes were incubated at 39 °C for 5 min, then inverted 10–15 times to mix, and briefly centrifuged. The tubes were immediately placed in Applied Biosystems™ QuantStudio™ 7 Flex Real-Time PCR System to incubate at 39 °C for 15 min. The fluorescence signal was monitored in real time and measured every 20 s (cycle) using the fluorophore (FAM) channel. A positive control using *M. hapla* extract (one J2 per reaction tube) and negative control without any nematode DNA were included in each run. Two or three replicates of each variant across several runs were performed for sensitivity and specificity experiments.

### 4.5. LF-RPA Assay

Two TwistAmp® nfo probes were designed according to the manufacturer's instructions and tested in the same conditions. Only one probe (Probe-hapla-nfo1) was selected for the assay based on the best visualization results. The LF-RPA assay products were accomplished using the TwistAmp® nfo kit (TwistDx, Cambridge, UK). The reaction mixture for each RPA assay was prepared according to the manufacturer's instructions: the lyophilized reaction pellets were suspended in 29.5 μL of the rehydration buffer, 2.1 μL of each forward and reverse primers (10 μM), 0.6 μL of the probe (10 μM), 1 μL of the DNA template or nematode extract and 12.2 μL of distilled water. Magnesium acetate in a volume of 2.5 μL was added to the lid of each tube, the lids were carefully closed, and the tubes were inverted 10–15 times and briefly centrifuged. The reaction tubes were incubated at 39 °C for 4 min, then inverted 10–15 times to mix, briefly centrifuged and returned to the incubator block at 39 °C for 16 min. The tubes were placed in the freezer to stop the reaction. For visual analysis with Milenia® Genline Hybridetect-1 strips (Milenia Biotec GmbH, Giessen, Germany), testing solution containing 48 μL of HybriDetect assay buffer and 12 μL of the sample RPA product was prepared in a 0.5 mL PCR tube. Ten μL of the testing solution was placed directly onto the sample area of the dipstick. Dipsticks were placed upright into 100 μL of the assay buffer and visual results were observed within 5 min. The amplification product was indicated by the development of a colored test line, and/or a separate control line to confirm that the system worked properly (Figures 4B and 5). Two or three replicates of each variant were performed for sensitivity and specificity experiments.

**Author Contributions:** Conceptualization, S.A.S.; methodology, S.A.S.; software, J.B. and S.A.S.; validation, J.B. and S.A.S.; formal Analysis, J.B. and S.A.S.; investigation, J.B. and S.A.S.; resources, J.B. and S.A.S.; data Curation, J.B. and S.A.S.; writing—original draft preparation, S.A.S.; writing—review and editing, J.B. and S.A.S.; visualization, J.B. and S.A.S.; supervision, S.A.S.; project administration, S.A.S.; funding acquisition, S.A.S. All authors have read and agreed to the published version of the manuscript.

**Funding:** This work was sponsored by the Specialty Crop Block Grant Program (USDA Project Number: AM190100XXXXG008; CDFA Grant Number: 19-0001-035-MU).

**Institutional Review Board Statement:** Not applicable.

**Informed Consent Statement:** Not applicable.

**Data Availability Statement:** The datasets generated during and/or analyzed during the currentstudy are available from the corresponding author on reasonable request.

**Acknowledgments:** The authors thank Álvarez-Ortega, S., Brito, J., Chizhov, V.N., Castillo, P., Inserra, R.N., Roberts, P. and Williamson V. for providing the root-knot nematode samples.

**Conflicts of Interest:** The authors declare no conflict of interest.

# References


1. Koenning, S.; Overstreet, C.; Noling, J.; Donald, P.; Becker, J.; Fortnum, B. Survey of crop losses in response to phytoparasitic nematodes in the United States for 1994. *J. Nematol.* **1999**, *31*, 587–618. [PubMed]
2. Perry, R.N.; Moens, M.; Starr, J.L. *Meloidogyne* species—A diverse group of novel and important plant parasites. In *Root-Knot Nematodes*; Perry, R.N., Moens, M., Starr, J.L., Eds.; CAB International: Wallingford, UK, 2009; pp. 1–17.
3. Alvarez-Ortega, S.; Brito, J.A.; Subbotin, S.A. Multigene phylogeny of root-knot nematodes and molecular characterization of *Meloidogyne nataliei* Golden, Rose & Bird, 1981 (Nematoda: Tylenchida). *Sci. Rep.* **2019**, *9*, 11788. [PubMed]
4. Williamson, V.M.; Caswell-Chen, E.P.; Westerdahl, B.B.; Wu, F.F.; Caryl, G. A PCR assay to identify and distinguish single juveniles of *Meloidogyne hapla* and *M. chitwoodi*. *J. Nematol.* **1997**, *29*, 9–15.
5. Zijlstra, C. Identification of *Meloidogyne chitwoodi*, *M. fallax* and *M. hapla* based on SCAR-PCR: A powerful way enabling reliable PCR-based techniques for *Meloidogyne* identification of populations or individuals that share common traits. *Eur. J. Plant Pathol.* **2000**, *106*, 283–290. [CrossRef]
6. Dong, K.; Dean, R.A.; Fortnum, B.A.; Lewis, S.A. Development of PCR primers to identify species of root-knot nematodes: *Meloidogyne arenaria*, *M. hapla*, *M. incognita* and *M. javanica*. *Nematropica* **2001**, *31*, 271–280.
7. Wishart, J.; Phillips, M.S.; Blok, V.C. Ribosomal intergenic spacer: A Polymerase Chain Reaction diagnostic for *Meloidogyne chitwoodi*, *M. fallax, and M. hapla*. *Phytopathology* **2002**, *92*, 884–892. [CrossRef] [PubMed]
8. Dong, L.L.; Yao, H.; Li, Q.S.; Song, J.Y.; Li, Y.; Luo, H.M.; Chen, S.L. Investigation and integrated molecular diagnosis of root-knot nematodes in *Panax notoginseng* root in the field. *Eur. J. Plant Pathol.* **2013**, *137*, 667–675. [CrossRef]
9. Watanabe, T.; Masumura, H.; Kioka, Y.; Noguchi, K.; Min, Y.Y.; Murakami, R.; Toyota, K. Development of a direct quantitative detection method for *Meloidogyne incognita* and *M. hapla* in andosol and analysis of relationship between the initial population of *Meloidogyne* spp. and yield of eggplant in an andosol. *Nematol. Res.* **2013**, *43*, 21–29. [CrossRef]
10. Hay, F.S.; Herdina Ophel-Keller, K.; Hartley, D.M.; Pethybridge, S.J. Prediction of potato tuber damage by root-knot nematodes using quantitative DNA assay of soil. *Plant Dis.* **2016**, *100*, 592–600. [CrossRef]
11. Sapkota, R.; Skantar, A.M.; Nicolaisen, M. A TaqMan real-time PCR assay for detection of *Meloidogyne hapla* in root galls and in soil. *Nematology* **2016**, *18*, 147–154. [CrossRef]
12. Gorny, A.M.; Wang, X.H.; Hay, F.S.; Pethybridge, S.J. Development of a species-specific PCR for qetection and quantification of *Meloidogyne hapla* in soil using the 16D10 root-knot nematode effector gene. *Plant Dis.* **2019**, *103*, 1902–1909. [CrossRef]
13. Peng, H.; Long, H.; Huang, W.; Liu, J.; Cui, J.; Kong, L.; Hu, X.; Gu, J.; Peng, D. Rapid, simple and direct detection of *Meloidogyne hapla* from infected root galls using loop-mediated isothermal amplification combined with FTA technology. *Sci. Rep.* **2017**, *7*, 44853. [CrossRef]
14. Piepenburg, O.; Williams, C.H.; Stemple, D.L.; Armes, N.A. DNA detection using recombination proteins. *PLoS Biol.* **2006**, *4*, e204. [CrossRef]
15. Daher, R.K.; Stewart, G.; Boissinot, M.; Bergeron, M.G. Recombinase polymerase amp lification for diagnostic applications. *Clin. Chem.* **2016**, *62*, 947–958. [CrossRef]
16. Zhang, Y.; Hu, J.Q.; Li, Q.M.; Guo, J.Q.; Zhang, G.P. Chapter 10—Detection of microorganisms using recombinase polymerase amplification with lateral flow dipsticks. *Methods Microbiol.* **2020**, *47*, 319–349. [CrossRef]
17. Miles, T.D.; Martin, F.N.; Coffey, M.D. Development of rapid isothermal amplification assays for detection of *Phytophthora* spp. in plant tissue. *Phytopathology* **2015**, *105*, 265–278. [CrossRef]
18. Lau, H.Y.; Wang, Y.L.; Wee, E.J.H.; Botella, J.R.; Trau, M. Field demonstration of a multiplexed point-of-care diagnostic platform for plant pathogens. *Anal. Chem.* **2016**, *88*, 8074–8081. [CrossRef]

19. Londoño, M.A.; Harmon, C.L.; Polston, J.E. Evaluation of recombinase polymerase amplification for detection of begomoviruses by plant diagnostic clinics. *Virol. J.* **2016**, *13*, 48. [CrossRef]

20. Cabada, M.M.; Malaga, J.L.; Castellanos-Gonzalez, A.; Bagwell, K.A.; Naeger, P.A.; Rogers, H.K.; Maharsi, S.; Mbaka, M.; White, A.C., Jr. Recombinase Polymerase Amplification compared to real-time Polymerase Chain Reaction test for the detection of *Fasciola hepatica* in human stool. *Am. J. Trop. Med. Hyg.* **2017**, *96*, 341–346. [CrossRef]

21. Li, T.-T.; Wang, J.-L.; Zhang, N.-Z.; Li, W.-H.; Yan, H.-B.; Li, L.; Jia, W.-Z.; Fu, B.-Q. Rapid and visual detection of *Trichinella* spp. using a lateral flow strip-based Recombinase Polymerase Amplification (LF-RPA) assay. *Front. Cell. Infect. Microbiol.* **2019**, *9*, 1. [CrossRef]

22. Jarvi, A.I.; Atkinson, E.S.; Kaluna, L.M.; Snook, K.A.; Steel, A. Development of a recombinase polymerase amplification (RPA-EXO) and lateral flow assay (RPA-LFA) based on the ITS1 gene for the detection of *Angiostrongylus cantonensis* in gastropod intermediate hosts. *Parasitology* **2020**, *4*, 1–8. [CrossRef] [PubMed]

23. Subbotin, S.A. Recombinase polymerase amplification assay for rapid detection of the root-knot nematode *Meloidogyne enterolobii*. *Nematology* **2019**, *21*, 243–251. [CrossRef]

24. Ju, Y.L.; Lin, Y.; Yang, G.G.; Wu, H.P.; Pan, Y.M. Development of recombinase polymerase amplification assay for rapid detection of *Meloidogyne incognita*, *M. javanica*, *M. arenaria*, and *M. Enterolobii*. *Eur. J. Plant Pathol.* **2019**, *155*, 1155–1163. [CrossRef]

25. Chi, Y.-K.; Zhao, W.; Ye, M.-D.; Ali, F.; Wang, T.; Qi, R.-D. Evaluation of recombinase polymerase amplification assay for detecting *Meloidogyne Javanica*. *Plant Dis.* **2020**. [CrossRef]

26. Cha, D.J.; Kim, D.S.; Lee, S.K.; Han, H.R. A new on-site detection method for *Bursaphelenchus xylophilus* in infected pine trees. *For. Pathol.* **2019**, *49*, e12503. [CrossRef]

27. Cha, D.; Kim, D.; Choi, W.; Park, S.; Han, H. Point-of-care diagnostic (POCD) method for detecting *Bursaphelenchus xylophilus* in pinewood using recombinase polymerase amplification (RPA) with the portable optical isothermal device (POID). *PLoS ONE* **2020**, *15*, e0227476. [CrossRef]

28. Song, Z.Q.; Yang, X.; Zhang, X.W.; Luan, M.B.; Guo, B.; Liu, C.N.; Pan, J.P.; Mei, S.Y. Rapid and visual detection of *Meloidogyne hapla* using recombinase polymerase amplification combined with a lateral flow dipstick (RPA-LFD) assay. *Plant Dis.* **2021**, in press. [CrossRef]

29. Peterson, D.J.; Vrain, T.C. Rapid identification of *Meloidogyne chitwoodi*, *M. hapla*, and *M. fallax* using PCR primers to amplify their ribosomal intergenic spacer. *Fundam. Appl. Nematol.* **1996**, *19*, 601–605.

30. Petersen, D.J.; Zijlstra, C.; Wishart, J.; Blok, V.; Vrain, T.C. Specific probes efficiently distinguish root-knot nematode species using signature sequences in the ribosomal intergenic spacer. *Fundam. Appl. Nematol.* **1997**, *20*, 619–626.

31. Lillis, L.; Siverson, J.; Lee, A.; Cantera, J.; Parker, M.; Piepenburg, O.; Lehman, D.A.; Boyle, D.S. Factors influencing Recombinase polymerase amplification (RPA) assay outcomes at point of care. *Mol. Cell. Probes* **2016**, *30*, 74–78. [CrossRef]

Article

# Species Diversity of Pin Nematodes (*Paratylenchus* spp.) from Potato Growing Regions of Southern Alberta, Canada

Maria Munawar [1], Dmytro P. Yevtushenko [1,*], Juan E. Palomares-Rius [2] and Pablo Castillo [2]

1. Department of Biological Sciences, University of Lethbridge, 4401 University Drive W, Lethbridge, AB T1K 3M4, Canada; maria.munawar@uleth.ca
2. Institute for Sustainable Agriculture (IAS), Spanish National Research Council (CSIC), Campus de Excelencia Internacional Agroalimentario, ceiA3, Avenida Menéndez Pidal s/n, 14004 Córdoba, Spain; palomaresje@ias.csic.es (J.E.P.-R.); p.castillo@csic.es (P.C.)
* Correspondence: dmyto.yevtushenko@uleth.ca
http://zoobank.org/urn:lsid:pub:39C84EDC-15ED-491E-9373-8876D34C35ED

**Abstract:** Pin nematodes (*Paratylenchus* spp.) are polyphagous parasitic species with a wide host range and geographical distribution; their diversity is unknown in the potato growing region of Alberta, Canada. The present study aims to provide morphological and molecular characterization of three pin nematode species, namely *P. neoprojectus*, *P. tateae*, and a new species, *Paratylenchus enigmaticus* sp. nov. All of them were recovered from the potato growing region of southern Alberta. The nematodes were isolated using the sieving and flotation-centrifugation method, and their morphology was assessed by light microscopy. Molecular characterization was performed using partial 18S, D2–D3 expansion domains of the 28S and ITS ribosomal genes. This study is the first report of molecular characterization of *P. tateae* and *P. neoprojectus*, being new records from southern Alberta, and two Spanish populations of *P. tateae* comprising the first report of this species in Europe. The phylogenetic analysis of the 18S, D2–D3 expansion domains of the 28S and ITS ribosomal DNA regions underscores the importance of using molecular data for accurate species identification and clarifies the status of *P. nanus* type B and *P. sheri*. Moreover, our findings will be useful to determine the impact of pin nematodes on potato production in future field research.

**Keywords:** *Paratylenchus tateae*; *Paratylenchus neoprojectus*; plant-parasitic nematode; integrative taxonomy; morphology; DNA sequencing; phylogeny; new record; new species

**Citation:** Munawar, M.; Yevtushenko, D.P; Palomares-Rius, J.E; Castillo, P. Species Diversity of Pin Nematodes (*Paratylenchus* spp.) from Potato Growing Regions of Southern Alberta, Canada. *Plants* **2021**, *10*, 188. https://doi.org/10.3390/plants10020188

Received: 26 December 2020
Accepted: 18 January 2021
Published: 20 January 2021

**Publisher's Note:** MDPI stays neutral with regard to jurisdictional claims in published maps and institutional affiliations.

## 1. Introduction

Potato is one of the most important crops in Canada, with Alberta ranking among the top provinces producing superior quality potatoes with the highest marketable yields [1]. To maintain high standards of potato production, Alberta's farmed fields are regularly surveyed and examined for the presence of pest species. Recent reports have described the incidence of plant-parasitic nematodes (PPN) in cultivated soils of Canada [2–4].

*Paratylenchus* species are commonly known as pin nematodes. The short stylet species feed ecto-parasitically; however, some species feed endo-parasitically by gaining entry into lateral roots [5–7]. Pin nematodes are amongst the most frequently occurring PPN in Canada [8], and previous studies have reported the association of pin nematodes with forages, turf grasses, legumes, and cereal crops of Eastern and Central Canada [2,9–13]. Biological studies have indicated that females of *P. projectus* Jenkins [14] lay 1–2 eggs/day, with an average life cycle of 30–38 days at 20–28 °C. Additionally, several *Paratylenchus* species have a persistent survival stage (mainly the fourth stage), which helps them to maintain inoculum levels during periods of adversity [15].

*Paratylenchus* species have a wide host-range, and several short stylet species, such as *P. bukowinensis* Micoletzky [16], *P. dianthus* Jenkins and Taylor [17], *P. hamatus* Thorne and Allen [18], *P. microdorus* Andrassy [19], *P. neoamblycephalus* Geraert [20], *P. shenzhenensis*

Wang, Xie, Li, Xu, Yu, and Wang [21] and *P. projectus*, cause varying degrees of damage to their hosts, including root injury and poor plant development, consequently decreasing yield and plant longevity [7,22].

Currently, the genus contains over 100 species, with only 11 reported in Canada [23,24]. *Paratylenchus* species are among the smallest PPN and this, together with their apparent similarities with other related species, makes them challenging to study and identify [25]. During a survey of potato fields, we isolated three *Paratylenchus* species. Preliminary examination revealed that all the species have advulval flaps, 4 lateral lines, and short stylets (<40 μm).

As several short stylet pin nematodes species are considered to be plant-pathogenic [22], we performed morphological/morphometrical and molecular studies on these *Paratylenchus* populations and identified them as *P. neoprojectus* Wu and Hawn [26], *P. tateae* Wu and Townshend [27], and a new *Paratylenchus* sp. that we named *P. enigmaticus* sp. nov. As the diversity of pin nematode species associated with potato growing areas of Alberta is largely unknown, the aims of the present work were to: (i) characterize the populations of *P. tateae*, *P. neoprojectus*, and *P. enigmaticus* sp. nov. found in potato growing areas of southern Alberta; (ii) update the pin nematode diversity record from Canada; (iii) study the phylogenetic relationship of these species with other pin nematodes. The results of this study will aid in distinguishing pathogenic forms from non-pathogenic species, and our findings will be useful in future field experiments to determine the impact of these PPN on potato production.

## 2. Results

### 2.1. Description of Female Paratylenchus neoprojectus Wu and Hawn

(Figures 1 and 2; Table 1) [26].

Body slender, ventrally arcuate with a bend in the middle of the body when heat relaxed; cuticle finely annulated; lateral field equidistant with four distinct lines; lip region rounded narrow, with anterior end flattened, continuous with the rest of the body; labial framework sclerotization weak; pharyngeal region typical paratylenchoid type; stylet rigid, straight; rounded stylet knobs; dorsal pharyngeal gland opening 5.0–6.0 μm behind stylet knobs; median pharyngeal bulb large elongate, bearing distinct large valves; isthmus short slender, surrounded by nerve ring; basal bulb pyriform, pharyngeal-intestinal valve bilobed; excretory pore situated at the level or middle of pharyngeal basal bulb. Hemizonid 1–2 annuli long situated just posterior to the excretory pore. The body slightly narrower posterior to vulva; ovary outstretched, well developed, in some specimens it reaches to the level of pharynx; spermatheca and crustaformeria well developed, the columnar arrangement of crustaformeria usually not discernable; spermatheca rounded; the vulva a transverse slit occupying half of the corresponding body width; vulval lips prominent, the anterior lip protrudes further than the posterior lip; vulval flaps present, but not prominent in fresh specimens; a small, rudimentary post uterine branch present along the ventral body wall; anus indistinct; tail slender, conoid, finely annulated, and gradually tapers to form a finely rounded terminus.

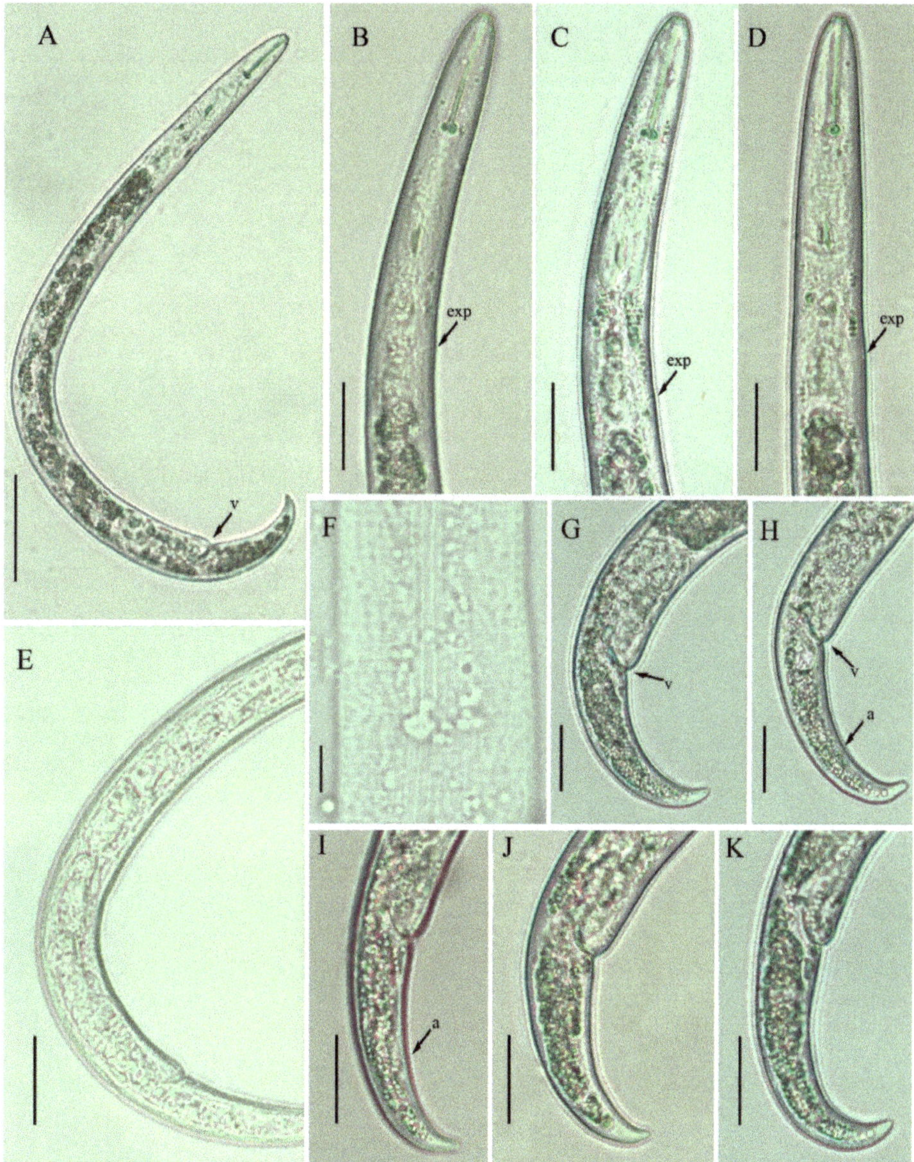

**Figure 1.** Light photomicrographs of *Paratylenchus neoprojectus* females. (**A**) Entire body; (**B–D**) pharyngeal regions; (**E**) posterior region with gonad; (**F**) lateral lines; (**G–K**) tails. Scale bars: (**A**) 50 μm; (**B–E**) 20 μm; (**F**) 5 μm; (**G–K**) 20 μm. Arrowheads: (a) Anus; (exp) excretory pore; (v) vulva.

**Table 1.** Morphometrics of *Paratylenchus neoprojectus* females and juveniles. All measurements are in μm and presented as mean ± standard deviation (range).

| Characters | Present Study | | Wu & Hawn [26] | * Van den Berg et al. [28] | |
|---|---|---|---|---|---|
| | Females | Juveniles | Females | Females | Juveniles |
| n | 11 | 4 | 76 | 17 | 4 |
| Body length | 383.5 ± 36.7 (330.0–434.0) | 342.0 ± 19.6 (322.0–365.0) | 327–405 | 359 (300–415) | 339.5 (299–390) |
| a | 24.0 ± 1.7 (21.0–26.0) | 22.3 ± 1.9 (20.5–24.3) | 18–26 | 22.1 (19.5–24.6) | 20.4 (17.7–22.9) |
| b | 3.8 ± 0.3 (3.3–4.3) | 3.9 ± 0.3 (3.5–4.1) | 3.8–4.6 | 3.9 (3.5–4.4) | 4.1 (3.7–4.7) |
| c | 14.6 ± 1.8 (12.1–18.5) | 12.8 ± 1.6 (11.0–15.0) | 14–16 | 15.3 (14–18.5) | 13.8 (12.3–18.9) |
| c′ | 2.7 ± 0.2 (2.3–3.0) | 2.3 ± 0.3 (1.9–2.6) | - | 2.4 (2.0–2.8) | 2.2 (1.7–2.5) |
| V | 84.4 ± 1.3 (82.0–85.8) | - | 82–85.7 | 84 (82.5–85) | - |
| Stylet percentage | 7.0 ± 0.8 (5.8–8.3) | - | - | 8 (6.8–9.3) | - |
| Lip height | 3.3 ± 0.4 (3.0–4.0) | - | - | 3.5 (3–4) | - |
| Lip width | 6.4 ± 0.4 (6.0–7.0) | - | - | 7 (6.5–7.5) | - |
| Stylet length | 25.3 ± 1.3 (25.0–29) | 13.3 ± 1.0 (12.0–14.0) | 28–31 | 28.5 (26–31) | 10 (3.5–14.5) |
| Median bulb length | 23.4 ± 1.6 (21.0–25.0) | - | - | - | - |
| Median bulb width | 9.3 ± 0.8 (8.0–11.0) | - | - | - | - |
| Anterior end to excretory pore | 79.1 ± 4.8 (70.0–85.0) | 75.0 ± 5.2 (70.0–80.0) | - | 77.5 (71–85) | 71 (65–78.5) |
| Pharynx length | 99.0 ± 4.2 (92.0–106.0) | 89.0 ± 6.2 (80.0–93.0) | 82–94 | 92 (85–110) | 83.5 (72.5–94.5) |
| Maximum body width | 16.0 ± 1.4 (13.5–18.0) | 15.4 ± 0.4 (15.0–15.8) | - | 16 (13–20) | - |
| Vulva body width | 13.6 ± 1.3 (12.0–15.0) | - | - | - | - |
| Anal body width | 9.7 ± 0.9 (8.0–11.2) | 11.7 ± 0.7 (10.7–12.4) | - | - | - |
| Distance from vulva to anus | 33.5 ± 5.8 (28.0–44.0) | - | 29–44 | 33.5 (26–44) | - |
| Distance from vulva to tail terminus | 60.0 ± 7.3 (50.0–72.0) | - | - | - | - |
| Tail length | 26.0 ± 2.9 (22.0–30.0) | 27.0 ± 3.5 (22.0–30.0) | 23–27 | 23.5 (17.5–29.5) | 23 (20.5–29.5) |

* Van den Berg et al. [28] represent the measurements of *P. nanus* type B. In this study, we refer this population as *P. neoprojectus*.

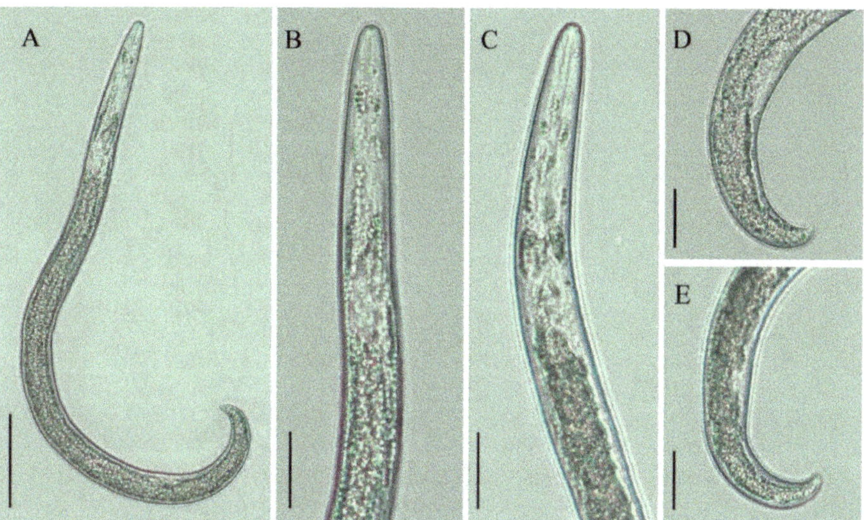

**Figure 2.** Light photomicrographs of *Paratylenchus neoprojectus* juvenile. (**A**) Entire body; (**B,C**) pharyngeal regions; (**D,E**) tails. Scale bars: (**A**) 50 µm; (**B–E**) 20 µm.

### 2.1.1. Juveniles

Only one juvenile form was detected. Individuals in this stage were similar in morphology to the adult females. However, they were characterized by the presence of weak stylet; pharynx components under-developed; genital primordium under-developed; anus indistinct; and a posterior body with a finely rounded terminus.

### 2.1.2. Remarks

*Paratylenchus neoprojectus* was originally described from Central Alberta, Canada in the rhizosphere of alfalfa [26]. Following the formal description, the species has appeared twice in the literature [23]. The first population was reported from India [29] without morphological characterization or illustrations; only morphometrics of adult females were provided. Since overlapping morphometrical characters are common in pin nematode species [25,28], the identification of this Indian population needs to be confirmed.

The second population was reported from Iran [30], and the illustrations showed the absence of a post uterine sac (vs. present in the original description), a broadly rounded tail terminus (vs. conically or finely rounded in the original description), and a short ovary (vs. an ovary that reaches to the pharyngeal basal bulb level in the original description). All these characters are not in agreement with the original description of *P. neoprojectus*, therefore a detailed re-evaluation based on integrative taxonomy is required to determine the exact status of this population.

In 2014, Van den Berg et al. [28] reported a detailed morphological and molecular characterization of several pin nematode species from the USA and South Africa. Based on their molecular data, the authors demonstrated that *P. nanus* has two sibling species type A and type B. Comparing the morphological, molecular, and morphometrical characteristics (Figures 1 and 2; Table 1), we conclude that *P. nanus* type B should be considered as *P. neoprojectus*. *Paratylenchus neoprojectus* and *P. nanus* are closely related species, but can be differentiated by the body shape (ventrally bent vs. open C-shape of *P. nanus*), position of the excretory pore (at the level or posterior to pharyngeal bulb vs. at level or anterior to pharyngeal bulb), ovary development (reaches the level of the pharyngeal basal bulb vs. short), presence of post uterine branch (vs. absent), and tail terminus morphology (conically or narrowly rounded vs. subacute to rounded, slightly indented). *Paratylenchus neoprojectus* is also close to *P. projectus* and can be differentiated from it by the lip region morphology (conical rounded vs. trapezoid), more posterior position of the excretory pore (vs. anterior), and tail terminus morphology (conically or narrowly rounded vs. often digitate terminus).

In the present study, the *P. neoprojectus* population from southern Alberta matches with the species' original description, except for minor differences in the body length; the southern Alberta population is slightly longer than the original one (330–434 vs. 327–405 µm).

### 2.1.3. Habitat and Locality

This population was found in the rhizosphere of *Chenopodium* sp. growing on the headland (uncultivated field margin) of a potato field, (latitude 49°48′40.5″ N; longitude—111°23′55.4″ W); Municipal District of Forty Mile County No. 8, Alberta, Canada.

### 2.2. Description of Female Paratylenchus tateae Wu and Townshend

(Figures 3–5; Table 2) [27].

Body slender, ventrally arcuate when heat relaxed; cuticle finely annulated; lateral field equidistant with four distinct lines; lip region conoid narrow, with anterior end flattened, continuous with the rest of the body; labial framework sclerotization weak; pharyngeal region, typical paratylenchoid type; stylet rigid, straight; stylet knobs, rounded; dorsal pharyngeal gland opening 4.5–6.0 µm behind stylet knobs; median pharyngeal bulb elongated, bearing distinct large valves; isthmus short slender, surrounded by nerve ring; basal bulb pyriform, pharyngeal-intestinal valve inconspicuous; excretory pore situated at the level of pharyngeal basal bulb or slightly anterior to it. Hemizonid 2–3 annuli long situated just anterior to excretory pore; body slightly narrower posterior to vulva; ovary outstretched, occasionally reflexed; spermatheca and crustaformeria not distinguishable in most of the specimens; in mature females, the spermatheca irregularly rounded without sperm; vulva a transverse slit occupying half of the corresponding body width; vulval lips prominent, the anterior lip protrudes further than the posterior lip; vulval flaps present, but not readily distinct in fresh specimens, observable in preserved specimens; a small, rudimentary post uterine branch present along the ventral body wall; anus indistinct; tail slender, conoid, finely annulated, and gradually tapers to form a finely pointed to rounded terminus, bluntly rounded terminus and tip with peg was observed in Spanish populations.

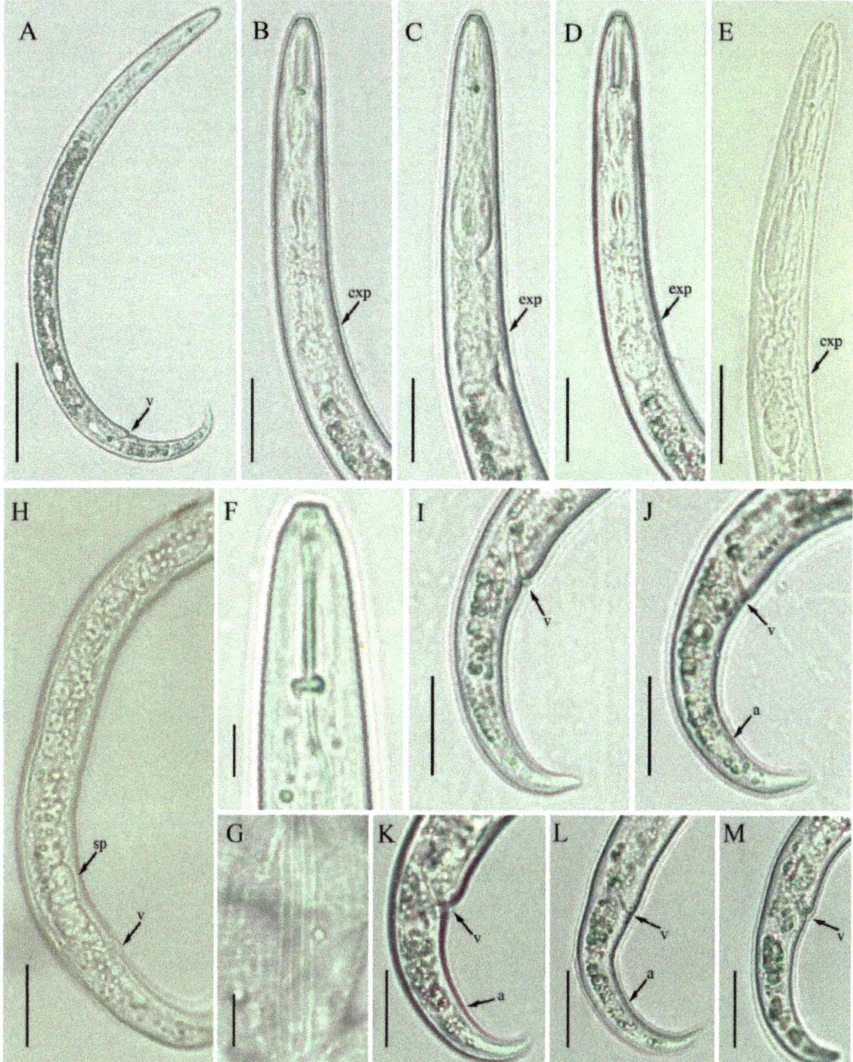

**Figure 3.** Light photomicrographs of *Paratylenchus tateae* female, Canadian population. (**A**) Entire body; (**B–E**) pharyngeal regions; (**F**) lip region; (**G**) lateral lines; (**H**) posterior region with gonad; (**I–M**) tails. Scale bars: (**A**) 50 μm; (**B–E**) 20 μm; (**F,G**) 5 μm; (**H–M**) 20 μm. Arrowheads: (a) Anus; (exp) excretory pore; (sp) spermatheca; (v) vulva.

**Table 2.** Morphometrics of Canadian and Spanish populations of *Paratylenchus lateae*. All measurements are in μm and presented as mean ± standard deviation (range).

| Characters | Canadian Populations | | | Spanish Populations | | Wu & Townshend [27] |
| --- | --- | --- | --- | --- | --- | --- |
| | Females | | Juveniles | Females | | Females |
| Populations | 091 | 041 | 091 | Ariza, Zaragoza | Alpera, Albacete | type population |
| n | 18 | 18 | 6 | 20 | 8 | 43 |
| Body length | 333.6 ± 33.7 (269.0–380.0) | 349.5 ± 25.4 (314.0–388.0) | 315.5 ± 26.4 (267.0–342.0) | 346.2 ± 25.8 (310.0–425.0) | 334.4 ± 14.3 (310.0–353.0) | 315–401 |
| a | 23.4 ± 1.4 (21.4–26.2) | 23.9 ± 1.9 (20.4–27.0) | 22.0 ± 1.9 (18.8–24.0) | 21.8 ± 1.6 (17.4–24.3) | 21.7 ± 1.7 (19.1–23.5) | 19–26 |
| b | 3.6 ± 0.3 (3.2–4.0) | 3.9 ± 0.3 (3.3–4.7) | 3.9 ± 0.2 (3.5–4.1) | 3.7 ± 0.2 (3.3–4.2) | 3.6 ± 0.2 (3.3–4.1) | 3.8–5.9 |
| c | 11.9 ± 1.1 (10.0–13.8) | 13.5 ± 1.4 (11.6–16.9) | 15.4 ± 1.9 (13.5–18.9) | 13.9 ± 1.9 (10.5–17.7) | 13.2 ± 1.8 (11.3–15.3) | 11.7–15.8 |
| c′ | 3.5 ± 0.4 (3.0–4.5) | 3.3 ± 0.3 (2.8–3.9) | 2.6 ± 0.1 (2.4–2.8) | 2.9 ± 0.4 (2.5–3.8) | 2.9 ± 0.2 (2.6–3.1) | - |
| V | 82.3 ± 1.2 (80.0–84.3) | 82.9 ± 0.9 (80.8–84.1) | - | 82.9 ± 1.4 (80.2–85.6) | 82.6 ± 1.4 (81.3–85.0) | 80.5–84.7 |
| Lip height | 2.6 ± 0.2 (2.0–3.0) | 2.8 ± 0.3 (2.0–3.0) | - | - | - | - |
| Lip width | 5.5 ± 0.2 (5.0–6.0) | 5.6 ± 0.4 (5.0–6.0) | - | 5.2 ± 0.4 (4.5–6.0) | 5.2 ± 0.4 (4.5–6.0) | - |
| Stylet length | 17.3 ± 0.9 (15.0–19.0) | 16.5 ± 0.9 (14.5–18.0) | 12.0 ± 1.1 (10.0–13.0) | 15.5 ± 0.4 (14.5–16.0) | 15.4 ± 0.4 (15.0–16.0) | 15–16.8 |
| Median bulb length | 21.9 ± 1.5 (19.4–24.2) | 20.6 ± 2.3 (16.0–24.0) | - | 18.2 ± 1.7 (15.5–22.0) | 17.4 ± 1.2 (16.0–19.0) | - |
| Median bulb width | 8.1 ± 0.6 (7.2–9.0) | 8.2 ± 0.8 (7.2–10.0) | - | 8.9 ± 0.6 (8.0–10.0) | 8.6 ± 0.5 (8.0–9.5) | - |
| Anterior end to excretory pore | 73.9 ± 3.8 (64.0–81.0) | 73.4 ± 5.3 (63.0–84.0) | 66.8 ± 4.8 (60.0–71.0) | 78.2 ± 6.0 (70.5–93.0) | 77.4 ± 3.8 (72.5–84.0) | 68–81 |
| Pharynx length | 91.7 ± 3.4 (83.0–98.0) | 90.2 ± 4.9 (82.0–98.0) | 80.3 ± 2.9 (76.0–83.0) | 93.1 ± 5.0 (85.5–103.0) | 92.1 ± 5.6 (85.5–102.0) | 77–89 |
| Maximum body width | 14.2 ± 1.2 (12.0–16.0) | 14.6 ± 0.8 (13.0–16.0) | 14.3 ± 0.6 (13.0–15.0) | 16.0 ± 1.9 (14.5–21.5) | 15.5 ± 1.3 (14.5–18.5) | - |
| Anal body width | 8.1 ± 0.8 (6.0–9.0) | 7.9 ± 0.6 (7.0–9.0) | 7.9 ± 0.2 (7.5–8.0) | 8.7 ± 0.4 (8.0–9.5) | 8.9 ± 0.9 (8.0–11.0) | - |
| Distance from vulva to anus | 30.6 ± 3.6 (26.0–39.0) | 33.5 ± 4.2 (27.0–43.0) | - | - | - | 28–41 |
| Distance from vulva to tail terminus | 58.8 ± 5.6 (52.0–70.6) | 59.6 ± 4.8 (51.0–67.0) | - | - | - | - |
| Tail length | 28.2 ± 3.0 (24.0–35.0) | 26.1 ± 2.5 (21.0–30.0) | 20.7 ± 1.7 (18.0–23.0) | 25.3 ± 3.3 (21.5–32.5) | 25.6 ± 2.9 (22.5–30.0) | 22–33 |

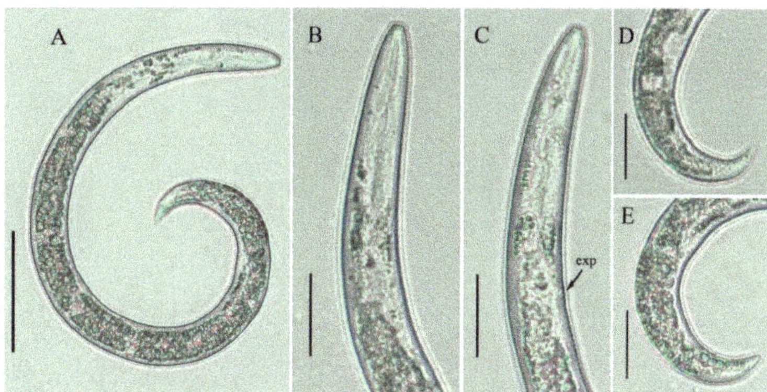

**Figure 4.** Light photomicrographs of *Paratylenchus tateae* juvenile, Canadian population. (**A**) Entire body; (**B,C**) pharyngeal regions; (**D,E**) tails. Scale bars: (**A**) 50 μm; (**B–E**) 20 μm.

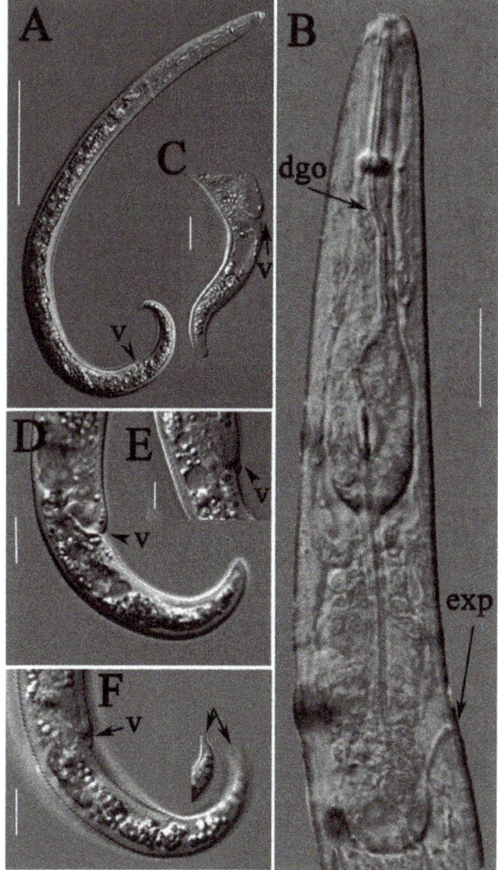

**Figure 5.** Light photomicrographs of *Paratylenchus tateae* female, Spanish population. (**A**) Entire body; (**B**) pharyngeal regions; (**C,D,F**) tails; (**E**) vulval region. Scale bars: (**A**) 50 μm; (**B–F**) 10 μm. Arrowheads: (dgo) Dorsal pharyngeal gland orifice; (exp) excretory pore; (v) vulva.

### 2.2.1. Juveniles

Only one juvenile form was detected. This stage of individuals was similar in morphology to the adult females. However, they were characterized by the presence of weak stylet; pharynx components under-developed; genital primordium under-developed; anus indistinct; posterior body with a finely pointed terminus.

### 2.2.2. Remarks

*Paratylenchus tateae* was originally described from Ontario, Canada, in the rhizosphere of several crops, such as corn, alfalfa, timothy, and white and red clover [27]. After the formal description, the species was reported twice in the literature [23], one of them reported in Saskatchewan [31], however Anderson and Kimpinski [32] collected samples from the same location and considered the Saskatchewan population as *P. labiosus*. The other population was described in India [29], and the author suggests that the Indian population differs from the Canadian population by smaller body length and a more posterior position of the vulva. Additionally, the description of the Indian population includes a rounded head, a disc-like lip region with prominent projecting submedian lobes, and the absence of a post uterine sac. All of these characteristics are contrary to the original description of *P. tateae*, which states the presence of a distinctive truncated lip region, weakly developed spermatheca, and a short, rudimentary post-uterine branch. Based on our current knowledge, we conclude that the Indian population presented by Bajaj [29] might not be *P. tateae*.

Morphologically and morphometrically, *P. tateae* is similar to *P. brevihastus* Wu [33]; the later species was also described in Ontario in the rhizosphere of alfalfa, blue violets, oats, red clover, and grasses. The only characters differentiating *P. tateae* from *P. brevihastus* are the absence of males and weakly developed spermatheca. We do not suggest synonymization here; we are in agreement with Van den Berg et al. [28], who stated that such actions should only be performed after careful molecular and morphological comparisons.

In the present study, we found two populations of *P. tateae* from southern Alberta, and two from Spain. All the populations match with the original description, except for minor differences in body length, as the Alberta population is slightly shorter than the original description (269–380 vs. 315–401 μm), while other characteristics are in the species variability range.

### 2.2.3. Habitat and Locality

Two *P. tateae* populations were found in the potato growing fields of the Municipal District of Taber, Alberta, Canada. The first field was located at latitude 49°46′55.8″ N, longitude—112°21′30.8″ W, whereas the second was located at latitude 49°47′48.5″ N, longitude—112°20′49.6″ W. Two *P. tateae* populations were found in Spain, in the rhizosphere of almond and wheat, at Ariza, Zaragoza province and Alpera, Albacete province, respectively.

*2.3. Description of Female Paratylenchus enigmaticus sp. nov.*

(Figures 6–8; Table 3).

http://zoobank.org/urn:lsid:zoobank.org:act:39C84EDC-15ED-491E-9373-8876D34C35ED.

Body slender, ventrally arcuate to form an open, C-shaped body habitus when heat relaxed; cuticle finely annulated; lateral field equidistant with four distinct lines, outer lines are more prominent than the inner ones; lip region conoid rounded, with anterior end flattened, continuous with the rest of the body; labial framework sclerotization weak; pharyngeal region typical paratylenchoid type; stylet rigid, straight; stylet knobs rounded; dorsal pharyngeal gland opening 4.0–6.0 μm behind stylet knobs; median pharyngeal bulb slender elongate, bearing distinct large valves; isthmus short slender, surrounded by nerve ring; basal bulb pyriform, pharyngeal-intestinal valve rounded; excretory pore situated at the level or anterior to pharyngeal basal bulb; hemizonid 1–2 annuli long situated immediately posterior to excretory pore; body slightly narrower posterior to vulva; ovary outstretched, well developed; spermatheca and crustaformeria well developed; spermatheca rounded; vulva a transverse slit occupying half of the corresponding body width; vulval lips prominent, the anterior lip is protruding further than the posterior lip; vulval flaps present, but not prominent in fresh specimens; a small rudimentary post uterine branch present along the ventral body wall; anus indistinct; the tail slender, conoid, finely annulated, and gradually tapers to form a rounded terminus.

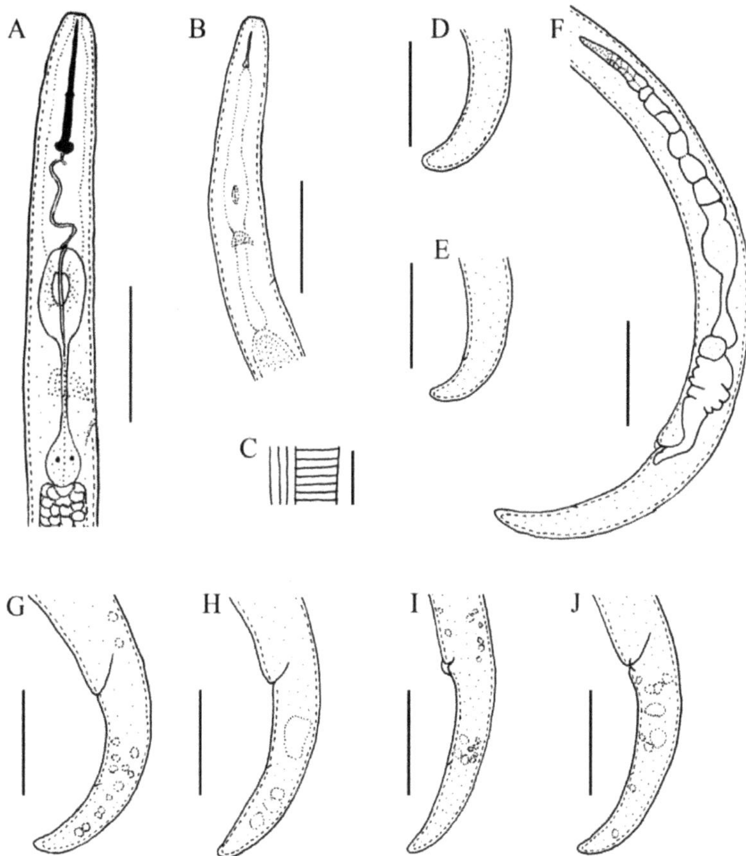

**Figure 6.** Line drawings of *Paratylenchus enigmaticus* sp. nov. (**A**) Pharyngeal region female; (**B**) pharyngeal region juvenile; (**C**) lateral field lines; (**D,E**) juvenile tails; (**F**) posterior region with genital branch; (**G–J**) female tails. Scale bars: (**A,B**) 20 µm; (**C**) 5 µm; (**D–J**) 20 µm.

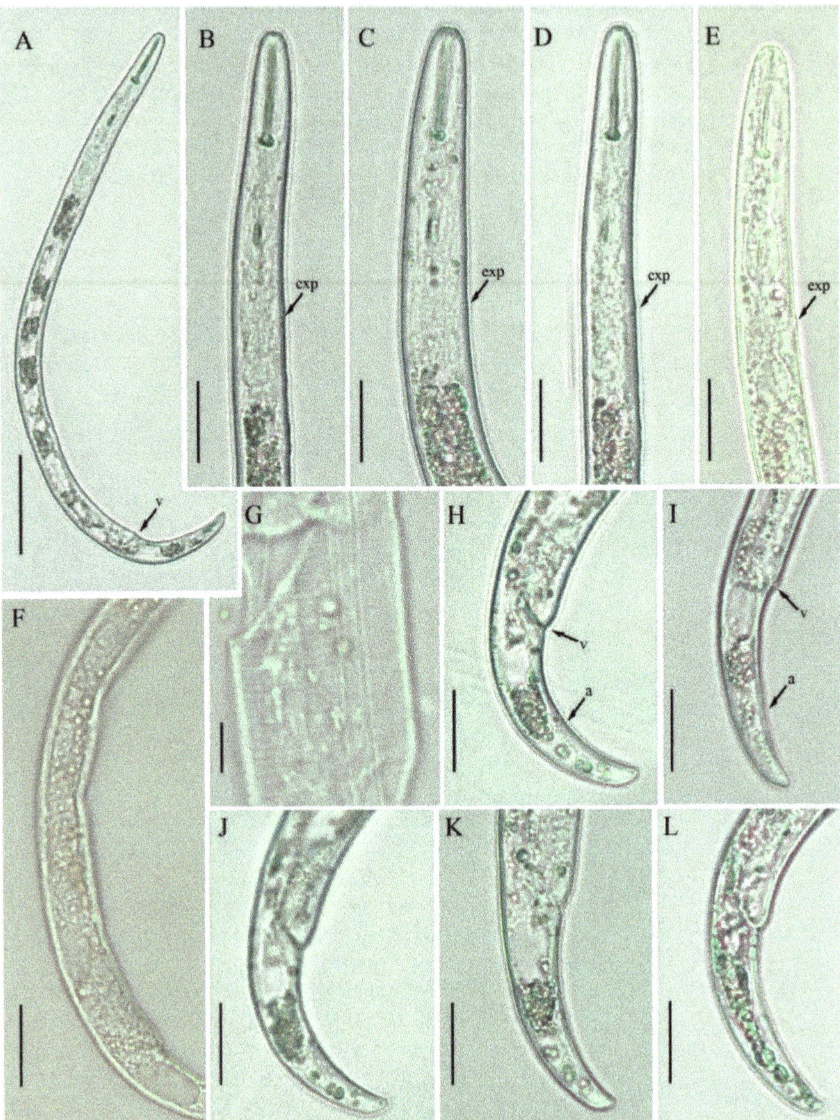

**Figure 7.** Light photomicrographs of *Paratylenchus enigmaticus* sp. nov. female. (**A**) Entire body; (**B–E**) pharyngeal regions; (**F**) posterior region with gonad; (**G**) lateral lines; (**H–L**) tails. Scale bars: (**A**) 50 µm; (**B–F**) 20 µm; (**G**) 5 µm; (**H–L**) 20 µm. Arrowheads: (a) Anus; (exp) excretory pore; (v) vulva.

**Table 3.** Morphometrics of Canadian and Belgian populations of *Paratylenchus enigmaticus* sp. nov. All measurements are in μm and presented as mean ± standard deviation (range).

| | Canadian Population | | | * Belgian Population Claerbout et al. [34] | | | | |
| | Holotype | Paratype | | | | | | |
| Characters | Female | Females | Juveniles | T1 | T2 | T3 | T4 | T5 |
| --- | --- | --- | --- | --- | --- | --- | --- | --- |
| n | | 11 | 5 | 10 | 10 | 10 | 10 | 10 |
| Body length | 372 | 382.7 ± 30.9 (343.0–431.0) | 344.3 ± 9.5 (331.0–357.0) | 365 ± 40 (308–465) | 335 ± 20 (302–360) | 365 ± 39 (313–422) | 358 ± 43 (300–411) | 328 ± 31 (293–368) |
| a | 24.6 | 25.7 ± 2.1 (21.7–28.7) | 23.8 ± 0.4 (23.1–24.4) | 24.2 ± 3.8 (14.9–27.6) | 24.3 ± 3.4 (19.3–27.2) | 26.7 ± 2.3 (22–29) | 23.7 ± 2.6 (18.5–27.5) | 23.2 ± 3.3 (18.1–28.1) |
| b | 3.9 | 4.1 ± 0.3 (3.7–4.7) | 4.2 ± 0.2 (3.9–4.4) | 3.7 ± 0.7 (2.7–4.6) | - | 3.4 ± 0.7 (2.5–4.9) | 3.2 ± 0.5 (2.8–4.2) | - |
| c | 15.7 | 15.4 ± 1.3 (12.9–17.5) | 14.9 ± 0.5 (14.4–15.7) | 15.0 ± 1.5 (12.3–17.2) | 14.9 ± 1.5 (13.2–17) | 14.9 ± 1.9 (12.7–17.8) | 14.8 ± 2.3 (13.7–19.8) | 13.0 ± 1.5 (10.1–15.7) |
| c' | 2.5 | 2.6 ± 0.3 (2.3–3.1) | 2.3 ± 0.3 (1.9–2.6) | - | - | - | - | - |
| V | 84.1 | 85 ± 0.9 (83.0–86.3) | - | 83.2 ± 2.1 (80.4–87.8) | 83.2 ± 2.1 (80–87) | 83.0 ± 1.5 (80–84) | 83.5 ± 0.9 (82.8–84.9) | 83.1 ± 2.1 (80.1–88) |
| Lip height | 3.1 | 3.0 ± 0.3 (2.6–3.6) | - | - | - | - | - | - |
| Lip width | 7.5 | 7.1 ± 0.4 (6.5–7.7) | - | - | - | - | - | - |
| Stylet length | 28.9 | 28.8 ± 1.1 (27.3–30.8) | 12.5 ± 0.9 (11.2–13.5) | 27.3 ± 1.3 (23.5–28.4) | 25.5 ± 1.6 (22.3–26.5) | 26.6 ± 1.5 (25.2–30.5) | 26.8 ± 1.3 (24.6–27.9) | 27.0 ± 1.5 (24.6–28.6) |
| Stylet percentage | 7.7 | 7.6 ± 0.5 (6.8–8.2) | - | 7.5 ± 0.9 (6.0–8.8) | 7.6 ± 0.7 (7.2–8.8) | 7.3 ± 0.7 (6.2–7.9) | 7.6 ± 0.8 (6.6–8.4) | 8.3 ± 0.5 (7.3–8.9) |
| Median bulb length | 21.2 | 20.4 ± 1.0 (18.5–21.3) | - | - | - | - | - | - |
| Median bulb width | 9.8 | 9.6 ± 1.1 (8.0–11.4) | - | - | - | - | - | - |
| Anterior end to excretory pore | 79 | 76.0 ± 4.2 (70.0–82.0) | 65.2 ± 2.8 (63.0–70.0) | - | - | - | - | - |
| Pharynx length | 95 | 93.8 ± 5.2 (83.0–100.0) | 81.6 ± 4.3 (76.0–88.0) | 100.7 ± 19.7 (75.2–137.7) | 88.0 ± 23.3 (42.9–105.8) | 109.9 ± 16.9 (83.3–123.5) | 114.7 ± 18.4 (84.6–125.7) | 120.4 ± 14.6 (95.0–144.0) |
| Maximum body width | 15.1 | 15.0 ± 1.2 (12.6–16.4) | 14.4 ± 0.3 (14.2–14.9) | - | - | - | - | - |
| Vulva body width | 12.7 | 13.1 ± 1.0 (11.4–14.7) | - | - | - | - | - | - |
| Anal body width | 9.5 | 9.7 ± 0.9 (7.7–10.6) | 10.2 ± 1.2 (8.8–11.7) | - | - | - | - | - |
| Distance from vulva to anus | 36 | 33.3 ± 4.0 (26.0–37.0) | - | - | - | - | - | - |
| Distance from vulva to tail terminus | 59.6 | 59.9 ± 3.1 (53.4–65.0) | - | - | - | - | - | - |
| Tail length | 23.6 | 24.9 ± 2.1 (22.0–29.0) | 23.2 ± 0.8 (22.0–24.0) | 24.4 ± 3.1 (21.7–30.8) | 22.6 ± 1.6 (20.3–26.2) | 24.6 ± 1.8 (21.0–26.1) | 24.5 ± 3.3 (21.2–23.7) | 25.4 ± 2.6 (22.0–30.0) |

* Belgian populations (T1–T5) represent measurement of females.

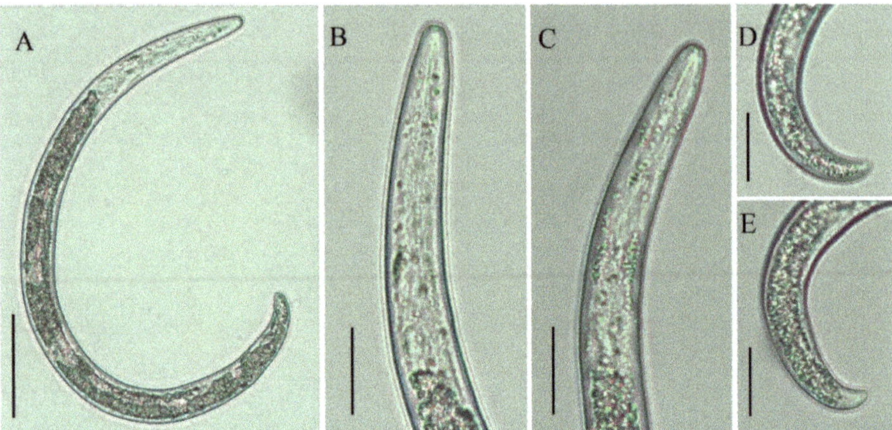

**Figure 8.** Light photomicrographs of *Paratylenchus enigmaticus* sp. nov. juvenile. (**A**) Entire body; (**B,C**) pharyngeal regions; (**D,E**) tails. Scale bars: (**A**) 50 µm; (**B–E**) 20 µm.

### 2.3.1. Juvenile

Only one form was detected. This stage of individuals was similar in morphology to the adult females. However, they were characterized by the presence of weak stylet; underdeveloped pharynx components; underdeveloped genital primordium; indistinct anus; and posterior body with a rounded terminus.

### 2.3.2. Diagnosis and Relationship

The new species is characterized by the presence of 4 lateral lines, advulval flaps, and a moderate stylet length of 28.8 (27.3–30.8) µm. The lip region is conoid rounded, with the anterior end flattened, continuous with the rest of the body. The excretory pore is situated at the level or anterior to the pharyngeal basal bulb. The spermatheca is rounded, and a small rudimentary post uterine branch is present. The tail conoid gradually tapers to form a rounded terminus.

Morphologically, the new species is close to *P. dianthus*, *P. neoprojectus*, *P. nanus* Cobb, [35] and *P. projectus*. The new species can be differentiated from *P. dianthus* by lip region morphology (conoid rounded vs. truncate), presence of small post uterine sac (vs. absent), tail terminus morphology (broadly rounded vs. finely rounded, rarely clavate, or sometimes digitate), and higher c′ value (3.5 (3.0–4.5) vs. 2.5). From *P. neoprojectus*, the new species can be differentiated by lip region morphology (conoid rounded vs. rounded), tail terminus morphology (broadly rounded vs. conically rounded), and position of excretory pore (at the level or anterior to pharyngeal bulb vs. at the level or middle of pharyngeal bulb). From *P. nanus* it differs by lip region morphology (conoid rounded vs. rounded), tail terminus morphology (broadly rounded vs. subacute to rounded, slightly indented), and shorter stylet length (28.8 (27.3–30.8) µm vs. 32–34 µm). From *P. projectus*, the new species differs by lip region morphology (conoid rounded vs. offset, conoid truncate, or trapezoid), presence of small post uterine sac (vs. absent), tail terminus morphology (broadly rounded vs. rounded dorsally sinuate), shorter stylet length (28.8 (27.3–30.8) µm vs. 25–37 µm), and higher c′ value (3.5 (3.0–4.5) vs. 2.7).

### 2.3.3. Remarks

The species was first found (but not described) in the glasshouse-grown lettuce from Belgium. The species causes damage to the root system, but this was not related to significant yield reduction in lettuce heads [34]. In the present study, same species was found in the potato growing region of southern Alberta. In the Belgian population, the authors noted the presence of a large proportion of pre-adults 51–96% and stated this might be due to soil disturbance [34]. The Canadian population also exhibits the same feature; the juveniles were observed in higher numbers than females. Morphological, molecular, and morphometrical comparisons indicate that the Canadian and the Belgian populations are conspecific, and in this study are described as *P. enigmaticus* sp. nov.

### 2.3.4. Type Habitat and Locality

*Paratylenchus enigmaticus* sp. nov. was found in a potato field (latitude 49°42′34.3″ N; longitude—112°3′54.1″ W); the municipal district of Taber, Alberta, Canada.

### 2.3.5. Etymology

The species name, *enigmaticus*, refers to the species identity remaining unresolved for several months.

### 2.3.6. Type Material

Holotype female, 9 paratypes females, and 2 juveniles (7 slides, numbers UL-DY1-01 to UL-DY1-07) and additional 5 slides containing females were deposited in the Nematode Collection of the University of Lethbridge, Alberta, Canada. Two females and three juveniles were deposited in the Nematode Collection of the Institute for Sustainable Agriculture, CSIC, Córdoba, Spain.

### *2.4. Molecular Characterization and Phylogenetic Analysis of Paratylenchus Populations from Canada and Spain*

The amplification of the D2–D3 expansion domains of the 28S rRNA, ITS region, and 18S rRNA genes of *Paratylenchus* populations yielded single fragments of ~1000 bp, 800 bp, and 800 bp, respectively. Ten new sequences from the D2–D3 expansion domains of the 28S rRNA gene, 11 from ITS, and two new sequences from the 18S rRNA gene were obtained in this study.

The D2–D3 expansion domains of the 28S rRNA sequences of *P. enigmaticus* sp. nov. (MW282760–MW282761) and *Paratylenchus* sp. T1–T5 (MN535542–MN535545) from Belgium showed no intraspecific variability (100% similarity) from each other. The sequence identities of *P. enigmaticus* sp. nov. with *Paratylenchus* sp. T1–T5 from Belgium, *P. tenuicaudatus* Wu [36] (KU291239, from Iran), and *P. tateae* (MW282754–MW282759) were 99% (1 bp difference and 0 indels), 95% (38 bp difference and 1 indel), and 99% (3–4 different nucleotides and 0 indels), respectively. Similarly, the D2–D3 sequences of *P. tateae* from Canada and Spain showed low intraspecific variability (99% similarity). The sequence identities of *P. tateae* with *P. sheri* Raski [37] (MN088374, from Iran), and *P. similis* Khan, Prasad, Mathur [38] (MN088375, from Iran) were 99% (differed in 5 nucleotides and 0 indels) and 98% (differed in 16 bp and 0 indels). *Paratylenchus neoprojectus* (MW282762–MW282763) sequences obtained in this study differs in 0–7 nucleotides and 0 indels (99–100% similarity) from sequences of *P. neoprojectus* (=*P. nanus* type B) from USA (KF242201, MH790252, MH6722687, MH237651), South Korea (KY468900, KY468899, KF242199, KY468901) and South Africa (KF242200, KF242198). Finally, Canadian *P. neoprojectus* sequence differs in 10 nucleotides and 0 indels (98% similarity) from a short 542 bp sequence of *P. coronatus* Colbran [39] (MK506808) from Iran.

The ITS sequences of Canadian and Spanish populations of *P. tateae* MW282766–MW282771) showed lower intraspecific variability at 99% similarity with 3 different nucleotides and 1–2 indels. The ITS sequences of *P. neoprojectus* (MW282775–MW282776) and *P. enigmaticus* sp. nov. showed low intraspecific variability with 4 and 1–11 different nucleotides, respectively, and 0–3 indels (98–99% similarity). The ITS sequences of *P. enigmaticus* sp. nov. (MW282772–MW282774) and *Paratylenchus* sp. T1–T5 from Belgium (MN535542–MN535545) are very similar, with 97% similarity (16–17 nucleotides difference, 4 indels), whereas the other close species, i.e., *P. hamatus* (KF242253, KF242246), *P. tenuicaudatus* (KF24226, KF242261), and *Paratylenchus* sp. SAS (KF242243) from the USA showed 90–91% (60–71 nucleotides difference, 13–18 indels) similarity with *P. enigmaticus* sp. nov.

The *P. neoprojectus* sequence of the Canadian population differs in 4–25 nucleotides and 0–7 indels (97–99% similarity) from sequences of *P. neoprojectus* (=*P. nanus* type B) from USA (MH236098), South Korea (MN710514, MN710515, KY468905, KY468904), and South Africa (KF242264, KF242263). The molecular information in the NCBI database regarding the 18S rRNA gene of pin nematode species is insufficient to calculate the sequence identities for this marker because few sequences have been deposited and there are not many molecular differences between species.

Phylogenetic relationships among *Paratylenchus* species inferred from analyses of the D2–D3 expansion domains of 28S rRNA, ITS region, and partial 18S rRNA sequences using BI are shown in Figures 9–11, respectively. The phylogenetic trees generated from the three nuclear markers, included 89, 81, and 50 sequences, with 680, 875, and 1610 nucleotides, respectively.

The D2–D3 expansion domains of the 28S rRNA phylogenetic tree of *Paratylenchus* spp. showed two main clades, one highly supported (PP = 1.00), including the three species described in this study, and another weakly supported (PP = 0.51), including several *Paratylenchus* spp.; most of them with a longer stylet (>40 μm; Figure 9). The *P. enigmaticus* sp. nov. clustered together in a highly supported subclade (PP = 1.00) with sequences of *Paratylenchus* sp. T1–T5 from Belgium, and was well separated (PP = 0.98) from *Paratylenchus* sp. A (AY780945) from California, USA (Figure 9). Moreover, *P. neoprojectus* clustered together in a highly supported subclade (PP = 1.00) with sequences of *P. neoprojectus* (=*P. nanus* type B) and *P. coronatus* (MK506808). It is also noted that the sequence of *P. sheri* (MN088374) provided by Mirbabaei et al. [40] grouped with the Canadian and Spanish populations of *P. tateae*. The molecular identities suggest that this sequence belongs to *P. tateae* instead of *P. sheri*. The morphological and molecular details associated with the *P. sheri* sequence suggest a possible error in the sequencing. It is therefore recommended to use the same specimen for morphological and molecular studies. Consequently, we consider MN088374 as *P. tateae* in our study.

The 50% majority rule consensus ITS BI tree also shows 2 clades, one representing short stylet species, including the three species described in this study, and the second containing mostly long stylet species (Figure 10). Likewise, the D2–D3 expansion domains of the 28S rRNA tree, *P. enigmaticus* sp. nov. grouped with *Paratylenchus* sp. T1–T5 from Belgium (PP = 1.00), and shares a clade with *P. hamatus*, *P. tenuicaudatus*, and *Paratylenchus* sp. SAS. Canadian and Spanish populations of *P. tateae* grouped with several populations of *P. neoprojectus* (PP = 0.91).

Finally, the phylogenetic relationships of *Paratylenchus* species inferred from analysis of partial 18S rRNA gene sequences shows two clades that are well defined (Figure 11), but several subclades that do not resolve well in the clade include *P. enigmaticus* sp. nov. (MW282764) and *P. neoprojectus* (MW282765).

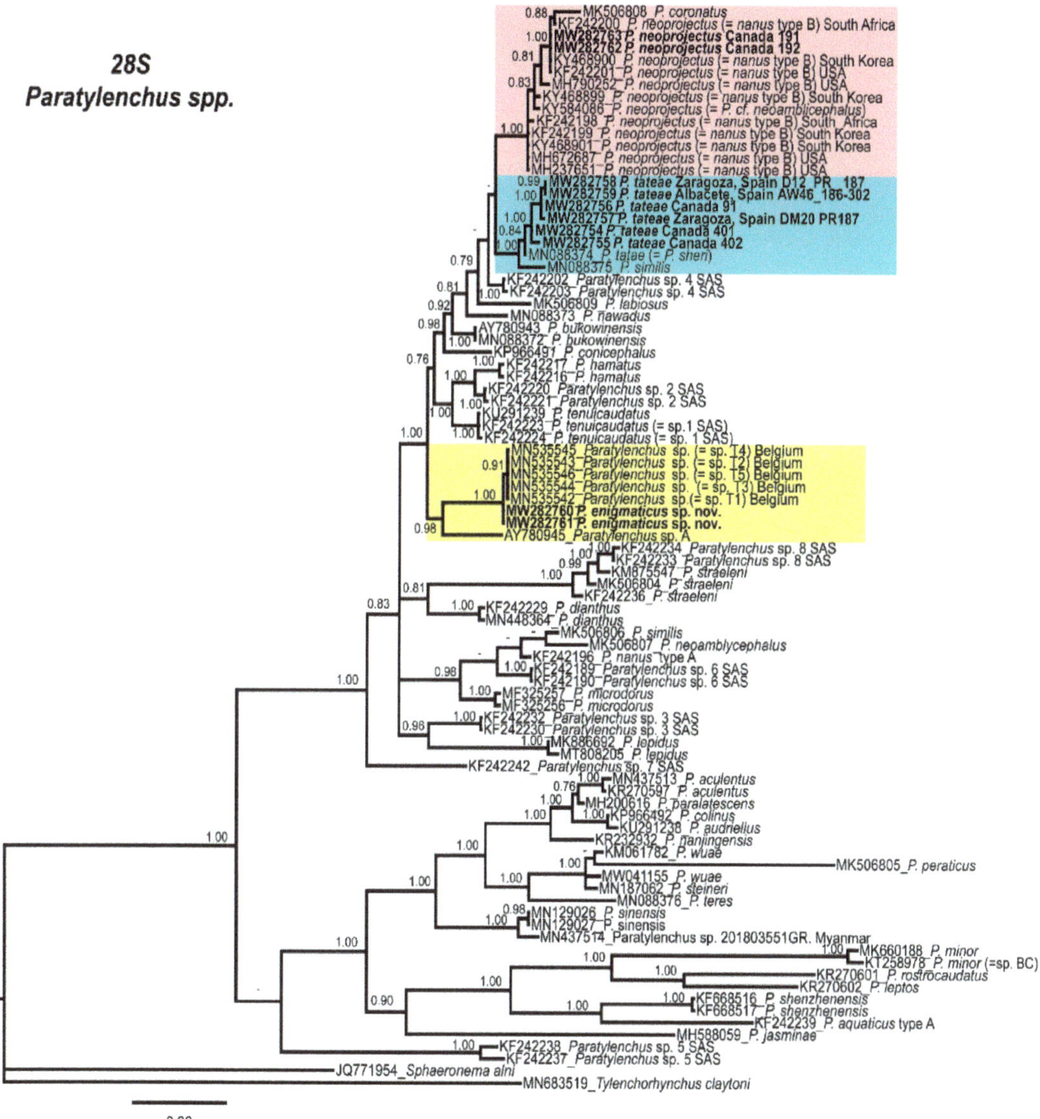


**Figure 9.** Phylogenetic relationships within the genus *Paratylenchus*. Bayesian 50% majority rule consensus tree as inferred from the D2–D3 expansion domains of the 28S rRNA sequence alignment under the general, time-reversible model of sequence evolution with correction for invariable sites and a gamma-shaped distribution (GTR + I+ G). Posterior probabilities of more than 0.70 are given for appropriate clades. Newly obtained sequences in this study are shown in bold. The scale bar indicates expected changes per site.

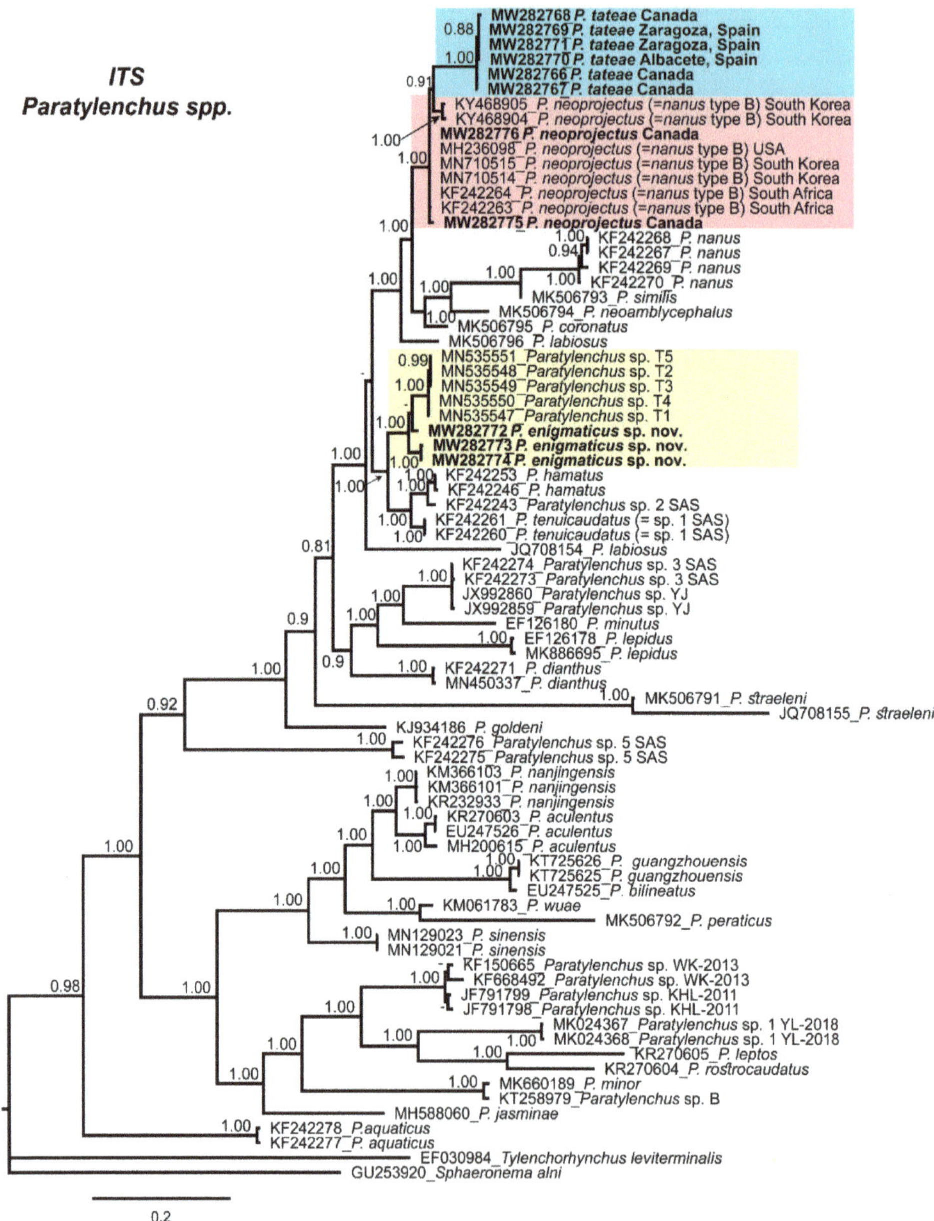



**Figure 10.** Phylogenetic relationships within the genus *Paratylenchus*. Bayesian 50% majority rule consensus tree as inferred from ITS rRNA sequence alignment under the general, time-reversible model of sequence evolution with correction for invariable sites and a gamma-shaped distribution (GTR + I+ G). Posterior probabilities greater than 0.70 are given for the corresponding clades. Newly obtained sequences in this study are shown in bold. The scale bar indicates expected changes per site.

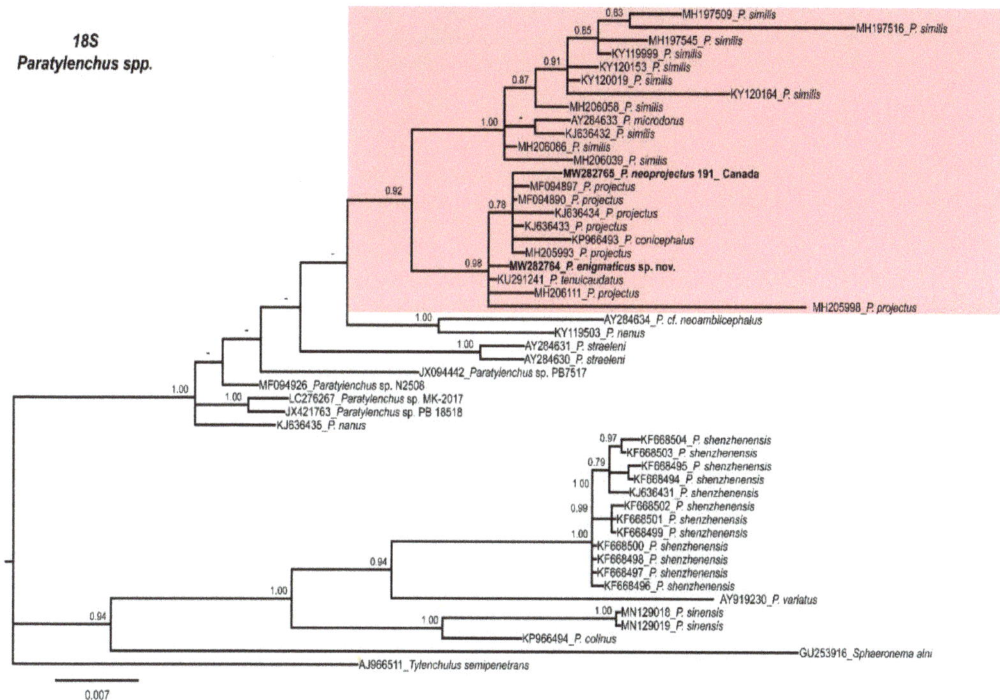



**Figure 11.** Phylogenetic relationships within the genus *Paratylenchus*. Bayesian 50% majority rule consensus tree as inferred from the partial 18S rRNA sequence alignment under the general, time-reversible model of sequence evolution with correction for invariable sites and a gamma-shaped distribution (GTR + I+ G). Posterior probabilities greater than 0.70 are given for appropriate clades. Newly obtained sequences in this study are shown in bold. The scale bar indicates expected changes per site.

*Paratylenchus* is a large genus that comprises short and long stylet species [23]. The majority of short stylet species are considered pathogenic and cause significant damage to their host plants [22]. So far, six short stylet species from Canada have been reported, namely *P. brevihastus*, *P. labiosus*, *P. neoprojectus*, *P. projectus*, *P. tateae*, and *P. tenuicaudatus*. All of these are Canadian native species except *P. projectus*, which is a cosmopolitan species known to have a global distribution [23].

Morphological identification of *Paratylenchus* species is difficult because of their variable characters and overlapping morphometrical values. Stylet length, number of lateral lines, and presence/absence of vulva flaps are considered to be robust characters for species differentiation; however, body length, tail length and shape, position of excretory pore, and ratios of c, c′ were concluded to be unreliable for species separation [25,41,42]. As the majority of *Paratylenchus* species presents a limited selection of differences in morphology, several nematologists have attempted to synonymize morphologically similar species. For example, Brzeski [43] synonymized *P. tateae*, *P. labiosus*, and *P. italiensis* with *P. similis*, because of their similar morphology and overlapped morphometrical values. Ghaderi et al. [25] accepted the synonymization of *P. similis* and *P. tateae*; however, with the availability of molecular data, the same authors [23] rejected the change and referred to both species as valid taxa, and also commented that several populations of *P. similis* may indeed be *P. tateae*. Bahmani et al. [44] also presented a detailed argument on the validity of *P. labiosus*, which was supported by molecular data in Mirbabaei et al. [40].

The possible presence of species complexes in pin nematodes was highlighted by Van den Berg et al. [28] and Mirbabaei et al. [40]. We are in agreement with the authors that similar appearances and overlapping morphometrical characters may present difficulties in ascertaining species status. Nevertheless, such morphological complexes can be resolved using molecular data. Several taxonomic issues have been successfully addressed with molecular studies, such as the validity and differentiation of *Radopholoides* from *Hoplotylus* and *Radopholus* [45], the transfer of *Tylaphelenchus jiaae* to the genus *Pseudaphelenchus* as *P. jiaae* [46], the revision and species synonymization in *Laimaphelenchus* [47], the species delimitation in members of Criconematoidea [48–51], and the resolution of the cryptic diversity and species complexes in Longidoridae [52–54].

Our phylogenetic analysis of D2–D3 expansion domains of the 28S rRNA also indicates that the status of *P. nanus* type B [28] and *P. sheri* [40] need detailed revision. By comparing all the available molecular and morphometric data from both species, it is evident that *P. nanus* type B is a population of *P. neoprojectus* and *P. sheri* is a population of *P. tateae*. Additionally, our *P. enigmaticus* sp. nov. appears conspecific with the Belgian population (T1–T5). It is notable that molecular data not only resolve the taxonomic issues, but also aids in eliminating the propagation of redundant data.

In the literature, several studies have outlined a wide host range [55–57] and survival abilities of pin nematodes [58,59]. Biologically, the final juvenile stage of certain species of pin nematode constitutes the highest portion of the total population. Rhoades and Linford [58] and Wood [15] refer to this stage as a resistant non-feeding stage which is more capable of withstanding desiccation and sudden freezing than the younger and adult stages.

The Canadian and Belgian populations of *P. enigmaticus* sp. nov. have a higher proportion of juveniles than adults, whereas *P. tateae* and *P. neoprojectus* have higher quantities of females than juveniles. It appears that *P. enigmaticus* sp. nov. has a resistant stage; however, the presence of such a stage needs confirmation through further study.

There are limited data regarding the prevalence of pin nematodes in the potato growing areas of southern Alberta and other parts of Canada. Thus far, *P. labiosus* and *P. projectus* are the only species detected in the potato growing areas of Prince Edward Island and New Brunswick [13,32,60,61]. In the present study, we identified *P. neoprojectus*, *P. tateae*, and *P. enigmaticus* sp. nov. in southern Alberta, along with *P. tateae* populations from Spain, using an integrative taxonomical approach. Our study also underscores the importance of using molecular data for accurate species identification and clarifying the status of *P. nanus* type B and *P. sheri*.

Lower densities of identified species in the samples suggest that these are mild parasitic species and, as of yet, do not behave as potential pests. However, pin nematodes have a reputation of building high population densities in short periods, and, under favorable circumstances, can be a threat to their hosts [22,34]. Indeed, a higher incidence of root lesion nematodes (*Pratylenchus* spp.) in southern Alberta has been reported by Forge et al. [4]. Having that in mind, the densities of pin nematodes are worth monitoring as some species can penetrate roots through existing entry points and may aggravate the plant damage. Therefore, further studies are required to assess species-specific yield losses and thresholds.

## 3. Materials and Methods

### 3.1. Isolation and Morphological/Morphometrical Studies

Nematodes were extracted from soil samples using the modified Cobb sieving and flotation-centrifugation method [62]. For preliminary examinations, fresh nematodes were transferred to the drop of distilled water, heat relaxed at 60 °C for 30–45 s, and observed under the Zeiss Axioskope 40 microscope. Permanent mounts were prepared as described in Seinhorst [63] and De Grisse [64]. Light micrographs of the mounted specimens were acquired using a Zeiss Axioskope 40 microscope equipped with a Zeiss Axiocam 208 camera (Carl Zeiss Microscopy, Jena, Germany). Standard morphometrical characters were selected based on previously published studies [25,28,57,65]. Measurements were made using ZEN blue 3.1 imaging software (Carl Zeiss Microscopy).

### 3.2. DNA Extraction, PCR and Sequencing

Nematode DNA was prepared according to Maria et al. [65]. Three sets of DNA primers (Integrated DNA Technologies, Coralville, IA, USA) were used in the PCR analyses to amplify the nucleotide sequences of the partial 18S, D2–D3 expansion domains of the 28S rRNA and ITS of ribosomal genes, including 5.8S rRNA and both ITS regions (ITS1 and ITS2) (rRNA). The partial 18S rRNA region was amplified with 1813F and 2646R primers [66]. The D2–D3 expansion domains of the 28S rRNA regions were amplified using 28–81F and 28–1006rev primers [67], and the ITS region was amplified using F194 [68] and AB28 primers [69]. The ribosomal gene cluster (whole rDNA cistron) is a multicopy, tandem repeated array in the genome. Each repeat is transcribed as a single rRNA precursor and cleaved, leading to the mature small subunit rRNA (SSU), the mature 5.8S rRNA, and the mature large subunit rRNA (LSU). The SSU is separated from the 5.8S rRNA by the first internal transcribed spacer (ITS1), and the second internal transcribed spacer (ITS2) is located between the 5.8S rRNA and the LSU [70]. A nice scheme of these repeats and the position of many of the primers used by nematologists could be found in Carta and Li [71]. The PCR conditions were as described in Holterman et al. [66,67] and in Ferris et al., [68]. Amplified PCR products were resolved by electrophoresis in 1% agarose gels and visualized by staining with GelRed (Biotium, Fremont, CA, USA). Amplified DNA fragments were purified using an E.Z.N.A Gel Extraction kit (Omega Biotek, Norcross, GA, USA), following manufacturer's instructions, ligated into the pJET1.2 vector (Thermo Fisher Scientific, Mississauga, ON, Canada), and introduced into *Escherichia coli* DH5α competent cells (Thermo Fisher Scientific). The presence of the PCR-derived inserts in the plasmids from transformed *E. coli* cells was confirmed by PCR. Plasmid DNA was isolated and purified using E.Z.N.A Plasmid DNA minikit I (Omega Biotek), according to the manufacturer's instructions, and sent to Genewiz, Inc for DNA sequencing (South Plainfield, NJ, USA). DNA sequences were aligned using the Bioedit sequence alignment tool and compared for similarities with all known nematode species sequences in the GenBank database.

### 3.3. Phylogenetic Analyses

Sequenced genetic markers from the nematodes examined in the present study (after discarding primer sequences and ambiguously aligned regions) and several pin nematode sequences obtained from GenBank were used in the phylogenetic reconstruction. Outgroup taxa for each dataset were selected based on previously published studies [57]. Multiple sequence alignments of the newly obtained and published sequences were made using the FFT-NS-2 algorithm of MAFFT V.7.450 [72]. Sequence alignments were visualized with BioEdit [73] and manually edited using Gblocks ver. 0.91b [74] in the Castresana Laboratory server (http://molevol.cmima.csic.es/castresana/Gblocks_server.html) using options for a less stringent selection (minimum number of sequences for a conserved or a flanking position: 50% of the number of sequences +1; maximum number of contiguous non-conserved positions: 8; minimum length of a block: 5; allowed gap positions: With half).

Phylogenetic analyses of the sequence datasets were conducted based on Bayesian inference (BI) using MRBAYES 3.2.7a [75]. The best-fit model of DNA evolution was calculated with the Akaike information (AIC) of JMODELTEST V.2.1.7 [76]. The best-fit model, base frequency, proportion of invariable sites, substitution rates and gamma distribution shape parameters in the AIC were used for phylogenetic analyses. BI analyses were performed under a general time reversible model, with a proportion of invariable sites and a rate of variation across sites (GTR + I + G) for the partial 18S rRNA, D2–D3 expansion domains of the 28S rRNA, and ITS region sequences. These BI analyses were run separately per dataset with four chains for $2 \times 10^6$ generations. The Markov chains were sampled at intervals of 100 generations. Two runs were conducted for each analysis. After discarding burn-in samples of 20% and evaluating convergence, the remaining samples were retained for more in-depth analyses. The topologies were used to generate a 50% majority-rule consensus tree. Posterior probabilities (PP) are given on appropriate clades.

Trees from all analyses were edited using FigTree software V.1.4.4 (http://tree.bio.ed.ac. uk/software/figtree/).

**Author Contributions:** Conceptualization, M.M. and D.P.Y.; methodology, M.M., D.P.Y., J.E.P.-R., and P.C.; software, M.M., J.E.P.-R., and P.C.; validation, M.M., D.P.Y., P.C. and J.E.P.-R.; formal analysis, M.M. and J.E.P.-R.; investigation, M.M., D.P.Y. and J.E.P.-R.; resources, D.P.Y., P.C.; data curation, M.M. and J.E.P.-R.; writing—original draft preparation, M.M., D.P.Y. and J.E.P.-R., writing—review and editing, M.M., D.P.Y., P.C. and J.E.P.-R.; visualization, M.M. and J.E.P.-R.; supervision and project administration, D.P.Y.; funding acquisition, D.P.Y. and P.C. All authors have read and agreed to the published version of the manuscript.

**Funding:** This work was supported by the Potato Early Dying Complex project funded by the University of Lethbridge Research Operating Fund, and the Canadian Potato Early Dying Network project funded by the Canadian Agri-Science Cluster for Horticulture 3 grant to D.P.Y., in collaboration with the Potato Growers of Alberta, McCain Foods Canada Ltd., Cavendish Farms Corp. and Lamb Weston Inc.; and by the Spanish Ministry of Science, Innovation and Universities, grant number "RTI2018-095925-A-I00" to P.C. and J.E.P.-R.

**Institutional Review Board Statement:** Not applicable.

**Informed Consent Statement:** Not applicable.

**Data Availability Statement:** The datasets generated during and/or analyzed during the current study are available from the corresponding author on reasonable request.

**Acknowledgments:** We thank potato growers in Alberta, Canada, for providing access to their fields, and Mariana Vetrici (University of Lethbridge, AB, Canada) for the collection of soil samples. We also thank Carolina Cantalapiedra-Navarrete (Institute for Sustainable Agriculture (IAS), CSIC, Spain), for the excellent technical assistance in molecular analyses.

**Conflicts of Interest:** The authors declare no conflict of interest.

## References


1. Statistics Canada. Table 32-10-0358-01 Area, production and farm value of potatoes. *Stat. Can.* **2020**. [CrossRef]
2. Potter, J.; McKeown, A. Fluctuations of populations of the pin nematode *Paratylenchus projectus* under selected potato management practices. *Phytoprotection* **2002**, *83*, 147–155. [CrossRef]
3. Mahran, A.; Tenuta, M.; Shinners-Carnelly, T.; Mundo-Ocampo, M.; Daayf, F. Prevalence and species identification of *Pratylenchus* spp. in Manitoba potato fields and host suitability of 'Russet Burbank'. *Can. J. Plant Pathol.* **2010**, *32*, 272–282. [CrossRef]
4. Forge, T.A.; Larney, F.J.; Kawchuk, L.M.; Pearson, D.C.; Koch, C.; Blackshaw, R.E. Crop rotation effects on *Pratylenchus neglectus* populations in the root zone of irrigated potatoes in southern Alberta. *Can. J. Plant Pathol.* **2015**, *37*, 363–368. [CrossRef]
5. Loof, P.A.A. *Paratylenchus projectus, CI.H., Descriptions of Plant-Parasitic Nematodes*; Set 5, No. 71; Commonwealth Agricultural Bureau: St Albans, UK, 1975.
6. Siddiqi, M.R. *Tylenchida: Parasites of Plants and Insects*, 2nd ed.; CABI Publishing: Wallingford, UK, 2000; p. 833.
7. Wang, K.; Li, Y.; Xie, H.; Wu, W.J.; Xu, C.L. Pin nematode slow decline of *Anthurium andreanum*, a new disease caused by the pin nematode *Paratylenchus shenzhenensis*. *Plant Dis.* **2016**, *100*, 940–945. [CrossRef] [PubMed]
8. Pereira, G.F. Survey of Plant-Parasitic Nematodes in Pulse Crop Fields of the Canadian Prairies. Master's Thesis, University of Manitoba, Winnipeg, MB, Canada, 2018.
9. Townshend, J.L.; Eggens, J.L.; McCollum, N.K. Occurrence and population densities of nematodes associated with forage crops in eastern Canada. *Can. Plant Dis. Surv.* **1973**, *53*, 131–136.
10. Townshend, J.L.; Potter, J.W.; Marks, C.F.; Loughton, A. The pin nematode, *Paratylenchus projectus*, in rhubarb in Ontario. *Can. J. Plant Sci.* **1973**, *53*, 377–381. [CrossRef]
11. Townshend, J.L.; Potter, J.W. Evaluation of forage legumes, grasses, and cereals as hosts of forage nematodes. *Nematologica* **1976**, *22*, 196–201. [CrossRef]
12. Townshend, J.L.; Cline, R.A.; Driks, V.A.; Marks, C.F. Assessment of turfgrasses for the management of *Pratylenchus penetrans* and *Paratylenchus projectus* in orchards. *Can. J. Plant Sci.* **1984**, *64*, 355–360. [CrossRef]
13. Kimpinski, J. Nematodes associated with potato in Prince Edward Island and New Brunswick. *Ann. App. Nematol.* **1987**, *1*, 17–19.
14. Jenkins, W.R. *Paratylenchus projectus*, new species (Nematoda: Criconematidae), with a key to the species of *Paratylenchus*. *J. Wash. Acad. Sci.* **1956**, *46*, 296–298.
15. Wood, F.H. Biology and host range of *Paratylenchus projectus* Jenkins, 1956 (Nematoda: Criconematidae) from a sub-alpine tussock grassland. *N. Z. J. Agric. Res.* **1973**, *16*, 381–384. [CrossRef]
16. Micoletzky, H. Die Freilebenden Erd-Nematoden. *Archiv für Naturgeschichte Berlin A* **1921**, *87*, 1–650.

17. Jenkins, W.R.; Taylor, D.P. *Paratylenchus dianthus*, n. sp. (Nematoda, Criconematidae), a parasite of carnation. *Proc. Helminthol. Soc. Wash.* **1956**, *23*, 124–127.

18. Thorne, G.; Allen, M.W. *Paratylenchus hamatus* n. sp. and *Xiphinema index* n. sp., two nematodes associated with fig roots, with a note on Paratylenchus anceps Cobb. *Proc. Helminthol. Soc. Wash.* **1950**, *17*, 27–35.

19. Andrássy, I. Neue und wenig bekannte nematoden aus Jugoslawien. *Ann. Hist.-Nat. Musei Natl. Hung.* **1959**, *51*, 259–275.

20. Geraert, E. The genus *Paratylenchus*. *Nematologica* **1965**, *11*, 301–334. [CrossRef]

21. Wang, K.; Xie, H.; Li, Y.; Xu, C.L.; Yu, L.; Wang, D.W. *Paratylenchus shenzhenensis* n. sp. (Nematoda: Paratylenchinae) from the rhizosphere soil of *Anthurium andreanum* in China. *Zootaxa* **2013**, *3750*, 167–175. [CrossRef]

22. Ghaderi, R. The damage potential of pin nematodes, *Paratylenchus* Micoletzky, 1922 *sensu lato* spp. (*Nematoda: Tylenchulidae*). *J. Crop Prot.* **2019**, *8*, 243–257.

23. Ghaderi, R.; Geraert, E.; Karegar, A. *The Tylenchulidae of the World; Identification of the Family Tylenchulidae (Nematoda: Tylenchida)*, 2nd ed.; Academia Press: Ghent, Belgium, 2016; p. 453.

24. Yu, Q.; Ye, W.; Powers, T. Morphological and Molecular Characterization of *Gracilacus wuae* n. sp. (Nematoda: Criconematoidea) associated with Cow Parsnip (*Heracleum maximum*) in Ontario, Canada. *J. Nematol.* **2016**, *48*, 203–213.

25. Ghaderi, R.; Kashi Nahanji, L.; Karegar, A. Contribution to the study of the genus *Paratylenchus* Micoletzky, 1922 *sensu lato* (Nematoda: Tylenchulidae). *Zootaxa* **2014**, *3841*, 151–187. [CrossRef] [PubMed]

26. Wu, L.Y.; Hawn, E.J. *Paratylenchus neoprojectus* n. sp. (Paratylenchinae: Nematoda) from alfalfa fields in Alberta, Canada. *Can. J. Zool.* **1975**, *53*, 1841–1843. [CrossRef]

27. Wu, Y.L.; Townshend, J.L. *Paratylenchus tateae* n. sp. (Paratylenchinae, Nematoda). *Can. J. Zool.* **1973**, *51*, 109–111. [CrossRef]

28. Van den Berg, E.; Tiedt, L.R.; Subbotin, S.A. Morphological and molecular characterisation of several *Paratylenchus* Micoletzky, 1922 (Tylenchida: Paratylenchidae) species from South Africa and USA, together with some taxonomic notes. *Nematologica* **2014**, *16*, 323–358. [CrossRef]

29. Bajaj, H.K. On the species of *Paratylenchus* Micoletzky (Nematoda: Criconematina) from Haryana, India. *Indian J. Nematol.* **1987**, *17*, 318–326.

30. Ghaderi, R.; Karegar, A. Some species of *Paratylenchus* (Nematoda: Tylenchulidae) from Iran. *Iran. J. Plant Pathol.* **2013**, *49*, 137–156, (In Persian with English Abstract).

31. Raski, D.J. Revision of the genus *Paratylenchus* Micoletzky, 1922 and descriptions of new species. Part II of three parts. *J. Nematol.* **1975**, *7*, 274–295.

32. Anderson, R.V.; Kimpinski, J. *Paratylenchus labiosus* n. sp. (Nematoda: Paratylenchidae) from Canada. *Can. J. Zool.* **1977**, *55*, 1992–1996. [CrossRef]

33. Wu, L.Y. *Paratylenchus brevihastus* n. sp. (Nematoda: Criconematidae). *Can. J. Zool.* **1962**, *40*, 391–393. [CrossRef]

34. Claerbout, J.; Vandevelde, I.; Venneman, S.; Kigozi, A.; de Sutter, N.; Neukermans, J.; Bleyaert, P.; Bert, W.; Hofte, M.; Viaene, N. A thorough study of a *Paratylenchus* sp. in glasshouse-grown lettuce: Characterisation, population dynamics, host plants and damage threshold as keys to its integrated management. *Ann. Appl. Biol.* **2020**, *178*, 62–79. [CrossRef]

35. Cob, N.A. Notes on *Paratylenchus*, a genus of nemas. *J. Wash. Acad. Sci.* **1923**, *13*, 251–257.

36. Wu, L.Y. *Paratylenchus tenuicaudatus* n. sp. (Nematoda: Criconematidae). *Can. J. Zool.* **1961**, *39*, 163–165. [CrossRef]

37. Raski, D.J. Paratylenchoides gen. n. and two new species (Nematoda: Paratylenchidae). *Proc. Helminthol. Soc. Wash.* **1973**, *40*, 230–233.

38. Khan, E.; Prasad, S.K.; Mathur, V.K. Two new species of the genus *Paratylenchus* Micoletzky, 1922 (Nematoda: Criconematidae) from India. *Nematologica* **1967**, *13*, 79–84. [CrossRef]

39. Colbran, R.C. Studies of plant and soil nematodes. 10. *Paratylenchus coronatus* n. sp. (Nematoda: Criconematidae), a pin nematode associated with citrus. *Qld. J. Agric. Anim. Sci.* **1965**, *22*, 277–279.

40. Mirbabaei, H.; Eskandari, A.; Ghaderi, R.; Karegar, A. On the synonymy of *Trophotylenchulus asoensis* and *T. okamotoi* with *T. arenarius*, and intra-generic structure of Paratylenchus (Nematoda: Tylenchulidae). *J. Nematol.* **2019**, *51*, 1–14. [CrossRef]

41. Brzeski, M.W.; Hanel, L. Paratylenchinae: Evaluation of diagnostic morpho-biometrical characters of females in the genus *Paratylenchus* Micoletzky, 1922 (Nematoda: Tylenchulidae). *Nematology* **2000**, *2*, 253–261. [CrossRef]

42. Akyazi, F.; Felek, A.F.; Cermak, V.; Cudejkova, M.; Foit, J.; Yildiz, S.; Hanel, L. Description of *Paratylenchus* (*Gracilacus*) *straeleni* (De Coninck, 1931) Oostenbrink, 1960 (Nematoda: Criconematoidea, Tylenchulidae) from hazelnut in Turkey and its comparison with other world populations. *Helminthologia* **2015**, *52*, 270–279. [CrossRef]

43. Brzeski, M.W. Paratylenchinae: Morphology of some known species and descriptions of *Gracilacus bilineata* sp. n. and *G. vera* sp. n. (Nematoda: Tylenchulidae). *Nematologica* **1995**, *41*, 535–565. [CrossRef]

44. Bahmani, J.; Barooti, S.; Ghaderi, R. On occurrence of *Paratylenchus labiosus* Anderson & Kimpinski, 1977 (Nematoda: Tylenchulidae) in Iran, with discussion on the validity of the species. *J. Crop Prot.* **2014**, *3*, 273–281.

45. Maria, M.; Gu, J.; Fang, Y.; He, J.; Castillo, P.; Li, H. *Radopholoides japonicus* n. sp. (Nematoda: Pratylenchidae) found in rhizosphere soil associated with *Podocarpus macrophyllus* from Japan. *Nematologica* **2017**, *19*, 1095–1105. [CrossRef]

46. Fang, E.; Li, H.; Maria, M.; Bert, W. Description of *Pseudaphelenchus zhoushanensis* n. sp. (Tylenchina: Aphelenchoididae) found in the wood of *Pinus thunbergii* at Zhoushan Islands, Zhejiang Province, China. *Nematologica* **2016**, *18*, 1151–1164. [CrossRef]

47. Pedram, M.; Pourhashemi, M.; Hosseinzadeh, J.; Koolivand, D. Comments on taxonomic status and host association of some *Laimaphelenchus* spp. (Rhabditida: Aphelenchoidea). *Nematologica* **2018**, *20*, 483–489. [CrossRef]

48. Powers, T.; Harris, T.; Higgins, R.; Mullin, P.; Sutton, L.; Powers, K. MOTUs, morphology and biodiversity estimation. A case study using nematodes of the suborder Criconematina and a conserved 18S DNA barcode. *J. Nematologica* **2011**, *43*, 35–48.

49. Powers, T.; Bernard, E.C.; Harris, T.; Higgins, R.; Olson, M.; Lodema, M.; Mullin, P.; Sutton, L.; Powers, K.S. COI haplotype groups in *Mesocriconema* (Nematoda: Criconematidae) and their morphospecies associations. *Zootaxa* **2014**, *3827*, 101–146. [CrossRef]

50. Powers, T.; Bernard, E.; Harris, T.; Higgins, R.; Olson, M.; Olson, S.; Lodema, M.; Matczyszyn, J.; Mullin, P.; Sutton, L.; et al. Species discovery and diversity in *Lobocriconema* (Criconematidae: Nematoda) and related plant-parasitic nematodes from North American ecoregions. *Zootaxa* **2016**, *4085*, 301–344. [CrossRef]

51. Powers, T.; Harris, T.; Higgins, R.; Mullin, P.; Powers, K. An 18S rDNA Perspective on the Classification of Criconematoidea. *J. Nematologica* **2017**, *49*, 236–244. [CrossRef]

52. Gutiérrez-Gutiérrez, C.; Palomares-Rius, J.E.; Cantalapiedra-Navarrete, C.; Landa, B.B.; Esmenjaud, D.; Castillo, P. Molecular analysis and comparative morphology to resolve a complex of cryptic *Xiphinema* species. *Zool. Scr.* **2010**, *39*, 483–498. [CrossRef]

53. Gutiérrez-Gutiérrez, C.; Cantalapiedra-Navarrete, C.; Decraemer, W.; Vovlas, N.; Prior, T.; Palomares-Rius, J.E.; Castillo, P. Phylogeny, diversity, and species delimitation in some species of the *Xiphinema americanum*-group complex (Nematoda: Longidoridae), as inferred from nuclear and mitochondrial DNA sequences and morphology. *Eur. J. Plant Pathol.* **2012**, *134*, 561–597. [CrossRef]

54. Palomares-Rius, J.E.; Cantalapiedra-Navarrete, C.; Castillo, P. Cryptic species in plant-parasitic nematodes. *Nematologica* **2014**, *16*, 1105–1118. [CrossRef]

55. Čermák, V.; Renčo, M. The family Paratylenchidae Thorne, 1949 in the rhizosphere of grass and woody species in Europe: A review of the literature. *Helminthol.* **2010**, *47*, 139–146. [CrossRef]

56. Esmaeili, M.; Heydari, R.; Castillo, P.; Ziaie Bidhendi, M.; Palomares-Rius, J.E. Molecular characterization of two known species of *Paratylenchus* Micoletzky, 1922 from Iran with notes on the validity of *Paratylenchus audriellus* Brown, 1959. *Nematologica* **2016**, *18*, 591–604. [CrossRef]

57. Maria., M.; Miao, W.; Castillo, P.; Zheng, J. A new pin nematode, *Paratylenchus sinensis* n. sp. (Nematoda: Paratylenchinae) in the rhizosphere of white mulberry from Zhejiang Province, China. *Eur. J. Plant Pathol.* **2020**, *156*, 1023–1039. [CrossRef]

58. Rhoades, H.L.; Linford, M.B. Molting of pre-adult nematodes of the genus *Paratylenchus* stimulated by root diffusates. *Science* **1961**, *130*, 1476–1477. [CrossRef]

59. Fisher, J.M. Effect of temperature and host on *Paratylenchus neoamblycephalus* and effect of the nematode on the host. *Aust. J. Agric. Res.* **1967**, *18*, 921–929. [CrossRef]

60. Wu, Y.L. *Paratylenchus projectus* (Paratylenchinae: Nematoda) and some closely related species. *Can. J. Zool.* **1975**, *53*, 1875–1881. [CrossRef]

61. Kimpinski, J.; Smith, E.M. Nematodes in potato soils in New Brunswick. *Can. Plant Dis. Surv.* **1988**, *68*, 147–148.

62. Jenkins, W.R. A rapid centrifugal-flotation technique for separating nematodes from soil. *Plant Dis. Rep.* **1964**, *48*, 692.

63. Seinhorst, J.W. A rapid method for the transfer of nematodes from fixative to anhydrous glycerin. *Nematologica* **1959**, *4*, 67–69. [CrossRef]

64. De Grisse, A.T. Redescription ou modifications de quelques techniques utilisées dans l'étude des nématodes phytoparasitaires. Mededelingen van de Faculteit Landbouwwetenschappen Rijksuniversiteit. *Gent* **1969**, *34*, 351–369.

65. Maria, M.; Powers, T.O.; Tian, Z.; Zheng, J. Distribution and description of criconematids from Hangzhou, Zhejiang Province, China. *J. Nematol.* **2018**, *50*, 183–206.

66. Holterman, M.; van der Wurff, A.; van den Elsen, S.; van Megen, H.; Holovachov, O.; Bakker, J.; Helder, J. Phylum wide analysis of SSU rDNA reveals deep phylogenetic relationships among nematodes and accelerated evolution toward crown clades. *Mol. Biol. Evol.* **2006**, *23*, 1792–1800. [CrossRef]

67. Holterman, M.; Rybarczyk, K.; van Den Elsen, S.; van Megen, H.; Mooyman, P.; Peña- Santiago, R.; Bongers, T.; Bakker, J.; Helder, J. A ribosomal DNA-based framework for the detection and quantification of stress-sensitive nematode families in terrestrial habitats. *Mol. Ecol. Resour.* **2008**, *8*, 23–34. [CrossRef]

68. Ferris, V.R.; Ferns, J.M.; Faghihi, J. Variation in spacer ribosomal DNA in some cyst-forming species of plant-parasitic nematodes. *Fundam. Appl. Nematol.* **1993**, *16*, 177–184.

69. Curran, J.; Driver, F.; Ballard, J.W.O.; Milner, R.J. Phylogeny of *Metarhizium*: Analysis of ribosomal DNA sequence data. *Mycol. Res.* **1994**, *98*, 547–552. [CrossRef]

70. Maroteaux, L.; Herzog, M.; Soye-Gobillard, M.O. Molecular organization of dinogflagellate ribosomal DNA: Evolutionary implications of the deduced 5.8 S rRNA secondary structure. *Biosystems* **1985**, *18*, 307–319. [CrossRef]

71. Carta, L.; Li, S. Improved 18S small subunit rDNA primers for problematic nematode amplification. *J. Nematol.* **2018**, *50*, 533–542. [CrossRef]

72. Katoh, K.; Rozewicki, J.; Yamada, K.D. MAFFT online service: Multiple sequence alignment, interactive sequence choice and visualization. *Brief. Bioinform.* **2019**, *20*, 1160–1166. [CrossRef]

73. Hall, T.A. BioEdit: A user-friendly biological sequence alignment editor and analysis program for windows 95/98/NT. *Nucleic Acids Symp. Ser.* **1999**, *41*, 95–98.

74. Castresana, J. Selection of conserved blocks from multiple alignments for their use in phylogenetic analysis. *Mol. Biol. Evol.* **2000**, *17*, 540–552. [CrossRef]

75. Ronquist, F.; Huelsenbeck, J.P. MRBAYES 3: Bayesian phylogenetic inference under mixed models. *Bioinformatics* **2003**, *19*, 1572–1574. [CrossRef]
76. Darriba, D.; Taboada, G.L.; Doallo, R.; Posada, D. jModelTest 2: More models, new heuristics and parallel computing. *Nat. Methods* **2012**, *9*, 772. [CrossRef]

*Article*

# Morphological and Molecular Characterization of *Pratylenchus dakotaensis* n. sp. (Nematoda: Pratylenchidae), a New Root-Lesion Nematode Species on Soybean in North Dakota, USA [†]

Zafar A. Handoo [1,*], Guiping Yan [2], Mihail R. Kantor [1], Danqiong Huang [2], Intiaz A. Chowdhury [2], Addison Plaisance [2], Gary R. Bauchan [3] and Joseph D. Mowery [3]

[1]  Mycology and Nematology Genetic Diversity and Biology Laboratory, USDA, ARS, Northeast Area, Beltsville, MD 20705, USA; mihail.kantor@usda.gov

[2]  Department of Plant Pathology, North Dakota State University, Fargo, ND 58108, USA; guiping.yan@ndsu.edu (G.Y.); danqiong.huang85@yahoo.com (D.H.); intiaz.chowdhury@ndsu.edu (I.A.C.); addison.plaisance@ndsu.edu (A.P.)

[3]  Electron & Confocal Microscopy Unit, USDA, ARS, Northeast Area, Beltsville, MD 20705, USA; gary.bauchan@usda.gov (G.R.B.); joseph.mowery@usda.gov (J.D.M.)

*  Correspondence: zafar.handoo@usda.gov

†  http://zoobank.org/urn:lsid:zoobank.org:pub:F89CA839-1A5B-4A27-BA53-8F8D6633E89C.

**Abstract:** Root-lesion nematodes (*Pratylenchus* spp.) of the genus *Pratylenchus* Filipjev, 1936, are among the most important nematode pests on soybean (*Glycine max* (L.) Merr.), along with soybean cyst and root-knot nematodes. In May 2015 and 2016, a total of six soil samples were collected from a soybean field in Walcott, Richland County, ND and submitted to the Mycology and Nematology Genetic Diversity and Biology Laboratory (MNGDBL), USDA, ARS, MD for analysis. Later, in 2019, additional nematodes recovered from a greenhouse culture on soybean originally from the same field were submitted for further analysis. Males, females, and juveniles of *Pratylenchus* sp. were recovered from soil and root samples and were examined morphologically and molecularly. DNA from single nematodes were extracted, and the nucleotides feature of three genomic regions targeting on the D2–D3 region of 28S rDNA and ITS rDNA and mitochondrial cytochrome oxidase subunit I (*COX1*) gene were characterized. Phylogeny trees were constructed to ascertain the relationships with other *Pratylenchus* spp., and polymerase chain reaction-restriction fragment length polymorphism (PCR-RFLP) was performed to provide a rapid and reliable differentiation from other common *Pratylenchus* spp. Molecular features indicated that it is a new, unnamed *Pratylenchus* sp. that is different from morphologically closely related *Pratylenchus* spp., including *P. convallariae*, *P. pratensis*, *P. fallax*, and *P. flakkensis*. In conclusion, both morphological and molecular observations indicate that the North Dakota isolate on soybean represents a new root-lesion nematode species which is named and described herein as *Pratylenchus dakotaensis* n. sp.

**Keywords:** D2/D3; description; *Glycine max*; lesion nematode; molecular; morphology; morphometrics; phylogeny; *Pratylenchus dakotaensis* n. sp.; soybean; ITS; *COX1* gene; PCR-RFLP

**Citation:** Handoo, Z.A.; Yan, G.; Kantor, M.R.; Huang, D.; Chowdhury, I.A.; Plaisance, A.; Bauchan, G.R.; Mowery, J.D. Morphological and Molecular Characterization of *Pratylenchus dakotaensis* n. sp. (Nematoda: Pratylenchidae), a New Root-Lesion Nematode Species on Soybean in North Dakota, USA. *Plants* **2021**, *10*, 168. https://doi.org/10.3390/plants10010168

Received: 7 January 2021
Accepted: 14 January 2021
Published: 17 January 2021

**Publisher's Note:** MDPI stays neutral with regard to jurisdictional claims in published maps and institutional affiliations.

## 1. Introduction

The genus *Pratylenchus* Filipjev, 1936, is one of the most important nematode genera in terms of the economic impact they have on crops [1,2]. Currently, the genus contains approximately 100 species [3–5], with new species being described very frequently. The root-lesion nematodes are ranked as the third most important group of plant-parasitic nematodes after root-knot and cyst nematodes [2] in terms of economic loss in agriculture and horticulture. Frederick and Tarjan [6] published a compendium of the *Pratylenchus* genus in 1989 in which they reported 89 species. In 1989, Handoo and Golden [7] also published a key and compendium to 63 valid species, including an update of the work done by different workers on the genus. The plants reported as hosts for the genus are very large. For example, one species, *Pratylenchus penetrans*, has been reported to have more than 400 plants as hosts [8].

On soybean, root-lesion nematodes are one of the most damaging nematodes that feed on the soybean roots [9]. Two species, *Pratylenchus brachyurus* and *P. penetrans*, have been reported to cause damage to the roots of soybean plants [10]. For example, soybean plant growth was suppressed by *Pratylenchus brachyurus* nematodes, with a negative correlation being reported between the number of nodes on the main stem and the number of nematodes at planting [11]. Despite the fact that nearly 100 different species of *Pratylenchus* have been described to date, only 5 have been reported in North Dakota, namely the *P. agilis* [12], *P. neglectus*, *P. scribneri* [13,14], and 2 new species of *Pratylenchus* [15,16].

The objective of this study was to describe one of these two new species using light microscopy (LM) and scanning electron microscopy (SEM) observations and assess the diagnostic values of their morphological and molecular characters. The morphometric details of females and males were recorded and compared to closely related species. Also, the molecular details using ITS, 28S, and *COX1* sequences were obtained and compared to the existing information in GenBank. PCR-RFLP was performed to rapidly and reliably differentiate it from other important *Pratylenchus* spp. species.

## 2. Materials and Methods

Nematode suspensions extracted from soil samples were sent to the MNGDBL, Beltsville, MD in May of 2015 and 2016. The origin of the soil samples was a field cultivated with soybean in Walcott, Richland County, ND. Nematodes were extracted from soil using the sugar centrifugal flotation method [17]. Each sample contained between 125 and 2000 root-lesion nematodes per kg soil [15]. In 2019, infested soil samples from the same field were planted to soybean cultivar Barnes in a greenhouse room at 22 °C. After 15 weeks of growth, the plants were harvested, and root-lesion nematodes were extracted from both the roots and soil using the Whitehead tray method [18]. Additional nematodes recovered from the greenhouse culture on soybean were submitted to the MNGDBL, Beltsville, MD for further analysis.

### 2.1. Morphological Examination

Females and males were recovered from the root and soil samples using the Whitehead tray method extraction method [18]. Nematodes were fixed in 3% formaldehyde and processed with glycerin by the formalin glycerin method [19,20]. Photomicrographs of females and males were made with an automatic 35 mm camera attached to a compound microscope with an interference contrast system, and light microscopic images of fixed nematodes were taken on a Nikon Eclipse Ni compound microscope using a Nikon DS-Ri2 camera. Measurements were made with an ocular micrometer on a Leica WILD MPS48 Leitz DMRB compound microscope. All measurements are in micrometers unless otherwise stated.

For the Low-Temperature Scanning Electron Microscopy (LT-SEM), nematodes were observed using the techniques described by Carta et al. [21], Kantor et al. [22], and Handoo et al. [23].

### 2.2. DNA Extraction, PCR, and Sequencing

DNA was extracted from a single individual nematode using the Proteinase K method [24]. Briefly, the chopped nematode pieces were transferred into a 0.5 mL sterile Eppendorf tube containing 2 µL of 10 × PCR buffer with $MgCl_2$, 2 µL of 600 µg/mL Proteinase K (Roche, Indianapolis, Indiana), and 6 µL of distilled $ddH_2O$. Tubes were incubated at −20 °C for at least 30 min followed by 65 °C for 1 h and 95 °C for 10 min. DNA samples from 3 nematode individuals were prepared, which represented 3 biological replicates.

Nucleotide sequences of D2D3 fragment of 28S rDNA and ITS rDNA regions from ribosomal DNA and *COX1* (cytochrome oxidase subunit I) gene from mitochondrial DNA were obtained by either direct sequencing using purified PCR products or T-A cloning. For D2–

D3 region of 28S rDNA, the primer set of D2A (5′-ACAAGTACCGTGAGGGAAAGTTG-3′) and D3B (5′-TCGGAAGGA ACCAGCTAC TA-3′) was used [25]. For ITS rDNA, the primer set of 18S (5′-TTGATTACGTCCCTGCCCTTT-3′) and 26S (5′-TTTCACTCG CCGTTACTAAGG-3′) was used [26]. For *COX1* gene, the primer set of JB3 (5′-TTTTTTGGG CATCCTGAGGTTTAT-3′) and JB4.5 (5′-TAAAGAAAGAACATAATGAAAATG-3′) was used [27]. The PCR were set up on Bio-Rad T100 Thermal Cycler (Hercules, CA, USA) as recommended [25–27]. For direct sequencing (D2–D3 region of 28S rDNA and *COX1* gene), PCR products were purified using Bio-tek E.Z.N.A. Cycle-Pure Kit (Omega, Norcross, GA, USA) and then sent to Genscript for sequencing (Genscript, Piscataway, NJ, USA). For cloning and sequencing, target PCR products were segregated on a 1.0% agarose gel, purified using Gel Extraction Kit (Omega), and cloned into pGEM-T Vector using pGEM-T Vector System II Kit (Promega, Madison, WI, USA) according to the manufacturer's instructions. Plasmid DNA was then extracted from the white colonies grown on indicator plates containing X-gal and IPTG, using a PerfectPrep™ Spin Mini Kit (5 PRIME Inc., Gaithersburg, MD, USA), and sent to Genscript for sequencing. Three sequences were generated for each of the three target areasand the corresponding consensus sequences of D2–D3 region of 28S rDNA, ITS rDNA, and *COX1* gene were deposited into GenBank to obtain their accession numbers (MW290216.1 for D2–D3, MW290217.1 for ITS, and MW309316.1 for *COX1*).

### 2.3. Phylogenetic Analysis

Phylogenetic relationships among *Pratylenchus* spp. were analyzed using Maximum Likelihood (ML) method of MEGA7 software [28]. Available DNA sequences of the 28S rDNA, ITS rDNA, and *COX1* gene of other *Pratylenchus* spp. species were retrieved from the NCBI Nucleotide Database (https://www.ncbi.nlm.nih.gov/nucleotide) and aligned with corresponding sequences obtained in this study by MUSCLE software (v3.8.31) with default settings for the highest accuracy. The gaps and missing data were completely removed. Accordingly, the General Time Reversible model was selected using gamma distribution, with invariant sites and 5 gamma-distributed rate categories to account for rates and patterns. Finally, the phylogeny trees were constructed using maximum likelihood method with 1000 bootstrap replications.

### 2.4. PCR-RFLP Analysis

To differentiate different *Pratylenchus* spp., polymerase chain reaction-restriction fragment length polymorphism (PCR-RFLP) analysis was performed using ITS rDNA amplified by the primers 18S/26S [26,29] with restriction endonucleases *Hind III* and *Hha I*. The PCR was carried out in a 20 μl reaction comprising of 1.5 μL DNA template, 0.4 μM forward and reverse primers, 0.2 mM dNTP, 1.5 mM $MgCl_2$, 1 × Green GoTaq® Flexi buffer, and 1 U GoTaq® Flexi DNA Polymerase (Promega, Madison, WI, USA) with conditions of pre-denaturing at 94 °C for 3 min followed by 35 cycles of 94 °C for 1 min, 55 °C for 1 min, and 72 °C for 1 min, with a final extension at 72 °C for 10 min. The digestion was performed in 20 μL reaction mixtures containing 5U restriction enzyme, 1× RE buffer, 2 μg acetylated BSA, and 10 μl PCR products at 37 °C for 2 h. The digested fragments were separated in 2 % agarose gel at 100 volts (V) for 20 min. The gel was visualized under UV light and images were captured using an AlphaImager® Gel Documentation System (Proteinsimple Inc., Santa Clara, CA, USA).

## 3. Results

### 3.1. Systematics

Pratylenchus dakotaensis n.sp.

http://zoobank.org/urn:lsid:zoobank.org:pub:F89CA839-1A5B-4A27-BA53-8F8D663 3E89C

(Figures 1–4, Table 1).

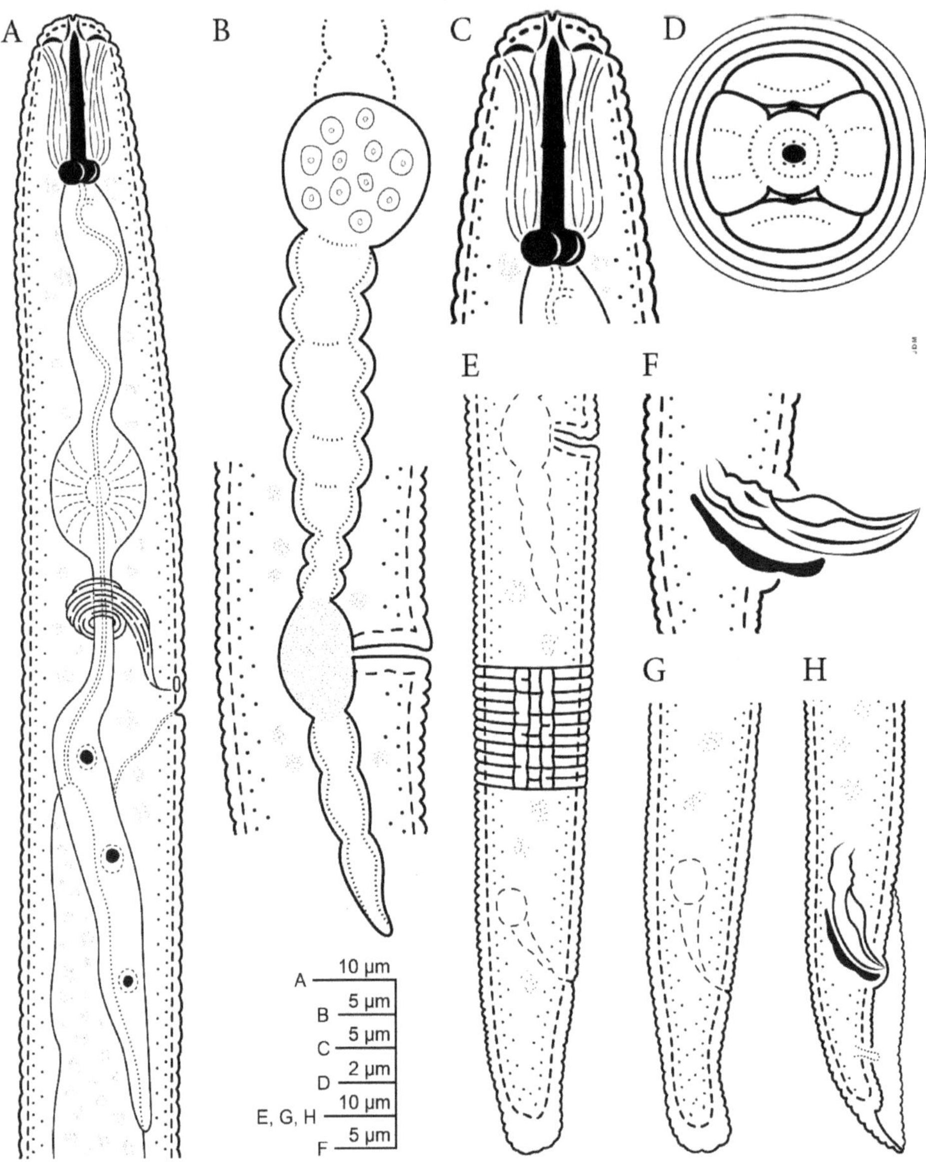

**Figure 1.** Line drawings of *Pratylenchus dakoaiensis* n. sp.: (**A**) Female pharyngeal region; (**B**) vulval region showing vulva, uterus, and spermatheca, (**C**) female lip region showing stylet; (**D**) details of the lip region showing the oral disc (*en face view*); (**E,G**) female tails with E showing lateral field with four lines; (**F,H**) male tails showing spicules and gubernaculum.

**Figure 2.** Photomicrographs of *Pratylenchus dakotaensis* n. sp.: (**A**) Female anterior end showing pharyngeal region; (**B**) female anterior end showing stylet; (**C**) female vulval area with arrow pointing the spermatheca; (**D**) female posterior end with arrows in black and white pointing to vulval and anal openings, respectively; (**E**) entire female with arrows in black and white pointing to vulval and anal openings; (**F**) female mid body showing lateral field with four lines; (**G,H,L**) female posterior ends showing tail variations and arrows pointing to anal areas; (**I,J**) male posterior ends showing spicule and bursa; (**K**) male posterior end with arrow pointing the phasmid.

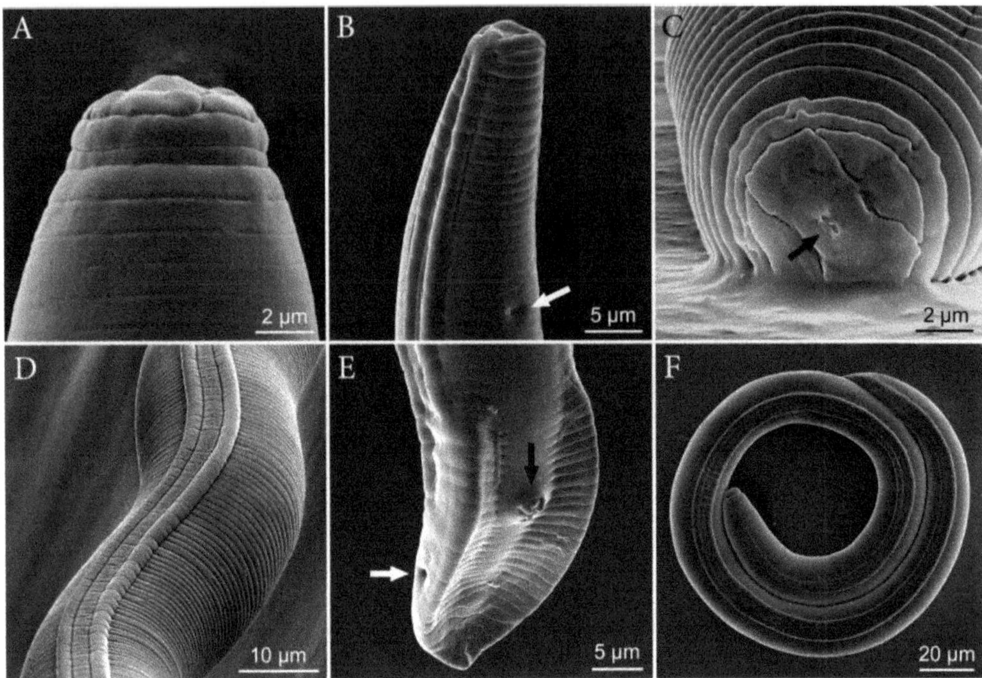

**Figure 3.** SEM images of *Pratylenchus dakotaensis* n. sp.: (**A**) Male specimen, head; (**B,C**) female posterior and anterior ends, (**B**) female posterior end, arrow showing anal opening, and (**C**) arrows showing oral opening; (**D**) female mid-body region showing lateral field; (**E**) male posterior end arrows in white and black showing cloaca opening and spicule, respectively; (**F**) whole specimen lateral field.

**Table 1.** Morphometrics of *Pratylenchus dakotaensis* n. sp. All measurements are in μm and in the form: mean ± standard deviation (s.d.) (range).

| Character | Holotype | Females | Males |
|---|---|---|---|
| n | | 22 | 7 |
| L | 552.0 | 484.5 ± 39.9 (390.0–555.0) | 445.7 ± 56.0 (355.0–502.0) |
| a | 27.6 | 23.4 ± 2.8 (20.8–29.8) | 23.7 ± 2.01 (20.8–25.2) |
| b | 4.2 | 4.0 ± 0.4 (3.2–4.8) | 4.06 ± 0.5 (3.2–4.8) |
| c | 22.0 | 20.2 ± 1.7 (16.8–24.1) | 20.0 ± 1.7 (16.7–21.3) |
| C' | 2.0 | 1.9 ± 0.3 (1.4–2.4) | 2.1 ± 0.17 (1.9–2.4) |
| Anal body width | 12.0 | 13.0 ± 2.0 (10.0–16.0) | 10.9 ± 0.7 (10.0–12.0) |
| V% | 80.0 | 80.2 ± 1.5 (78.0–83.0) | - |
| Maximum body width | 21.0 | 21.9 ± 2.5 (20.8–29.8) | 18.9 ± 1.5 (17.0–21.0) |
| Stylet length | 16.0 | 17.5 ± 0.3 (16.0–18.0) | 16.0 ± 0.3 (15.5–16.5) |
| Distance from head end to posterior end of esophageal glands | 130.0 | 118.8 ± 9.7 (110.0–140.0) | 109.3 ± 5.7 (101.0–115.0) |
| Tail length | 25.0 | 24.4 ± 2.4 (20.0–30.0) | 23.1 ± 1.8 (20.0–25.0) |
| Spicule length | - | - | 17.5 ± 0.82 (16.0–18.5) |
| Gubernaculum | - | - | 4.5 ± 0.5 (4.0–5.0) |

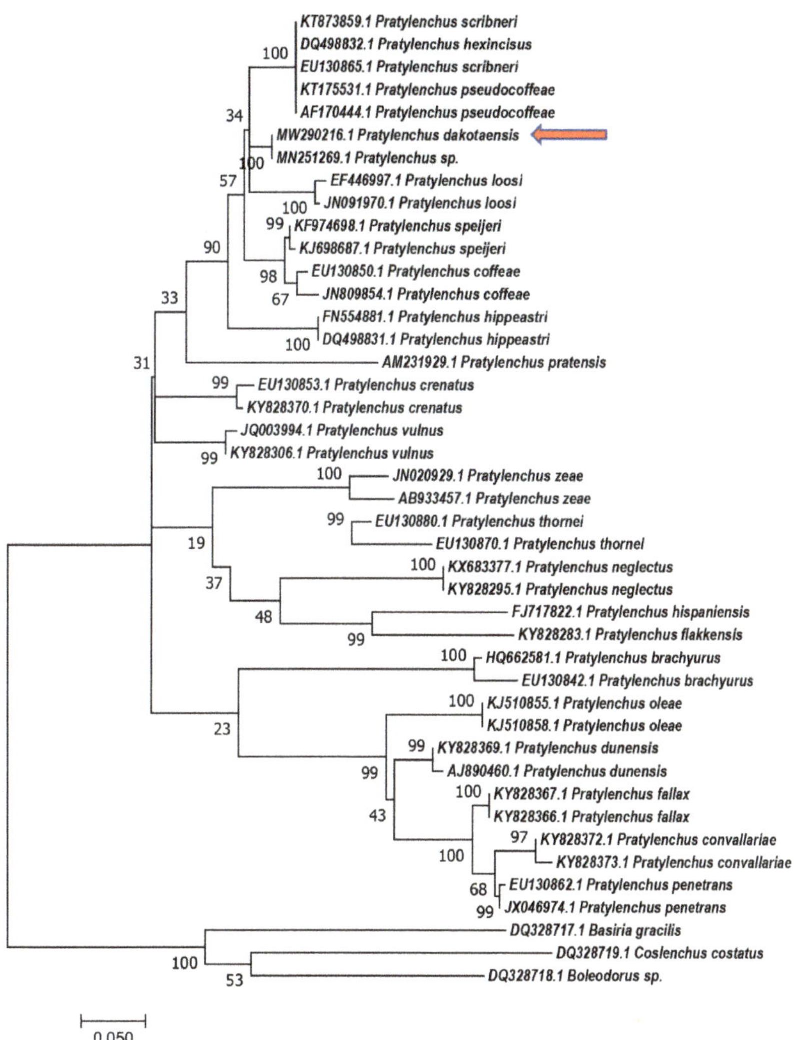


**Figure 4.** Phylogenetic relationships of *Pratylenchus dakotaensis* n. sp. (red arrow) from D2–D3 28S with related *Pratylenchus* spp. sequences from GenBank based on Maximum Likelihood analysis. Support values are given above branches. *Coslenchus costatus*, *Boleodorus* sp., and *Basiria gracilis* were served as outgroups.

### 3.1.1. Measurements

### 3.1.2. Description

*Female*: Slender and vermiform body, assuming straight or arcuate form when killed by gentle heat and tapering at both ends. Lateral field with four lines, with the outer two lines being areolated more so at tail region. Occasionally, additional oblique lines are noted in between the two inner lines. The lip region is flat to rounded or dome-shaped, slightly offset with the body contour and bearing three fine annuli. The en face view shows a divided face with rectangular subdorsal and subventral lips fused with oral disc in a dumbbell- to dome-shaped pattern that is separated from lateral lip sectors by three almost straight, often obscure incisures forming an obtuse angle. The stylet is short and robust with rounded knobs. The distance of the dorsal pharyngeal gland orifice to the stylet base is

2–3 μm. The procorpus is generally cylindrical but narrows near the middle and at junction with median bulb. The median bulb is muscular and rounded to slightly oval-shaped with cuticularized valve plates. The nerve ring encircles median part of isthmus. The excretory pore is located posterior to the nerve ring. The hemizonid is located at the two annuli anterior to excretory pore. The pharyngeal glands' nuclei are in tandem, elongate, and overlapping with the intestine ventrally. The reproductive system is monodelphic, prodelphic, with the ovary outstretched with single row of oocytes. The post-uterine sac is 18–20 μm long, and the vulva is located 78–83% of total body length from anterior end. The vulval lips are slightly protruding with no lateral flaps and epiptygma. The tail broad is conical, with 16–26 narrow irregularly annuli with terminal annuli usually wider than other tail annuli. The tail terminus is distinctly crenate/annulated with rounded to truncate- or clavate-shaped. The phasmids are prominently located at approximately the middle of the tail.

*Male*: Males are common and are similar to females, including the lip region, except for the sexual dimorphism. The stylet slightly is shorter than females, measuring 15.5 μm to 16.5 μm long. The lateral fields have four incisures, with the outer two lines mostly areolated. The reproductive system is composed of a single testis, which is anteriorly outstretched. The spicules and gubernaculum are ventrally curved, measuring 16–18.5 μm and 4–5 μm, respectively. The tail is short, bluntly rounded to pointed. The bursa encircle the entire tail. The ventral surface of the bursa is coarsely annulated. The phasmids are prominent.

### 3.2. Type Host and Locality

*Pratylenchus dakotaensis* n. sp is associated with roots and around soil from a soybean field in Richland County, ND. The global positioning coordinates for Richland County are 43.188221° N and 124.390174° W.

### 3.3. Type Material

Holotype (female): Slide T-740t, deposited in the United States Department of Agriculture Nematode Collection, Beltsville, MD, USA. Paratypes (Females, and Males): Same data and repository as holotype, Slides T-7153p to T-7158p. Additional females on slide numbers T-7159p at University of California, Riverside, CA, USA, and T-7160p at Fera, Plant Pest Disease Cultures and Collections, York, United Kingdom.

### 3.4. Diagnosis and Relationships

*Pratylenchus dakotaensis* n. sp. is characterized by a combination of the following morphological features in females: Slender, vermiform body, assuming straight or arcuate form, lateral field with four lines, with the outer two lines being areolated; the lip region is flat to rounded or dome-shaped, slightly offset with the body contour and bearing three fine annuli; the en face view shows a divided face with a rectangular subdorsal and subventral lips fused with the oral disc in a dumbbell- to dome-shaped pattern; the stylet is short and robust, with rounded knobs; the vulva is located at 78–83% of total body length from anterior end; the vulval lips are slightly protruding with no lateral flaps and epiptygma; the tail is broad and conical, with 16–26 narrow irregularly annuli, and the terminal annuli usually wider than the other tail annuli; the tail terminus is distinctly crenate/annulated with rounded to truncate- or clavate-shaped. Males are common; their stylet is slightly shorter than females; the spicules and gubernaculum are ventrally curved, measuring 16–18.5 μm and 4–5 μm, respectively; the tail is short, bluntly rounded to pointed; and the bursa encircle the entire tail.

*Pratylenchus dakotaensis* n. sp. is morphologically closely related to *Pratylenchus convallariae*, *P. pratensis*, and *P. fallax*. Sequence (GenBank accession No. MW290216, 702 bp) from the 28S D2–D3 had less than 94.2% similarity with these three species. In addition, it had 100% identity with *Pratylenchus* sp. (MN251269) from Lafayette County, Wisconsin and 98.6% identity with *P. scribneri* (MG925218) from Ohio, USA. The ITS sequence (GenBank accession No. MW290217, 1226 bp) of *P. dakotaensis* had less than 93.1% similarity with other *Pratylenchus* spp. including *P. convallariae*, *P. pratensis*, *P. fallax*, *P. scribneri*, and many isolates of an unknown *Pratylenchus* sp. Sequence (GenBank accession No. MW309316, 419 bp) from *COX1* gene had 97.5% identity with five isolates of a *Pratylenchus* sp. from Atchison County, Kansas, USA and less than 84.6% identity with other *Pratylenchus* spp. Thus, the sequence data did not support *P. convallariae*, *P. pratensis*, or *P. fallax*. Another morphologically closely related species is *P. flakkensis*, but *P. dakotaensis* differs from *P. flakkensis* in several morphological characters, with a high head, three head annuli, slight longer stylet in females, higher vulva percentage, and longer spicule in males. Accordingly, both morphological and molecular observations with the known and abovementioned closely related species indicate that the North Dakota isolate on soybean represents a new root-lesion nematode species, which is described herein as *Pratylenchus dakotaensis* n. sp.

### 3.5. Etymology

The species name was derived from North Dakota, the geographic origin.

### 3.6. Molecular Analysis

Phylogenetic relationships based on the D2–D3 region of 28S rDNA, ITS rDNA, and *COX1* gene were generated using the Maximum Likelihood method using corresponding nucleotides from *Pratylenchus* species (Figures 4–6). In the tree constructed using the D2–D3 region of 28S rDNA, which is considered as the most evolutionally conserved region, *P. dakotaensis* was more likely closely related with *P. scribneri*, *P. hexincisus*, *P. pseudocoffeae P. loosi*, *P. speijeri*, *P. coffeae*, and *P. hippeastri* (ML = 90), compared with those morphological closely related species, including *P. convallariae*, *P. pratensis*, *P. fallax*, and *P. flakkensis*. Similarly, in the trees constructed using ITS rDNA and *COX1* gene, *P. dakotaensis* was also clustered with those closely related species in the tree of 28S rDNA.

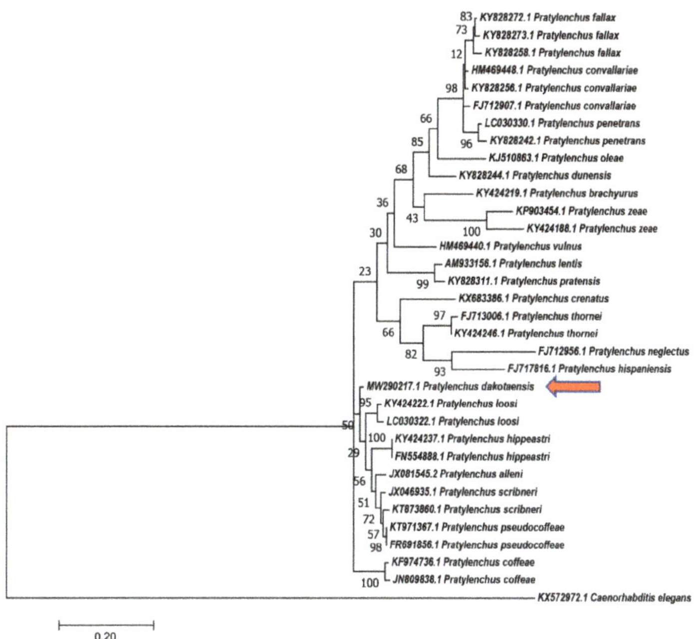

**Figure 5.** Phylogenetic relationships of *Pratylenchus dakotaensis* n. sp. (red arrow) from ITS rDNA with related *Pratylenchus* spp. sequences from GenBank based on Maximum Likelihood analysis. Support values are given above branches. *Caenorhabditiss elegans* was served as an outgroup.

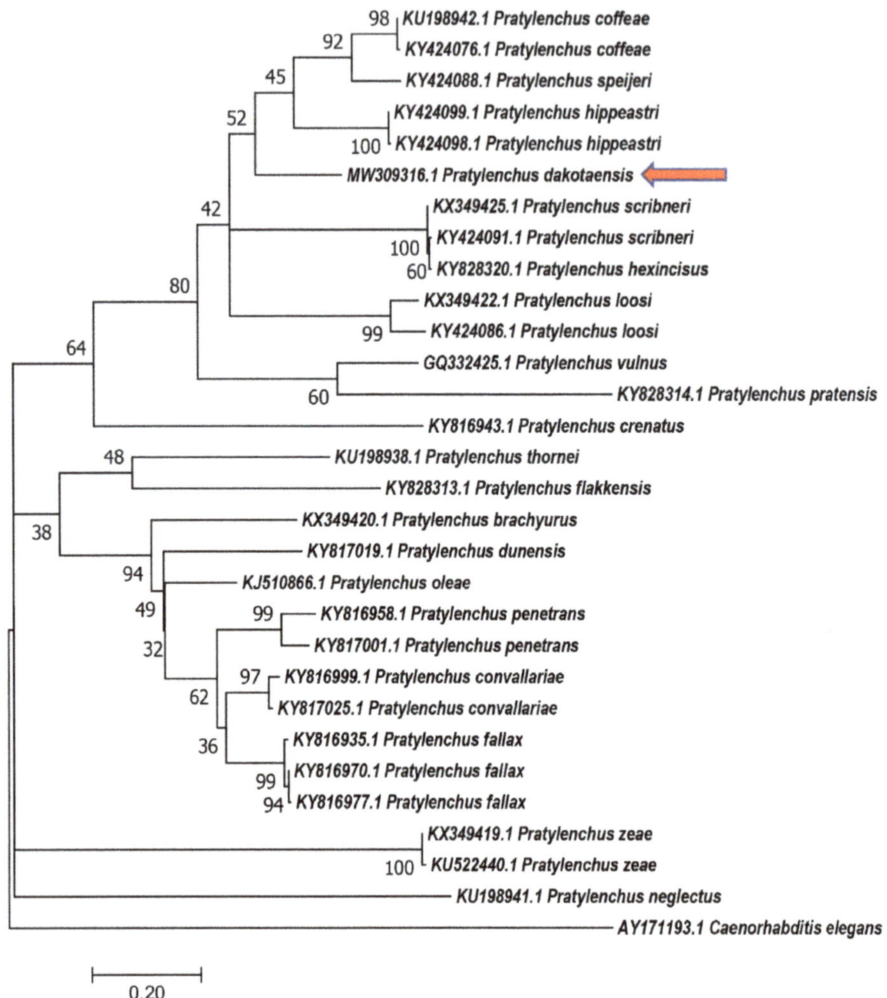



**Figure 6.** Phylogenetic relationships of *Pratylenchus dakotaensis* n. sp. (red arrow) from partial cytochrome oxidase subunit I (*COX1*) gene with related *Pratylenchus* spp. sequences from GenBank based on Maximum Likelihood analysis. Support values are given above branches. *Caenorhabditiss elegans* was served as an outgroup.

The RFLP analysis using the ITS rDNA region was performed to distinguish *P. dakotaensis* n. sp. from other common, important root-lesion nematode species (Figure 7). The results revealed that PCR products from the ITS region with two digestion enzymes (*Hind III* and *Hha I*) generated the same banding pattern for nine samples from the field infested with *P. dakotaensis* n. sp. but different banding patterns from *P. scribneri*, *P. neglectus*, and *P. penetrans*, which are the major *Pratylenchus* species in the region.

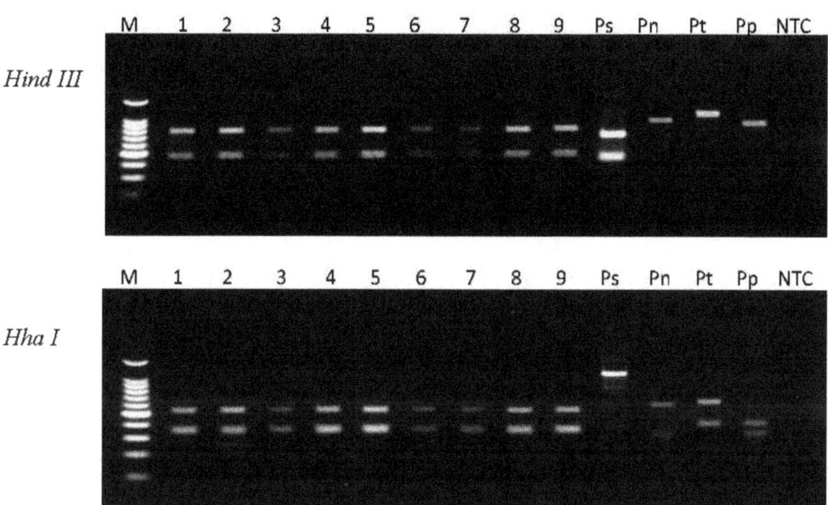

**Figure 7.** PCR-RFLP of *Pratylenchus* spp. using *Hind III* and *Hha I* enzymes. PCR products were amplified with primers rDNA2-V/PnSeqR targeted on ITS rDNA. The letter M refers to 100 bp DNA ladder from Promega, lanes 1–9 were nematode DNA extracted from single individuals isolated from the field having infestation of *Pratylenchus dakotaensis* n. sp., Ps refers to DNA from *P. scribneri* (ND), Pn refers to DNA from *P. neglectus* (ND), Pt refers to DNA from *P. thornei* (OR), Pp refers to DNA from *P. penetrans* (MN), and NTC refers to no-template control using ddH$_2$O instead of nematode DNA in the PCR reaction.

## 4. Discussion

Based on the molecular results obtained using the 28S D2–D3 primers, the North Dakota population had less than 94.2% similarity with morphologically closely related *Pratylenchus* spp., including *P. convallariae*, *P. pratensis*, *P. fallax*, and *P. flakkensis*. After analyzing the molecular data obtained by sequencing the ITS region, less than 93.1% similarity with *P. convallariae*, *P. pratensis*, and *P. fallax* was observed. The sequence from the *COX1* gene had less than 84.6% identity with other *Pratylenchus* spp., except 97.5% identity with five isolates of an undefined *Pratylenchus* sp. Looking at the morphometric data, the population of *P. dakotaensis* is similar to *P. flakkensis*. Despite the similarities between the two, several differences have been observed, such a high head in the North Dakota population, three head annuli instead of two, slightly longer stylet in females, higher vulva percentage, and longer spicule in males. In conclusion, combining all morphological and molecular data and observations with the known and abovementioned closely related species indicates that the North Dakota isolate on soybean represents a new root-lesion nematode species, described here as *Pratylenchus dakotaensis* n. sp. Interestingly, the 28S D2–D3 sequence of an unknown *Pratylenchus* sp. from Wisconsin, USA (GenBank accession No. MN251269) showed 100% identity with this new species. The specimens from Wisconsin and North Dakota need to be compared thoroughly to determine whether the Wisconsin population belongs to *Pratylenchus dakotaensis* n.sp.

**Author Contributions:** Conceptualization, Z.A., M.K., G.Y., D.H.; methodology Z.H., G.Y., and M.K.; software, M.K., J.M., G.B.I.C, G.Y., D.H.; validation, Z.A., G.Y.; formal analysis, Z.H.; investigation, G.Y., Z.H., D.H.; resources, G.Y., M.K., Z.H; data curation, Z.H., M.K.; writing—original draft preparation, Z.H., G.Y., M.K., D.H.; writing—review and editing, Z.H., G.Y., M.K., D.H., J.M., I.A.C., A.P.; visualization, Z.H., G.Y.; supervision, Z.H., G.Y.; project administration, Z.H., G.Y.; funding acquisition, G.Y. All authors have read and agreed to the published version of the manuscript.

**Funding:** We thank the North Dakota Soybean Research Council (NDSRC) for funding this research. This research was also supported by United States Department of Agriculture (USDA)-National Institute of Food and Agriculture (NIFA) Hatch Multistate project number ND02233.

**Data Availability Statement:** The data presented in this study are openly available in Plants at https://doi.org/10.3390/plants10010168.

**Acknowledgments:** Mihail Kantor was supported in part by an appointment to the Research Participation Program at the Mycology and Nematology Genetic Diversity and Biology Laboratory, USDA, ARS, Northeast Area, Beltsville, MD, administered by the Oak Ridge Institute for Science and Education through an interagency agreement between the U.S. Department of Energy and USDA-ARS. To Stephen Rogers of USDA-ARS, MNGDBL we are grateful for technical assistance. Mention of trade names or commercial products in this publication is solely for purpose of providing specific information and does not imply recommendation or endorsement by the U.S. Department of Agriculture. USDA is an equal opportunity provider and employer.

**Conflicts of Interest:** The authors declare no conflict of interest.

# References


1. Sasser, J.N.; Freckman, D.W. A world perspective on nematology; the role of the society. In *Vistas on Nematology*; SON: Hyatsville, MD, USA, 1987; pp. 7–14.
2. Castillo, P.; Vovlas, N. *Pratylenchus (Nematoda: Pratylenchidae): Diagnosis, Biology, Pathogenicity and Management*; Nematology Monographs and Perspectives; Brill: Leiden, The Netherlands, 2007; Volume 6, pp. 1–530.
3. Geraert, E. *The Pratylenchidae of the World: Identification of the Family Pratylenchidae (Nematoda: Tylenchida)*; Academia Press: Ghent, Belgium, 2013.
4. Handoo, Z.A.; Skantar, A.M.; Kantor, M.R.; Hafez, S.L.; Hult, M.N. Molecular and morphological characterization of the amaryllis lesion nematode, *Pratylenchus hippeastri* (Inserra et al., 2007), from California. *J. Nematol.* **2020**, *52*. [CrossRef] [PubMed]
5. Qing, X.; Bert, W.; Gamliel, A.; Bucki, P.; Duvrinin, S.; Alon, T.; Braun Miyara, S. Phylogeography and molecular species delimitation of *Pratylenchus capsici* n. sp., a new root lesion nematode in Israel on pepper (*Capsicum annuum*). *Phytopathology* **2019**, *109*, 847–858. [CrossRef] [PubMed]
6. Frederick, J.J.; Tarjan, A.C. A compendium of the genus Pratylenchus Filipjev, 1936 (Nemata: Pratylenchidae). *Rev. Nématologie* **1989**, *12*, 243–256.
7. Handoo, Z.A.; Golden, A.M. A key and compendium to the species of Pratylenchus (lesion nematodes). *J. Nematol.* **1989**, *21*, 202–218. [PubMed]
8. Vieira, P.; Maier, T.R.; Eves-van den Akker, S.; Howe, D.K.; Zasada, I.; Baum, T.J.; Eisenback, J.D.; Kamo, K. Identification of candidate effector genes of *Pratylenchus penetrans*. *Mol. Plant Pathol.* **2018**, *19*, 1887–1907. [CrossRef] [PubMed]
9. Lewis, S.A.; Drye, C.E.; Saunders, J.A.; Shipe, E.R.; Halbrendt, J.M. Plant-parasitic nematodes on soybean in South Carolina. *J. Nematol.* **1993**, *25*, 890.
10. Schmitt, D.P.; Barker, K.R. Damage and reproductive potentials of *Pratylenchus brachyurus* and *P. penetrans* on soybean. *J. Nematol.* **1981**, *13*, 327.
11. Koenning, S.R.; Schmitt, D.P.; Barker, K.R. Influence of selected cultural practices on winter survival of *Pratylenchus brachyurus* and subsequent effects on soybean yield. *J. Nematol.* **1985**, *17*, 464.
12. Thorne, G.; Malek, R.B. *Nematodes of the Northern Great Plains. Part I. Tylenchida (Nemata: Secernentea)*; Technical Bulletin 31. Paper 1; South Dakota State University: Brookings, SD, USA, 1968.
13. Yan, G.P.; Plaisance, A.; Huang, D.; Liu, Z.; Chapara, V.; Handoo, Z.A. First report of the root-lesion nematode *Pratylenchus neglectus* on wheat (*Triticum aestivum*) in North Dakota. *Plant Dis.* **2016**, *100*, 1794. [CrossRef]
14. Huang, D.; Yan, G. Specific detection of the root-lesion nematode *Pratylenchus scribneri* using conventional and real-time PCR. *Plant Dis.* **2017**, *101*, 359–365. [CrossRef]
15. Yan, G.; Plaisance, A.; Huang, D.; Chowdhry, I.A.; Handoo, Z.A. First report of the new root-lesion nematode *Pratylenchus sp.* on Soybean in North Dakota. *Plant Dis.* **2017**, *101*, 1554. [CrossRef]
16. Yan, G.P.; Plaisance, A.; Huang, D.; Handoo, Z.A.; Chitwood, D.J. First Report of a New, Unnamed Lesion Nematode *Pratylenchus sp.* Infecting Soybean in North Dakota. *Plant Dis.* **2017**, *101*, 1555. [CrossRef]
17. Jenkins, W.R. A rapid centrifugal flotation technique for separating nematodes from soil. *Plant Dis. Rep.* **1968**, *48*, 692.
18. Whitehead, A.G.; Hemming, J.R. A comparison of some quantitative methods of extracting small vermiform nematodes from soil. *Ann. Appl. Biol.* **1965**, *55*, 25–38. [CrossRef]
19. Golden, A.M. Preparation and Mounting Nematodes for Microscopic Observations. In *Plant Nematology Laboratory Manual*; Zuckerman, M., Mai, W.F., Krusberg, L.R., Eds.; University of Massachusetts Agricultural Experiment Station: Amherst, MA, USA, 1990; pp. 197–205.
20. Hooper, D.J. Handling, Fixing, Staining, and Mounting nematodes. In *Laboratory Methods for Work with Plant and Soil Nematodes*, 5th ed.; Southey, J.F., Ed.; Her Majesty's Stationery Office: London, UK, 1970; pp. 39–54.

21. Carta, L.K.; Handoo, Z.A.; Li, S.; Kantor, M.; Bauchan, G.; McCann, D.; Gabriel, C.K.; Yu, Q.; Reed, S.E.; Koch, J.; et al. Beech Leaf Disease Symptoms Caused by Newly Recognized Nematode Subspecies *Litylenchus crenatae mccannii* (Anguinata) Described from *Fagus grandifolia* in North America. *For Pathol.* **2020**, *50*, e12580. [CrossRef]

22. Kantor, M.; Handoo, Z.A.; Skantar, A.M.; Hult, M.N.; Ingham, R.E.; Wade, N.M.; Ye, W.; Bauchan, G.R.; Mowery, J.D. Morphological and molecular characterisation of Punctodera mulveyi n. sp. (Nematoda: Punctoderidae) from a golf course green in Oregon, USA, with a key to species of Punctodera. *Nematology* **2020**. [CrossRef]

23. Handoo, Z.; Kantor, M.; Carta, L. Taxonomy and Identification of Principal Foliar Nematode Species (Aphelenchoides and Litylenchus). *Plants* **2020**, *9*, 1490. [CrossRef]

24. Kumari, S.; Subbotin, S.A. Molecular characterization and diagnostic of stubby root and virus vector nematodes of the family *Truchodoridae* (Nametode: Triplonchida) using ribosomal RAN genes. *Plant Pathol.* **2012**, *61*, 1021–1031. [CrossRef]

25. Subbotin, S.A.; Sturhan, D.; Chizhov, V.N.; Vovlas, N.; Baldwin, J.G. Phylogenetic analysis of Tylenchida Thorne, 1949 as inferred from D2 and D3 expansion fragments of the 28S rRNA gene sequences. *Nematology* **2006**, *8*, 455–474. [CrossRef]

26. Vrain, T.C.; Wakarchuk, D.A.; Levesque, A.C.; Hamilton, R.I. Intraspecific rDNA restriction fragment length polymorphism in the *Xiphinema americanum* group. *Fundam. Appl. Nematol.* **1992**, *15*, 563–573.

27. Derycke, S.; Vanaverbeke, J.; Rigaux, A.; Backeljau, T.; Moens, T. Exploring the use of Cytochrome Oxidase c Subunit 1 (COI) for DNA barcoding of free-living marine nematodes. *PLoS ONE* **2010**, *5*, e13716. [CrossRef] [PubMed]

28. Kumar, S.; Stecher, G.; Tamura, K. MEGA7: Molecular evolutionary genetics analysis version 7.0 for bigger datasets. *Mol. Biol. Evol.* **2016**, *33*, 1870–1874. [CrossRef] [PubMed]

29. Waeyenberge, L.; Ryss, A.; Moens, M.; Pinochet, J.; Vrain, T.C. Molecular characterization of 18 *Pratylenchus* species using rDNA restriction fragment length polymorphism. *Nematology* **2000**, *2*, 135–142. [CrossRef]

*Article*

# Morphostatic Speciation within the Dagger Nematode *Xiphinema hispanum*-Complex Species (Nematoda: Longidoridae)

Antonio Archidona-Yuste [1,*], Ruihang Cai [2,3], Carolina Cantalapiedra-Navarrete [2], José A. Carreira [4], Ana Rey [5], Benjamín Viñegla [4], Gracia Liébanas [4], Juan E. Palomares-Rius [2] and Pablo Castillo [2]

[1] Department of Ecological Modelling, Helmholtz Centre for Environmental Research—UFZ, Permoserstrasse 15, 04318 Leipzig, Germany

[2] Instituto de Agricultura Sostenible (IAS), Consejo Superior de Investigaciones Científicas (CSIC), Avda. Menéndez Pidal s/n, 14004 Córdoba, Spain; ruihangcai@163.com (R.C.); carocantalapiedra@hotmail.com (C.C.-N.); palomaresje@ias.csic.es (J.E.P.-R.); p.castillo@csic.es (P.C.)

[3] Laboratory of Plant Nematology, Institute of Biotechnology, College of Agriculture and Biotechnology, Zhejiang University, Hangzhou 310058, China

[4] Departamento de Biología Animal, Biología Vegetal y Ecología, Universidad de Jaén, Campus Las Lagunillas, 37724 Jaén, Spain; jafuente@ujaen.es (J.A.C.); bvinegla@ujaen.es (B.V.); gtorres@ujaen.es (G.L.)

[5] Departamento de Biogeografía y Cambio Global, Museo Nacional de Ciencias Naturales-CSIC, José Abascal 2, 28006 Madrid, Spain; arey@mncn.csic.es

* Correspondence: antonio.archidona-yuste@ufz.de

http://zoobank.org:pub:6D60BC11-B301-42EF-9301-6E42A8E93B9C

Received: 23 October 2020; Accepted: 23 November 2020; Published: 26 November 2020

**Abstract:** Dagger nematodes of the genus *Xiphinema* include a remarkable group of invertebrates of the phylum Nematoda comprising ectoparasitic animals of many wild and cultivated plants. Damage is caused by direct feeding on root cells and by vectoring nepoviruses that cause diseases on several crops. Precise identification of *Xiphinema* species is critical for launching appropriate control measures. We deciphered the cryptic diversity of the *Xiphinema hispanum*-species complex applying integrative taxonomical approaches that allowed us to verify a paradigmatic example of the morphostatic speciation and the description of a new species, *Xiphinema malaka* sp. nov. Detailed morphological, morphometrical, multivariate and genetic studies were carried out, and mitochondrial and nuclear haploweb analyses were used for species delimitation of this group. The new species belongs to morphospecies Group 5 from the *Xiphinema* non*americanum*-group species. *D2-D3, ITS1*, partial *18S*, and partial *coxI* regions were used for inferring the phylogenetic relationships of *X. malaka* sp. nov. with other species within the genus *Xiphinema*. Molecular analyses showed a clear species differentiation not paralleled in morphology and morphometry, reflecting a clear morphostatic speciation. These results support the hypothesis that the biodiversity of dagger nematodes in southern Europe is greater than previously assumed.

**Keywords:** Bayesian inference; cryptic species; *coxI*; *D2-D3* expansion domains of *28S* rRNA-gene; integrative taxonomy; principal component analysis

---

## 1. Introduction

Plant-parasitic nematodes (PPN) are characterized by the presence of a stylet used for root tissue penetration, comprise about 15% of the total number of nematode species currently known, of which over 4100 species have been identified as PPN [1,2]. Annual crop losses caused by PPN are estimated to be about 8–15% of total crop production worldwide [3,4]. Accurate identification of PPN is essential for the selection of appropriate control measures against plant pathogenic species, as well as for a reliable method allowing distinction between species under quarantine or regulatory strategies and a better understanding of their implications in pest control and soil ecology [5,6]. PPN species have been defined historically based on morphological characteristics [7,8]. However, the adoption of molecular techniques in nematode taxonomy has revealed unexpected genetic diversity within species throughout the phylum Nematoda [9]. This has been especially accurate for the family Longidoridae, a large group of ectoparasitic nematodes feeding from the root tip zone to the hairy root region, and characterized by a substantial intra and interspecific homogeneity of the morphometric characters used for species discrimination [1,6,10,11]. Use of molecular data in species identification of dagger and needle nematodes over the last three decades has indicated that many widespread species actually comprise multiple genetically divergent and morphologically similar cryptic species [6,11–13]. Complexes of cryptic species often result from nonecological speciation in which diversification is not accompanied by apparent ecological or morphological separation in traditional quantitative traits [14].

The genus *Xiphinema* is one of the most diversified group species of longidorid nematodes with more than 280 valid species [5,6,11,15]. The ecological and phytopathological importance of this group of nematodes lies in its wide range of host plants and cosmopolitan distribution [5,11], but some species of this genus are vectors of several important plant viruses (genus *Nepovirus*, family Comoviridae) that cause significant damage to a wide range of crops [10,16]. Considering the great diversity of this group, the genus *Xiphinema* was divided into two different species groups [5,17,18]: (i) the *Xiphinema americanum*-group comprising a complex of about 60 species [15,17]; and (ii) the *Xiphinema* non*americanum*-group which comprises a complex of more than 220 species [5,6,19]. Later, this group was divided into eight morphospecies groups for helping identification [18]. However, some cryptic species and species complexes within *Xiphinema* have been recently revealed based on integrative taxonomical approaches, including morphometric multivariate methods, genetic analyses based on ribosomal and mitochondrial DNA (rDNA and mtDNA, respectively) and species delimitation (haplonet tools) [6,11,20,21]. A paradigmatic example of these species complexes comprises the *Xiphinema hispanum*-complex, *viz.* didelphic *Xiphinema* species from the Iberian Peninsula characterized by a rounded tail in females with or without an inconspicuous bulge projecting slightly ventrally and a uterus showing spiniform structures [22]. The cryptic diversity of this species complex has been deciphered by our team over the last ten years applying integrative taxonomical approaches that allowed us to verify these species as valid, and the recent description of a new species, *X. subbaetense* [11,20]. Recent studies on this species complex clearly separated three species (*X. adenohystherum*, *X. hispanum* and *X. subbaetense*) revealing high levels of genetic diversity within them that showed little morphological differentiation [11]. In new nematode surveys carried out in natural areas in the provinces of Málaga and Almería, Andalusia, southern Spain, we have detected nine unidentified *Xiphinema* isolates resembling *X. hispanum*-complex morphology. Detailed morphological and morphometrical observations using light microscopy indicated that these isolates appeared undistinguishable from *X. hispanum* complex species, a fact which prompted us to undertake comprehensive multivariate and genetic analyses, compared with previous reported data, to decipher this taxonomic conundrum.

Morphostatic evolution can be defined as genetic modifications, and even complete speciation events, which are not reflected in morphology, often being a result of nonadaptive radiation marked by the rapid proliferation of species without ecological differentiation [23,24]. Although no data have yet been specifically mentioned in Nematoda, morphostatic evolution seems not to be a rare phenomenon in longidorids based on the numerous complexes and cryptic species documented [6,11–13,15,20,25]. In Longidoridae, it is very common that molecular divergences among species are not reflected in morphological or morphometric traits, which conforms a morphostatic model of evolution with numerous cryptic species within this group [6,11,13,15,20,21,25,26].

In this context, we investigated (1) the existence of a new cryptic species within the *X. hispanum*-complex confirming a morphostatic speciation in this group using an integrative species delineation approach based on multivariate morphometric analysis and haplonet mitochondrial and nuclear haploweb tools; (2) a new species of the genus *Xiphinema* (*Xiphinema malaka* sp. nov.) described through integrative methods based on the combination of morphological, morphometric and molecular data; and (3) phylogenetic analyses based on *D2-D3* expansion domains of the *28S* rRNA gene, *ITS1*, the partial *18S* rRNA gene, and the partial mitochondrial *coxI* gene sequences to clarify the relationships of the new *Xiphinema* species.

## 2. Results

Species boundaries within the *Xiphinema* complex included in this research (Figure 1) were based on the integrative application of morphological, morphometric and molecular methods to unravel potential cryptic species diversity (Table 1). Species delimitation was carried out using two independent approaches based on morphometric (multivariate analysis) and molecular data using ribosomal and mitochondrial sequences (haplonet). Multivariate morphometric and haplonet methods were performed on the nine studied isolates including previous isolates from the *X. hispanum*-complex to verify species identifications. The integration of this procedure with the analysis of nematode morphology allowed us to verify *Xiphinema malaka* sp. nov. as a valid new species within the *X. hispanum* cryptic complex. Additionally, we maintained a consensus approach for the different species delimitation methods, including concordant results in phylogenetic trees inferred from nuclear and mitochondrial markers and/or different morphological or morphometric characteristics.

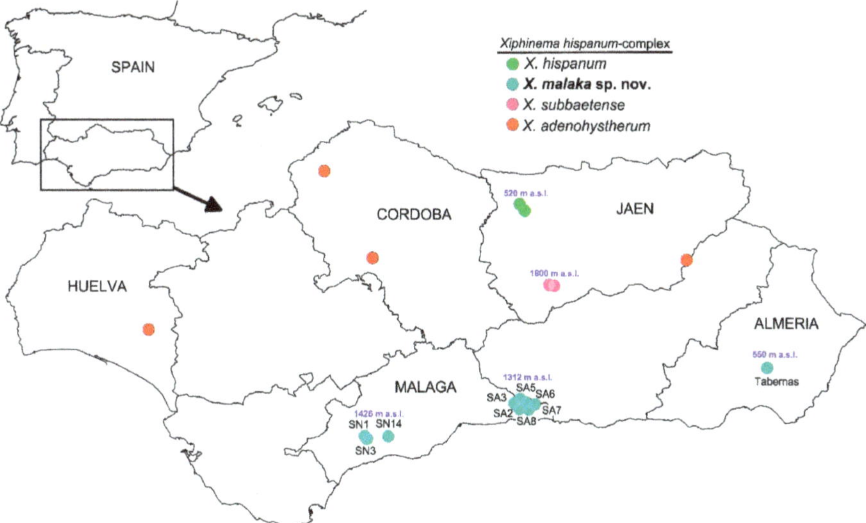


**Figure 1.** Geographic distribution of *Xiphinema hispanum*-complex species and locations of sampling sites of which the recovered isolates of the new species were characterized morphometrically and molecularly. Arrow indicates the location of Andalusia in the Iberian Peninsula.

**Table 1.** Isolates sampled for *Xiphinema malaka* sp. nov. from several localities of Málaga and Almería provinces (Southern Spain), and *Xiphinema adenohystherum* sequences used in this study.

| Sample Code—Host-Plant Locality | D2-D3 Haplotype | *coxI* Haplotype | D2-D3 | ITS1 | 18S | *coxI* |
|---|---|---|---|---|---|---|
| *Xiphinema malaka*sp. nov. | | | | | | |
| **Maritime Pine (*Pinus pinaster* Aiton)** | | | | | | |
| 1. SA03-DF91  Canillas de Albaida (Málaga) [a] | *D2-D3*-Hm3 | *coxI*-Hm2 | MT584052 | MT584088 | MT584086 | MT580263 |
| 2. SA03-DF92  Canillas de Albaida (Málaga) [a] | *D2-D3*-Hm10 | *coxI*-Hm2 | MT584053 | - | - | MT580264 |
| 3. SA03-DF15  Canillas de Albaida (Málaga) [a] | *D2-D3*-Hm10 | *coxI*-Hm2 | MT584054 | - | - | MT580265 |
| 4. SA03-DF90  Canillas de Albaida (Málaga) [a] | *D2-D3*-Hm10 | - | MT584055 | - | - | - |
| 5. SA03-DF93  Canillas de Albaida (Málaga) [a] | *D2-D3*-Hm10 | - | MT584056 | - | - | - |
| 6. SA03-DF94  Canillas de Albaida (Málaga) [a] | *D2-D3*-Hm10 | - | MT584057 | - | - | - |
| 7. SA02-DF42  Canillas de Albaida (Málaga) | *D2-D3*-Hm12 | - | MT584058 | MT584089 | - | |
| 8. SA02-AU62  Canillas de Albaida (Málaga) | *D2-D3*-Hm12 | - | MT584059 | - | - | |
| 9. SA05_DF16  Canillas de Albaida (Málaga) | *D2-D3*-Hm11 | - | MT584060 | MT584090 | - | |
| 10. SA05_DH96  Canillas de Albaida (Málaga) | *D2-D3*-Hm1 | - | MT584061 | - | - | - |
| 11. SA05_DH97  Canillas de Albaida (Málaga) | *D2-D3*-Hm2 | - | MT584062 | - | - | - |
| 12. SA05_DH98  Canillas de Albaida (Málaga) | *D2-D3*-Hm3 | - | MT584063 | - | - | - |
| 13. SA05_DH99  Canillas de Albaida (Málaga) | *D2-D3*-Hm3 | - | MT584064 | - | - | - |
| 14. SA06-DG12  Canillas de Albaida (Málaga) | *D2-D3*-Hm12 | - | MT584065 | MT584091 | - | - |
| 15. SA06-DG13  Canillas de Albaida (Málaga) | *D2-D3*-Hm5 | - | MT584066 | MT584092 | - | - |
| 16. SA06-DI01  Canillas de Albaida (Málaga) | *D2-D3*-Hm4 | - | MT584067 | - | - | - |
| 17. SA06-DI02  Canillas de Albaida (Málaga) | *D2-D3*-Hm5 | - | MT584068 | - | - | - |
| 18. SA06-DI03  Canillas de Albaida (Málaga) | *D2-D3*-Hm6 | - | MT584069 | - | - | - |
| 19. SA06-DI04  Canillas de Albaida (Málaga) | *D2-D3*-Hm4 | - | MT584070 | - | - | - |
| 20. SA07-AU46  Canillas de Albaida (Málaga) | *D2-D3*-Hm9 | *coxI*-Hm3 | MT584071 | MT584093 | MT584087 | MT580266 |
| 21. SA07-AU47  Canillas de Albaida (Málaga) | *D2-D3*-Hm9 | - | MT584072 | MT584094 | - | - |
| 22. SA07-AU48  Canillas de Albaida (Málaga) | *D2-D3*-Hm9 | - | MT584073 | - | - | - |
| 23. SA08-DF19  Canillas de Albaida (Málaga) | *D2-D3*-Hm2 | *coxI*-Hm1 | MT584074 | MT584095 | - | MT580267 |
| **Black pine (*Pinus nigra*Arnold)** | | | | | | - |
| 24. SN01-DI10  Igualeja (Málaga) | *D2-D3*-Hm7 | *coxI*-Hm5 | MT584075 | MT584096 | - | MT580268 |
| 25. SN01-DE88  Igualeja (Málaga) | *D2-D3*-Hm3 | *coxI*-Hm5 | MT584076 | - | - | MT580269 |
| 26. SN01-DF46  Igualeja (Málaga) | *D2-D3*-Hm3 | *coxI*-Hm6 | MT584077 | - | - | MT580270 |
| 27. SN03-DE93  Igualeja (Málaga) | *D2-D3*-Hm7 | *coxI*-Hm5 | MT584078 | MT584097 | - | MT580271 |

**Table 1.** *Cont.*

| Sample Code—Host-Plant Locality | | *D2-D3* Haplotype | *coxI* Haplotype | *D2-D3* | *ITS1* | *18S* | *coxI* |
|---|---|---|---|---|---|---|---|
| **Cork oak (*Quercus suber* L.)** | | | | | | | |
| 28. SN14-DF11 | Monda (Málaga) | *D2-D3*-Hm8 | - | MT584079 | MT584098 | - | - |
| 29. SN14-DI12 | Monda (Málaga) | *D2-D3*-Hm8 | - | MT584080 | - | - | - |
| 30. SN14-DI13 | Monda (Málaga) | *D2-D3*-Hm8 | - | MT584081 | - | - | - |
| 31. SN14-DI14 | Monda (Málaga) | *D2-D3*-Hm8 | - | MT584082 | - | - | - |
| 32. SN14-DI15 | Monda (Málaga) | *D2-D3*-Hm8 | *coxI*-Hm4 | MT584083 | - | - | MT580272 |
| 33. SN14-DI16 | Monda (Málaga) | *D2-D3*-Hm8 | *coxI*-Hm4 | MT584084 | - | - | MT580273 |
| **Yellow broom (*Cytisus scoparius* (L.) Link)** | | | | | | | |
| 34. RAMB-AO44 | Tabernas, Almería | *D2-D3*-Hm13 | *coxI*-Hm1 | MT584085 | MT584099 | - | MT580274 |
| ***Xiphinema adenohystherum*** | | | | | | | |
| **Grapevine (*Vitis vinifera* L.)** | | | | | | | |
| 406f | Villalba del Alcor, Huelva | - | - | - | MT584100 | - | - |
| **Wild olive (*Olea europaea* L. subsp *europaea* var. sylvestris)** | | | | | | | |
| AR139 | Aroche, Huelva | - | - | - | MT584101 | - | - |
| **European holly (*Ilex aquifolium* L.)** | | | | | | | |
| Z137f | Arevalo de la Sierra, Soria | - | - | - | MT584102 | - | - |

[a] Type locality (paratype specimens); (-) Not obtained or not performed.

### 2.1. Multivariate Morphometric Analysis

In principal component analysis (PCA), the first three components (sum of squares (SS) loadings > 1) accounted for 65.1% of the total variance in the morphometric characteristics of the *X. hispanum*-complex (Table 2). The eigenvalues for each character were used to interpret the biological meaning of the factors. First, the principal component 1 (PC1) was mainly dominated by a stylet with a high positive correlation (eigenvalue = 0.523). PC2 was mainly dominated by high negative correlation for the vulva position (eigenvalue = −0.547) as well as a high positive correlation for the a ratio (eigenvalue = 0.482) (Table 2). This component was, therefore, related with the overall nematode size and shape. Finally, PC3 was mainly dominated by the highest positive correlation found for the c′ ratio and lower, but also high, positive correlation for the hyaline region length (eigenvalues = 0.774 and 0.458, respectively). This component was then related with tail shape. Overall, these results suggest that all of the extracted PCs were related to the overall size and shape of nematode isolates. The results of the PCA were represented graphically in Cartesian plots in which isolates of the *X. hispanum*-complex were projected on the plane of the *x*- and *y*-axes, respectively, as pairwise combinations of components 1 to 3 (Figure 2). In the graphic representation of the *X. hispanum*-complex, and with the exception of *X. adenohystherum*, we observed that the specimens of all species were projected showing an expanded distribution along the plane for all the projected combinations of the components. One reason might be the wide morphometric variation detected in these species (Tables 3 and 4) [6,11]. As a consequence, we did not detect a clear separation amongst species within the *X. hispanum*-complex, all the specimens being projected at random for all the projected combinations. These patterns suggest a clear example of morphostatic speciation within the *X. hispanum*-complex. However, it should be noted that when projected on the plane of the combinations of PC1-2 and PC2-3, almost all specimens of *X. malaka* sp. nov. and *X. subbaetense* were separated among them (Figure 2). This graphical separation was shown by the projection of PC2 (dominated by the V and a ratios). This graphical separation is due to the variation found in the ratio a among these species, as pointed out below. A minimum spanning tree (MST) superimposed on the plot of the first three principal components showed the same patterns observed with PCA, that is, not clear separation amongst species within the *X. hispanum*-complex (Figure 2).

**Table 2.** Eigenvectors and SS loadings of factors derived from nematode morphometric characters for *Xiphinema hispanum*-complex (*Xiphinema malaka* sp. nov., *Xiphinema adenohystherum*, *Xiphinema hispanum*, *Xiphinema subbaetense*).

| | *Xiphinema hispanum*-Complex Principal Components [a] | | |
|---|---|---|---|
| **Character** [b] | **PC1** | **PC2** | **PC3** |
| Body length (L) | −0.382 | 0.032 | −0.003 |
| a | −0.081 | <u>0.483</u> | 0.262 |
| c′ | 0.009 | 0.137 | <u>0.774</u> |
| d | −0.371 | 0.444 | −0.100 |
| d′ | −0.318 | −0.384 | 0.086 |
| V | −0.170 | <u>−0.550</u> | −0.046 |
| Stylet | <u>−0.523</u> | −0.001 | −0.015 |
| Oa-gr | −0.440 | 0.198 | −0.320 |
| Hyaline region length | −0.334 | −0.256 | <u>0.458</u> |
| SS loadings | 1.67 | 1.35 | 1.13 |
| % of total variance | 30.80 | 20.21 | 14.10 |
| Cumulative % of total variance | 30.80 | 51.01 | 65.1 |

[a] Based on 41 female specimens of *Xiphinema malaka* sp. nov. from seven isolate samples, 25 female specimens of *Xiphinema subbaetense* from two isolate samples, eight female specimens of *Xiphinema adenohystherum* from a isolate sample, and 11 female specimens of *Xiphinema hispanum* from a isolate sample. Values of morphometric variables 1 to 3 (eigenvector > 0.458) are underlined. All isolates were molecularly identified and located at southern Spain. The c′ ratio was excluded by the multicollinearity test and then, it was not included in the multivariate analysis for the *Xiphinema hispanum*-complex; [b] Morphological and diagnostic characters according to Jairajpuri and Ahmad [7] with some inclusions. a = body length/maximum body width; c′ = tail length/body width at anus; d = anterior to guiding ring/body diam. at lip region; d′ = body diameter at guiding ring/body diameter at lip region; Oa-gr = Oral aperture-guiding ring distance; V = (distance from anterior end to vulva/body length) × 100.

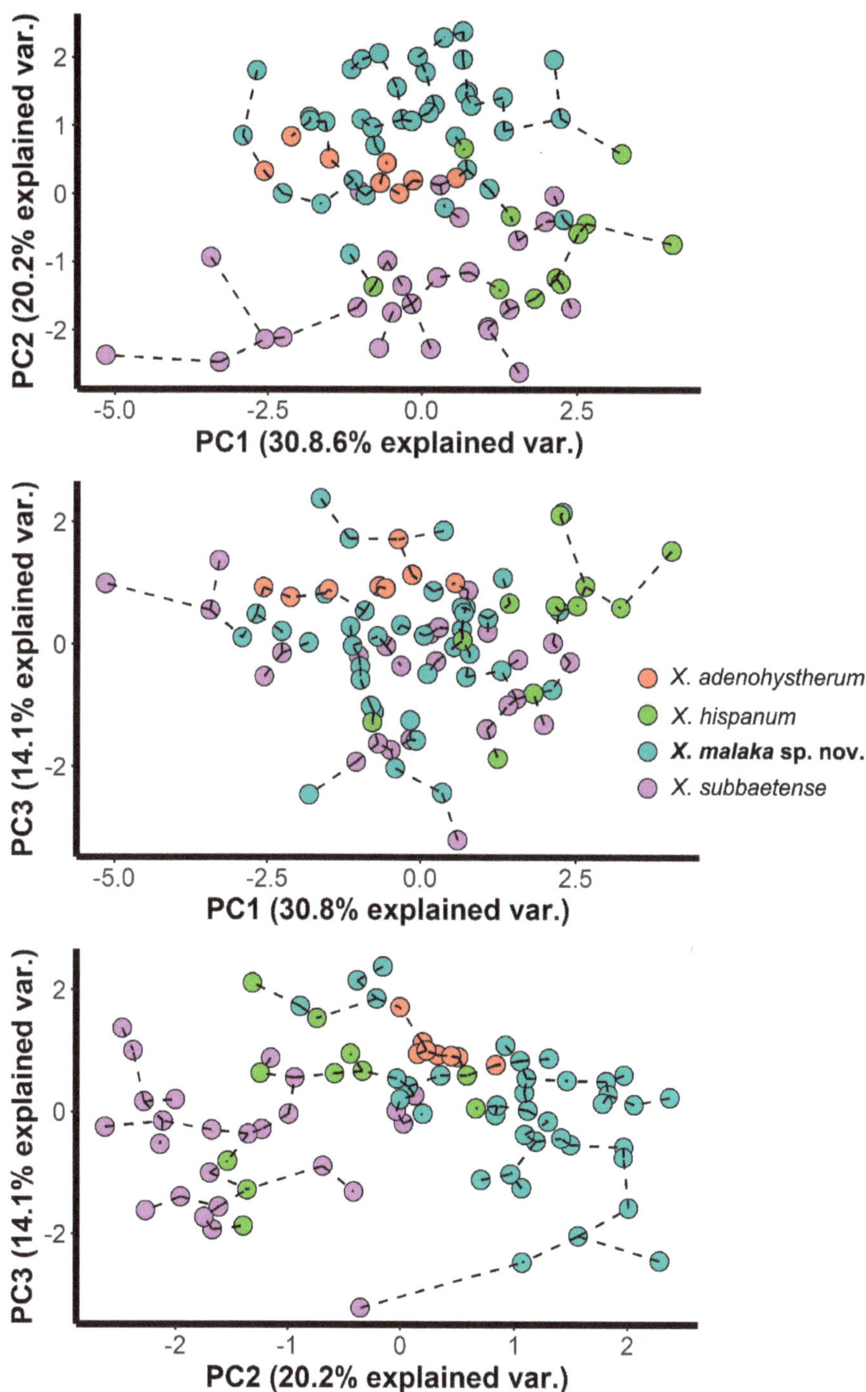



**Figure 2.** Principal component analysis on morphometric characters to characterize *Xiphinema hispanum*-complex with a superimposed minimum spanning tree (based on Euclidean distance).

**Table 3.** Morphometrics of paratypes for *Xiphinema malaka* sp. nov. from maritime pine (*Pinus pinaster* Aiton) at Canillas de Albaida (Málaga province) southern Spain [a].

| | | | Paratypes | | | |
|---|---|---|---|---|---|---|
| **Characters-Ratios** [b] | **Holotype** | **Females** | **J1** | **J2** | **J3** | **J4** |
| n | 1 | 20 | 7 | 4 | 4 | 6 |
| L (mm) | 4.3 | 4.2 ± 0.36 (3.51–4.90) | 1.25 ± 0.58 (1.13–1.31) | 1.81 ± 0.11 (1.66–1.92) | 2.19 ± 0.88 (2.11–2.33) | 3.11 ± 0.16 (2.86–3.31) |
| a | 73.6 | 74.2 ± 5.4 (65.6–82.2) | 48.8 ± 2.1 (45.3–51.5) | 43.2 ± 3.3 (40.8–48.1) | 56.9 ± 2.0 (53.5–58.7) | 64.8 ± 3.4 (62.1–70.0) |
| b | 8.8 | 8.0 ± 0.7 (6.7–9.3) | 5.1 ± 0.7 (4.4–6.1) | 6.0 ± 0.8 (5.3–7.0) | 5.9 ± 0.4 (5.5–6.3) | 6.6 ± 0.5 (6.0–7.2) |
| c | 117.4 | 110.6 ± 7.8 (97.3–126.3) | 20.5 ± 1.1 (18.6–22.4) | 28.5 ± 1.8 (26.8–31.0) | 39.9 ± 1.6 (37.4–41.5) | 67.4 ± 5.8 (60.2–75.4) |
| c′ | 0.9 | 0.9 ± 0.04 (0.9–1.0) | 3.5 ± 0.2 (3.2–3.8) | 2.4 ± 0.1 (2.3–2.6) | 2.0 ± 0.1 (1.8–2.1) | 1.3 ± 0.1 (1.2–1.4) |
| d | 8.9 | 8.4 ± 0.7 (6.9–9.3) | 5.4 ± 0.1 (5.3–5.6) | 7.6 ± 0.1 (7.5–7.6) | 7.3 ± 0.04 (7.3–7.4) | 8.4 ± 0.5 (8.1–9.2) |
| d′ | 2.7 | 2.6 ± 0.1 (2.4–2.9) | 2.0 ± 0.1 (1.9–2.1) | 2.68 ± 0.02 (2.67–2.70) | 2.4 ± 0.1 (2.3–2.5) | 2.7 ± 0.1 (2.6–2.8) |
| V | 50.1 | 50.2 ± 0.7 (49.2–51.5) | - | - | - | - |
| G1 | 13.7 | 13.6 ± 1.4 (11.1–15.8) | - | - | - | - |
| G2 | 12.5 | 11.6 ± 1.1 (9.9–13.5) | - | - | - | - |
| Odontostyle length | 143.0 | 140.4 ± 4.7 (131.0–148.5) | 64.1 ± 0.6 (63.5–65.0) | 76.3 ± 1.5 (75.0–78.0) | 93.6 ± 5.6 (84.0–98.0) | 117.8 ± 4.0 (115.0–125.5) |
| Odontophore length | 77.5 | 79.5 ± 1.9 (75.0–83.0) | 40.6 ± 0.7 (40.0–42.0) | 56.1 ± 1.5 (54.0–57.5) | 59.9 ± 2.2 (57.5–63.0) | 71.2 ± 2.0 (67.0–72.5) |
| Total stylet | 220.5 | 219.9 ± 5.8 (206.0–229.0) | 104.7 ± 1.1 (104.0–107.0) | 135.8 ± 3.2 (132.5–140.0) | 153.5 ± 5.9 (145.0–161.0) | 189.0 ± 3.8 (187.0–197.0) |
| Replacement odontostyle | - | - | 76.7 ± 2.4 (74.0–80.0) | 100.9 ± 4.5 (95.0–105.5) | 117.8 ± 8.8 (105.0–127.0) | 143.9 ± 2.9 (140.0–148.0) |
| Lip region width | 14.5 | 14.6 ± 0.4 (14.0–15.0) | 9.4 ± 0.3 (9.0–9.5) | 10.3 ± 0.4 (10.0–10.5) | 11.1 ± 0.3 (11.0–11.5) | 12.5 ± 0.4 (12.0–13.0) |
| Oral aperture-guiding ring | 129.0 | 124.9 ± 8.8 (96.0–135.0) | 51.6 ± 1.6 (50.0–54.0) | 75.1 ± 3.6 (71.5–80.0) | 81.80 ± 2.0 (80.0–84.0) | 103.6 ± 4.5 (98.0–110.0) |
| Tail length | 37.0 | 38.0 ± 2.3 (34.0–43.0) | 60.8 ± 1.9 (57.5–64.0) | 63.5 ± 1.7 (62.0–65.0) | 54.8 ± 2.2 (53.0–58.0) | 46.3 ± 2.6 (43.0–49.0) |
| J | 12.0 | 11.9 ± 1.1 (10.0–13.5) | 9.1 ± 0.6 (8.5–10.5) | 13.8 ± 1.3 (12.5–15.0) | 12.3 ± 0.3 (12.0–12.5) | 10.5 ± 0.9 (9.5–11.5) |

[a] Measurements are in μm and in the form: mean ± standard deviation (range); [b] a = body length/maximum body width; b = body length/pharyngeal length; c = body length/tail length; c′ = tail length/body width at anus; d = anterior to guiding ring/body diam. at lip region; d′= body diam. at guiding ring/body diam. at lip region; V = (distance from anterior end to vulva/body length) × 100; J = hyaline tail region length; G1 = (anterior genital branch length/body length) × 100; G2 = (posterior genital branch length/body length) × 100.

## 2.2. Mitochondrial Haplonet and Nuclear Haploweb Networks

Species delimitation using haplonet methods in *X. hispanum*-complex species contained 75 sequences (35 sequences from *X. malaka* sp. nov., four sequences from *X. adenohystherum*, 13 sequences from *X. hispanum*, and 23 sequences from *X. subbaetense*) with 13, 3, 4, and 3 different haplotypes and several heterozygous individuals, respectively (Table 1, Figure 3A). The TCS haplotype analysis inferred from the *D2-D3* region showed four well-differentiated haplogroups corresponding to four different main lineages (*X. adenohystherum*, *X. hispanum*, *X. malaka* sp. nov., and *X. subbaetense*) (Figure 3A). *Xiphinema malaka* sp. nov. comprised a higher diversity in Mountain Almijara (SA, with nine haplotypes) than that detected in Mountain Nieves (SN 2 haplotypes), one haplotype in Tabernas, and one haplotype (Hm3) jointly detected in SA and SN (Figure 3A).

**Table 4.** Morphometrics of *Xiphinema malaka* sp. nov. from several sampling points and localities at Málaga and Almería provinces, southern Spain[a].

| Host-plant Locality Sample Code | Málaga Province | | | | | | | | Almería Province |
|---|---|---|---|---|---|---|---|---|---|
| | Maritime Pine (Canillas de Albaida) SA2 | Maritime Pine (Canillas de Albaida) SA5 | Maritime Pine (Canillas de Albaida) SA6 | Maritime Pine (Canillas de Albaida) SA8 | | Black Pine (Iguaneja) SN1 | Black Pine (Iguaneja) SN3 | Cork oak (Monda) SN14 | Yellow Broom Tabernas AO44 |
| Characters-ratios [b] | Females | Females | Females | Male | Female | Female | Female | Females | J4 |
| n | 1 | 4 | 9 | 1 | 1 | 1 | 1 | 2 | 1 |
| L (mm) | 4.6 | 4.5 ± 0.4 (4.0–5.0) | 4.8 ± 0.2 (4.5–5.2) | 4.8 | 4.3 | 4.3 | 4.2 | (4.85, 4.92) | 3.08 |
| a | 82.8 | 85.4 ± 11.0 (76.0–99.8) | 87.2 ± 5.2 (79.6–96.5) | 109.5 | 76.1 | 72.1 | 70.9 | (82.0, 82.2) | 70.0 |
| b | 9.9 | 9.0 ± 1.1 (8.0–10.0) | 9.5 ± 0.7 (8.6–11.0) | 9.2 | 9.0 | 12.2 | 8.1 | (9.8, 10.3) | 7.0 |
| c | 123.1 | 132.8 ± 15.2 (115.3–151.2) | 164.7 ± 16.3 (133.2–178.6) | 170.2 | 133.5 | 98.3 | 113.9 | (103.2, 105.8) | 65.5 |
| c′ | 0.91 | 0.91 ± 0.03 (0.87–0.95) | 0.78 ± 0.06 (0.72–0.89) | 0.8 | 0.8 | 1.0 | 0.9 | (1.0, 1.1) | 1.3 |
| d | 8.5 | 8.2 ± 0.2 (8.0–8.4) | 8.3 ± 0.3 (7.9–8.9) | 7.6 | 7.6 | 8.1 | 9.4 | (7.4, 7.9) | 9.2 |
| d′ | 2.6 | 2.48 ± 0.01 (2.47–2.50) | 2.54 ± 0.13 (2.33–2.75) | 2.3 | 2.5 | 2.6 | 2.8 | (2.6, 2.7) | 2.8 |
| V or T | 49.2 | 49.4 ± 0.6 (48.5–50.0) | 49.8 ± 1.1 (47.8–51.5) | 62.0 | 47.1 | 52.2 | 50.0 | (51.8, 53.8) | - |
| G1 | 9.4 | 11.5 ± 0.8 (10.7–12.3) | 12.1 ± 2.0 (9.9–13.6) | - | 13.5 | 12.8 | 10.5 | (9.2, 10.7) | - |
| G2 | 8.9 | 10.0 ± 0.7 (9.5–10.8) | 11.4 ± 1.2 (10.1–12.3) | - | 13.0 | 11.7 | 10.2 | (8.1, 10.2) | - |
| Odontostyle length | 135.5 | 137.0 ± 4.2 (132.0–141.0) | 143.1 ± 2.2 (138.5–145.5) | 144.5 | 131.0 | 134.5 | 136.0 | (143.0, 149.0) | 125.5 |
| Odontophore length | 79 | 80.9 ± 0.9 (80.0–82.0) | 79.1 ± 2.7 (76.0–84.0) | 78.0 | 83.0 | 80.0 | 80.0 | (82.0, 88.0) | 71.5 |
| Total stylet | 215 | 217.9 ± 4.9 (212.5–222.0) | 222.2 ± 3.5 (215.5–227.5) | 225.5 | 214.0 | 215.0 | 216.0 | (231.0) | 197.0 |

**Table 4.** *Cont.*

| | Málaga Province | | | | | | | | Almería Province |
|---|---|---|---|---|---|---|---|---|---|
| Host-plant Locality Sample Code | Maritime Pine (Canillas de Albaida) SA2 | Maritime Pine (Canillas de Albaida) SA5 | Maritime Pine (Canillas de Albaida) SA6 | | Maritime Pine (Canillas de Albaida) SA8 | Black Pine (Iguaneja) SN1 | Black Pine (Iguaneja) SN3 | Cork oak (Monda) SN14 | Yellow Broom Tabernas AO44 |
| Characters-ratios [b] | Females | Females | Females | Male | Female | Female | Female | Females | J4 |
| Replacement odontostyle | - | - | - | - | - | - | - | - | 146 0 |
| Lip region width | 14.5 | 14.5 ± 0.4 (14.0–15.0) | 14.7 ± 0.5 (14.0–15.5) | 16.0 | 15.0 | 14.0 | 14.0 | (15.0, 15.5) | 14.0 |
| Oral aperture-guiding ring | 115 | 118.5 ± 2.1 (116.0–121.0) | 122.5 ± 3.8 (119.0–129.0) | 121.5 | 114.0 | 114.0 | 122.0 | (115.0, 118.0) | 110.0 |
| Tail length | 37 | 33.6 ± 1.1 (32.5–35.0) | 29.2 ± 2.9 (26.0–34.5) | 28.0 | 32.5 | 44.0 | 37.0 | (46.5, 47.0) | 47.0 |
| J | 12.0 | 10.3 ± 0.4 (10.0–10.5) | 10.5 ± 2.0 (8.0–13.0) | 8.0 | 8.5 | 12.0 | 11.0 | (13.0) | 12.0 |
| Spicules | - | - | - | 64.0 | - | - | - | - | - |
| Lateral guiding pieces | - | - | - | 26.5 | - | - | - | - | - |

[a] Measurements are in μm and in the form: mean ± standard deviation (range); [b] a = body length/maximum body width; b = body length/pharyngeal length; c = body length/tail length; c′ = tail length/body width at anus; d = anterior to guiding ring/body diam. at lip region; d′= body diam. at guiding ring/body diam. at lip region; V = (distance from anterior end to vulva/body length) × 100; J = hyaline tail region length; G1 = (anterior genital branch length/body length) × 100; G2 = (posterior genital branch length/body length) × 100.

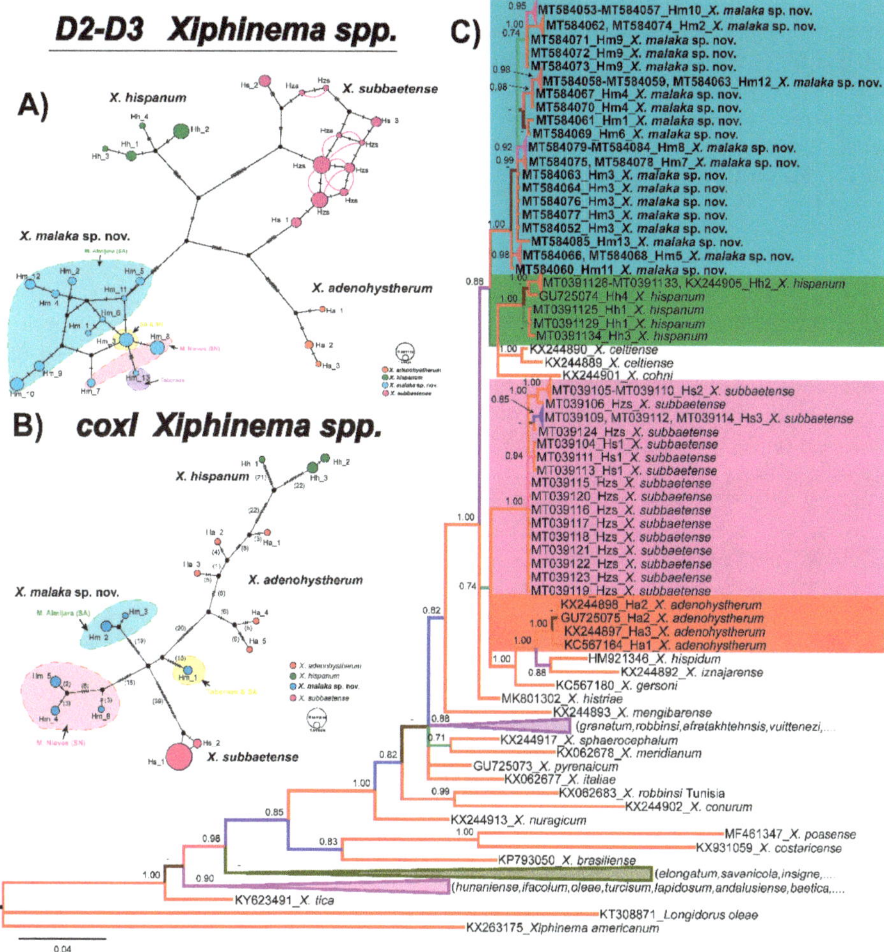



**Figure 3.** (**A**). Construction of *D2-D3* haploweb of *Xiphinema malaka* sp. nov. (**B**). *coxI* haplonet of *Xiphinema malaka* sp. nov. Coloured circles represent haplotypes and their diameter are proportional to the number of individuals sharing the same haplotype. Black short lines on the branches indicate the number of mutated positions in the alignment that separate each haplotype. Co-occurring haplotypes are enclosed in black dashes. (**C**). Phylogenetic relationships within the genus *Xiphinema*. Bayesian 50% majority rule consensus tree as inferred from D2 and D3 expansion domains of *28S* rRNA sequence alignment under the general time-reversible model of sequence evolution with correction for invariable sites and a gamma-shaped distribution (GTR + I + G) +. Posterior probabilities more than 0.70 are given for appropriate clades. Newly obtained sequences in this study are shown in bold. Scale bar = expected changes per site. Some branches were collapsed for improving readability of *Xiphinema* species. Abbreviations: Ha = *X. adenohystherum* haplotypes; Hh = *X. hispanum* haplotypes; Hm = *X. malaka* sp. nov. haplotypes; Hs = *X. subbaetense* haplotypes; Hzs = *X. subbaetense* heterozygous specimens. SA = Mountain of Almijara and Tejeda; SN = Mountain of Nieves.

However, in *coxI* haplonet (Figure 3B), six different haplotypes of *X. malaka* sp. nov. were detected, three in SN and three in SA. One from SA shared the same haplotype with the Tabernas isolate, and this haplotype kept a far molecular distance with the other two haplotypes from SA. It was worth noting that the number of *D2-D3* haplotypes of *X. malaka* sp. nov. was higher than *coxI* haplotypes (13 vs. 6), but there were more mutations between these *coxI* haplotypes than *D2-D3* haplotypes (Figure 3A,B). Besides, *X. subbaetense* also comprised more haplotypes in the *D2-D3* haplonet than the *coxI* haplonet (11 vs. 2); the situation of *X. hispanum*, *X. adenohystherum* were the same as previously described by Cai et al. [11].

## 2.3. Molecular Characterization

The amplification of *D2-D3* expansion domains of *28S* rRNA, *ITS1* rRNA, the partial *18S* rRNA, and partial *coxI* genes, yielded single fragments of ~900 bp, 1100 bp, 1800 bp, and 500 bp, respectively, based on gel electrophoresis. *D2-D3* for *X. malaka* sp. nov. (MT584052–MT584085) showed a low intraspecific variability with 1–7 different nucleotides and 0 indels (99% similarity). The molecular diversity of this marker within SA (1–7 nucleotides, 0 indels) and SN (2–3 nucleotides, 0 indels) isolates was similar among them and differed from the closest related species, *X. hispanum* (KX244905, MT039125–MT039134) by 20–21 different nucleotides and 1–2 indels (97% similarity), *X. subbaetense* (MT039104–MT039124) by 22–25 different nucleotides and 2–3 indels (97% similarity), and from *X. adenohystherum* (KC567164, KX244898, GU725075, KX244897) by 29–42 different nucleotides and 3 indels (96% similarity).

The *ITS1* region for *X. malaka* sp. nov. showed an intraspecific variability with 26–39 different nucleotides and 4–10 indels (96%–98% similarity). The molecular diversity of this marker within SA (18–24 nucleotides, 1–4 indels) and SN (26–29 nucleotides, 4 indels) isolates was also similar among them. *ITS1* for *X. malaka* sp. nov. (MT584088-MT584099) differed from the closest related species, *X. subbaetense* (MT026293–MT026295) by 132–136 different nucleotides and 28–29 indels (88% similarity), *X. hispanum* (GU725061) by 84–142 different nucleotides and 22–38 indels (87–90% similarity), and from *X. adenohystherum* (GU725063, MT584100–MT584102) by 133–139 different nucleotides and 40–45 indels (87–88% similarity).

For the *18S* rRNA, two new identical sequences for *X. malaka* sp. nov. (MT584086–MT584087) were obtained in this study and both of them showed very high similarity values with other accessions from *Xiphinema* spp. deposited in GenBank, being 98–99% similar. From the closet related species they differed by 1–2 nucleotides and 0 indels from *X. subbaetense* (MT039135–MT039140), *X. adenohystherum* (GU725084) by two nucleotides different and 0 indels, and *X. hispanum* (GU725083) by one nucleotide different and 0 indels. Finally, thirteen new *coxI* sequences for *X. malaka* sp. nov. (MT580263–MT580274) were deposited in GenBank in this study. This gene showed an intraspecific variability with 3–48 different nucleotides and 0 indels (88–99% similarity). The molecular diversity of this marker within SA (0–2 nucleotides, 0 indels) and SN (0–9 nucleotides, 0 indels) isolates was similar among them. *coxI* for *X. malaka* sp. nov. (MT580263–MT580274) differed from the closest related species, *X. subbaetense* (MT040280–MT010300) by 59–66 different nucleotides and 0 indels (82% similarity), *X. hispanum* (KY816614, MT040301-MT040305) by 51–78 different nucleotides and no indels (78–81% similarity), and from *X. adenohystherum* (KY816588–KY816592) by 58–65 different nucleotides and no indels (82–85% similarity).

## 2.4. Phylogenetic Relationships

Phylogenetic relationships among *Xiphinema* species inferred from analyses of *D2-D3* expansion domains of *28S* rRNA, *ITS1*, the partial *18S* rRNA and the partial *coxI* mtDNA gene sequences using BI are shown in Figures 3C, 4, 5 and 6, respectively. The phylogenetic trees generated with the nuclear and mitochondrial markers included 136, 49, 65 and 95 sequences with 747, 1106, 1547 and 372 positions in length, respectively (Figures 3C, 4, 5 and 6). The *D2-D3* tree of *Xiphinema* spp. showed a well-supported clade (PP = 1.00), including 10 species from morphospecies Groups 5 and 6, seven of them belonging to morphospecies Group 5 and three to Group 6, all of them reported from the Iberian peninsula, and included *X. malaka* sp. nov. (MT584052–MT584085). All other clades followed the same pattern as previous studies. *Xiphinema malaka* sp. nov. was phylogenetically related with *X. hispanum*, *X. celtiense* and *X. cohni* in a moderately supported clade (PP = 0.88), but clearly separate from all of them (Figure 3C).

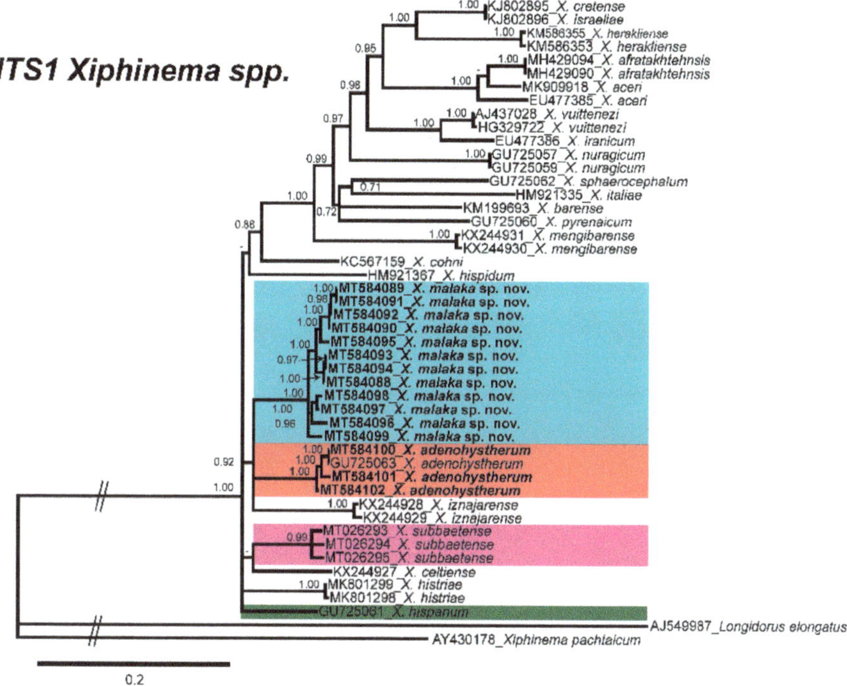


**Figure 4.** Phylogenetic relationships of *Xiphinema malaka* sp. nov. within the genus *Xiphinema*. Bayesian 50% majority-rule consensus trees as inferred from *ITS1* sequence alignments under transition model with a proportion of invariable sites and a rate of variation across sites (TIM2 + I + G). Posterior probabilities more than 70% are given for appropriate clades. Newly obtained sequences in this study are in bold letters, and each colour is associated with each species of the complex.

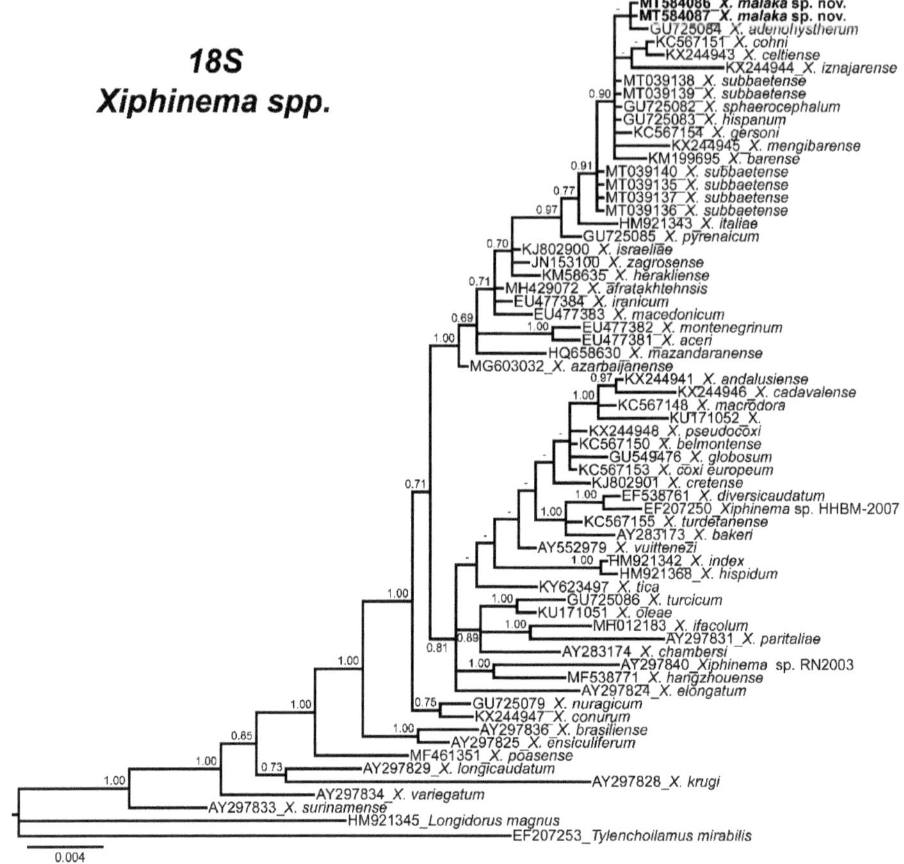



**Figure 5.** Phylogenetic relationships of *Xiphinema malaka* sp. nov. within the genus *Xiphinema*. Bayesian 50% majority-rule consensus trees as inferred from *18S* sequence alignments under the GTR + I + G model. Posterior probabilities more than 70% are given for appropriate clades. Newly obtained sequences in this study are in bold letters.

Difficulties were experienced with alignment of the ITS1 sequences due to scarce similarity. Thus, only related sequences were used for phylogeny. The 50% majority rule consensus *ITS1* BI tree showed several clades low to moderately supported (Figure 4). *Xiphinema malaka* sp. nov. was phylogenetically related to *X. adenohystherum* and *X. iznajarense* in a moderately supported clade (PP = 0.92), but clearly separate from all of them (Figure 4).

The 50% majority rule consensus *18S* rRNA gene BI tree showed several major clades (Figure 5). Phylogenetic inferences based on *18S* suggest that *X. malaka* sp. nov. was related to other species of the *X. hispanum*-complex in a moderately supported clade (PP = 0.91), together with other species such as *X. barense*, *X. celtiense*, *X. cohni*, *X. gersoni*, *X. iznajarense*, *X. mengibarense*, and *X. sphaerocephalum* (Figure 5).

Finally, the phylogenetic relationships of *Xiphinema* species inferred from analysis of partial *coxI* gene sequences showed several clades that were not well defined (Figure 6). *Xiphinema malaka* sp. nov. (MT580263-MT580274) was phylogenetically related to *X. hispanum*-complex species in a low supported clade (PP = 0.65), but clearly separate from all of them (Figure 6).

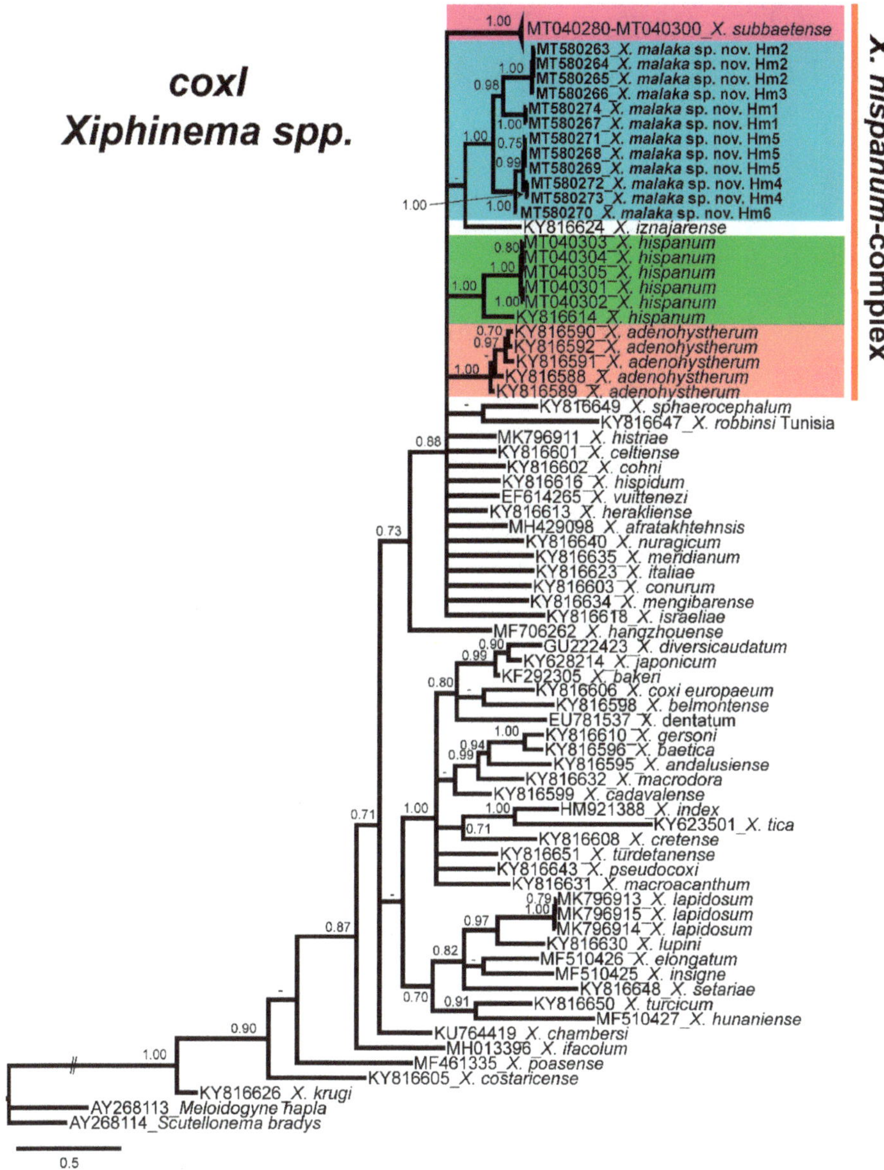



**Figure 6.** Phylogenetic relationships of *Xiphinema malaka* sp. nov. within the genus *Xiphinema*. Bayesian 50% majority-rule consensus trees as inferred from cytochrome c oxidase subunit I (*coxI*) mtDNA gene sequence alignments under the GTR + I + G model. Posterior probabilities more than 70% are given for appropriate clades. Newly obtained sequences in this study are in bold letters, and each colour is associated with each species of the complex.

*2.5. Morphology and Morphometry of Xiphinema malaka sp. nov*

***Xiphinema malaka* sp. nov.**

http://zoobank.org/urn:lsid:zoobank.org:act: BDBF964D-71E8-4E4F-B61C-50C5A7C51083

(Figures 7–10, Tables 3 and 4)

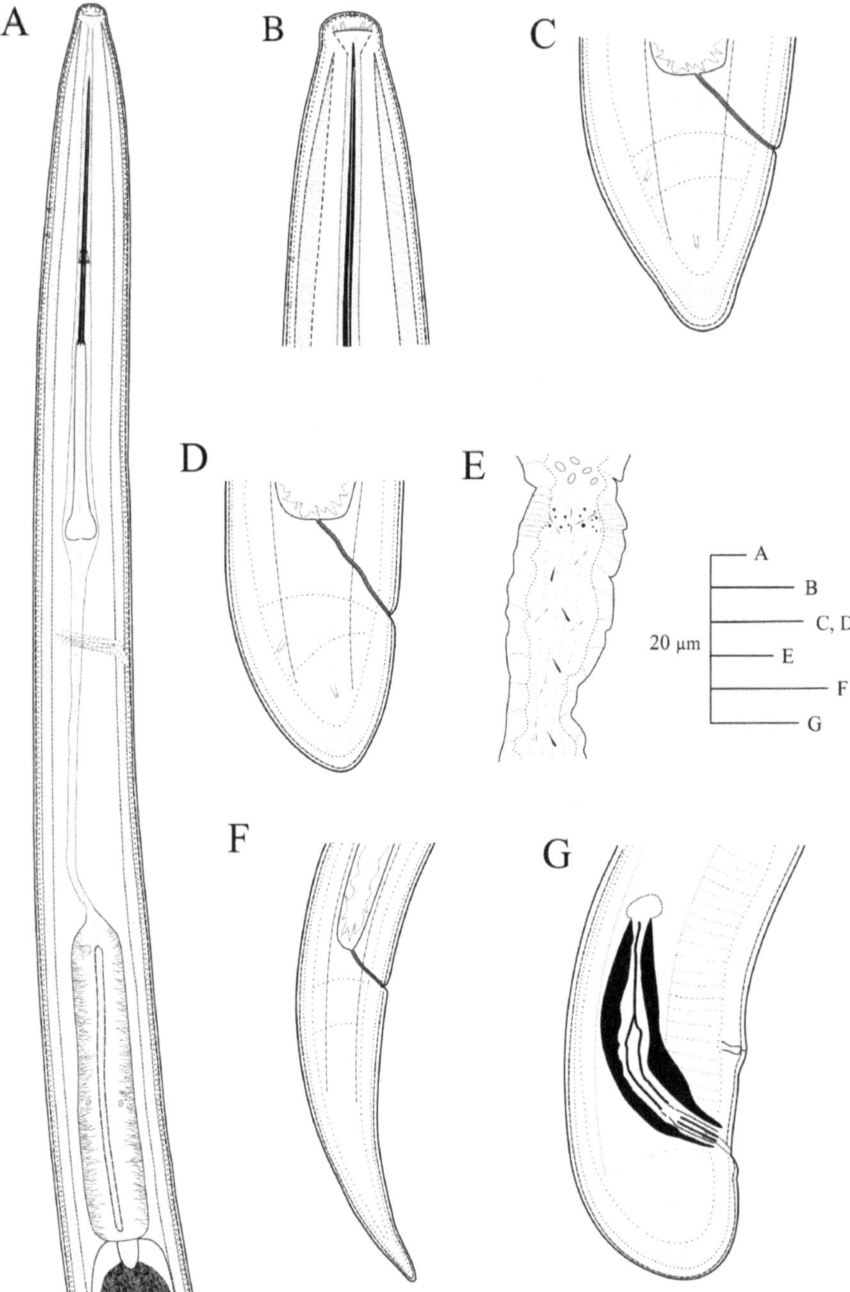



**Figure 7.** Line drawings of holotype for *Xiphinema malaka* sp. nov. (**A**), pharyngeal region; (**B**), detail of lip region; (**C,D**), female tails; (**E**), detail of uterine pseudo Z-differentiation.; (**F**), tail of first-stage juvenile (J1); (**G**), male tail.

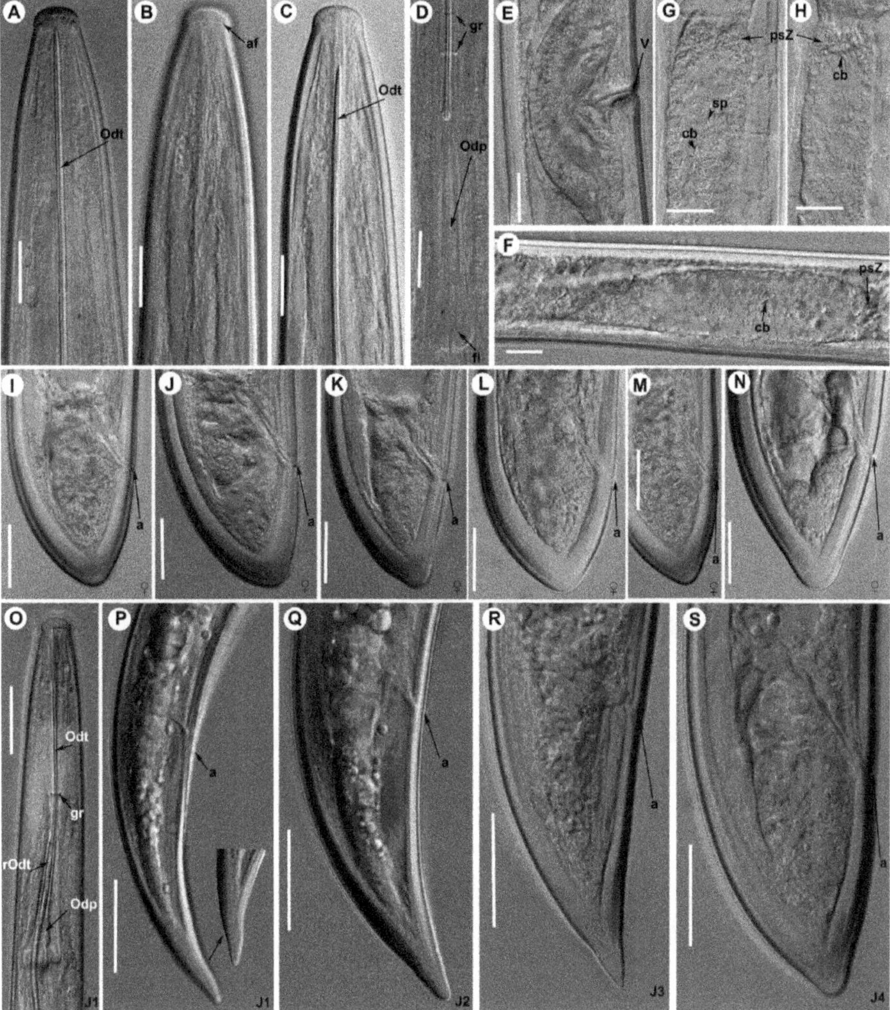

**Figure 8.** Light photomicrographs of *Xiphinema malaka* sp. nov. females holotype and paratypes: (**A**), anterior region holotype; (**B,C**) anterior regions paratypes; (**D**), detail of odontophore and guiding ring in holotype; (**E**), vulval region; (**F–H**), detail of female genital track showing Z-differentiation in holotype; (**I**), tail region of holotype; (**J–N**), tail region in paratypes; (**O**), detail of first-stage anterior region; (**P–S**), tail region of 1st, 2nd, 3rd and 4th stage juveniles. Abbreviations: a = anus; af = amphidial fovea; cb = crystalloid bodies; fl = odontophore flanges; gr = guiding ring; odp = odontophore; odt = odontostyle; psZ = pseudo-Z organ; rodt = replacement odontostyle; sp = spine; v = vulva. Scale bars: 20 μm.

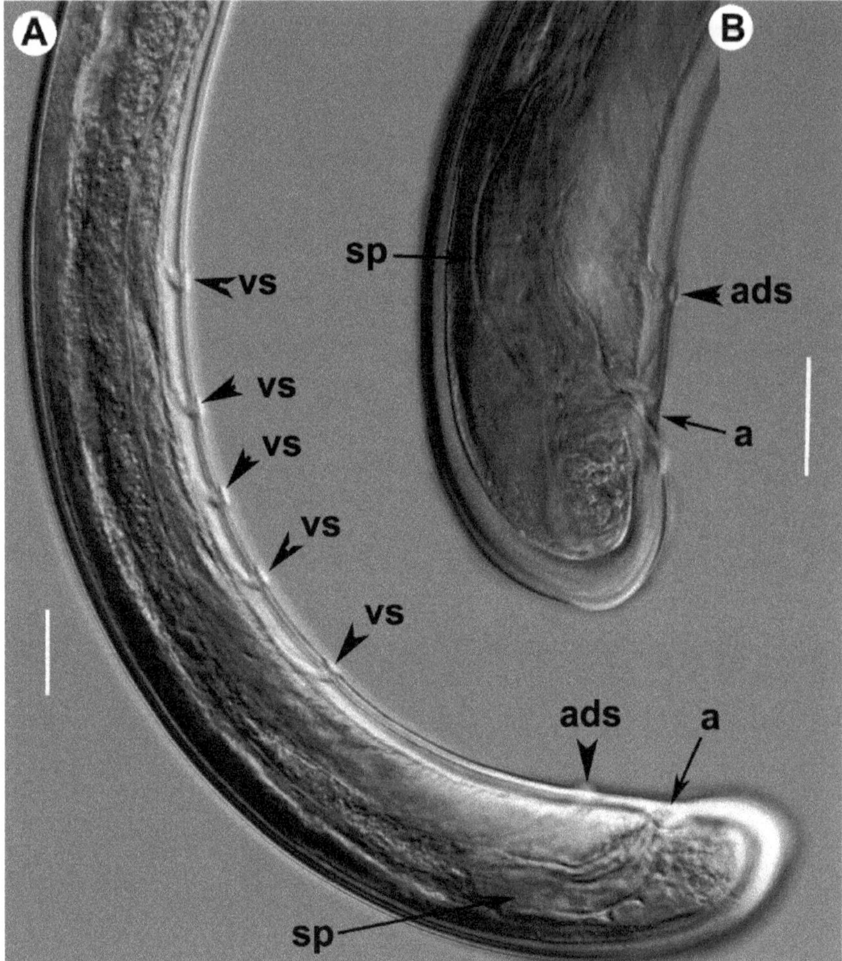



**Figure 9.** Light photomicrographs of *Xiphinema malaka* sp. nov. male: (**A**), posterior region; (**B**), detail of tail showing spicules. Abbreviations: a = anus; ads = adanal supplements; sp = spicules; vs = ventromedian supplements. Scale bars: 20 µm.

### 2.5.1. Material Examined

*Holotype.* Adult female was found in the rhizosphere of maritime pine (*Pinus pinaster* Aiton) at 1312 m a.s.l. from Canillas de Albaida, Málaga province, Spain (GPS: 36°52′21.81″ N; 3°55′41.00″ W) collected by A. Archidona-Yuste on 12 December 2019; mounted in pure glycerin and deposited in the nematode collection at Institute for Sustainable Agriculture (IAS) of Spanish National Research Council (CSIC), Córdoba, Spain (Slide number X-SA3-02).

*Paratypes.* Female and juvenile paratypes were collected from the same soil sample as the holotype (Table 3); mounted in pure glycerin and deposited in the Institute for Sustainable Agriculture (IAS) of the Spanish National Research Council (CSIC), Córdoba, Spain (Slide numbers X-SA3-03–X-SA3-08); one female at Istituto per la Protezione delle Piante (IPP) of Consiglio Nazionale delle Ricerche (C.N.R.), Sezione di Bari, Bari, Italy (X-SA3-011); one female at the USDA Nematode Collection (T-7474p).

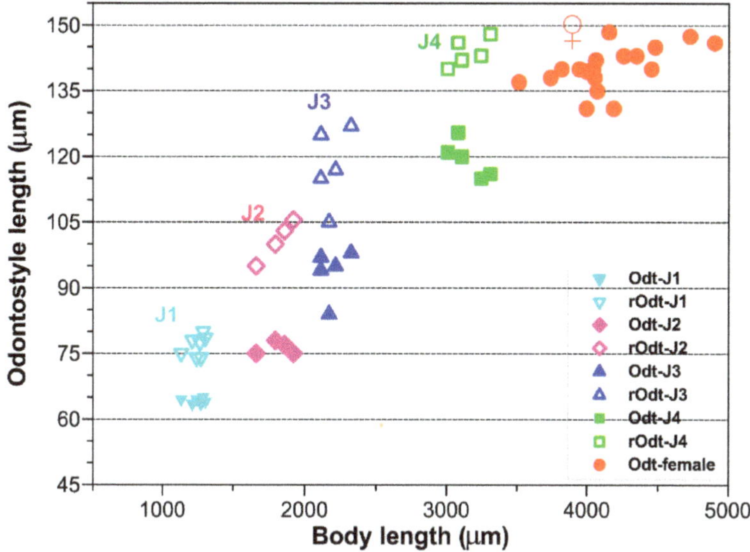



**Figure 10.** Relationship of body length to length of functional and replacement odontostyle (Ost and rOst, respectively) length in all developmental stages from first-stage juveniles (J1) to mature females of *Xiphinema malaka* sp. nov.

*Additional material examined.* Additional nematode isolates were studied and characterized from the rhizosphere of maritime pine, black pine, cork oak and yellow broom at several localities at Málaga and Almería provinces (Table 4). Morphometric measurements were taken for 62 individuals, 40 females, one male and 21 juveniles from J1 to J4 from several localities in Málaga province, Tables 3 and 4. Unfortunately, the scarce nematode isolate detected in the isolte of Tabernas (Almería) did not allow us to take measurements of adult females.

*Type locality.* Canillas de Albaida, Málaga province, Spain (GPS: 36°52′21.81″ N; 3°55′41.00″ W); 1254 m above sea level (a.s.l.) collected by A. Archidona-Yuste on 12 December 2019.

*Etymology.* The species epithet refers to the Phoenician word Malaka, the name of the province of Málaga where the species was found in several localities.

2.5.2. Diagnosis. *Xiphinema malaka* sp. nov.

Belongs to morphospecies Group 5 from the *Xiphinema* non*americanum*-group species [18]. It is characterized by a moderate long body (3.5–4.9 mm), assuming a J-shaped when heat-relaxed; lip region hemispherical, separate from the body contour by a depression, 14.0–15.0 μm wide; a relatively long odontostyle 131.0–148.5 μm; vulva located at 47.1–53.8% of body length; female reproductive system didelphic-amphidelphic having both branches about equally developed, pseudo Z-differentiation containing numerous small granular bodies, uterus tripartite with small crystalloid bodies and spines in low number and presence of prominent wrinkles in the uterine wall that may be confused with spiniform structures; female tail short convex-conoid on both sides, and bearing 3 caudal pores, ending in a rounded and broad terminus with a very small bulge at the end in some specimens; c′ ratio (0.9–1.0); male rare one individual out of 75 females. Four developmental juvenile stages were identified, the 1st-stage juvenile with tail elongate-conoid with characteristic subdigitate rounded terminus (c′ ratio 3.2–3.8). According to the polytomous key of Loof & Luc [18] and matrix codes sorted by Archidona-Yuste et al. [19], codes for the new species are (codes in parentheses are exceptions): A4-B23-C6-D6-E65-F4(5)-G3-H2-I3-J6-K2-L1. The DNA sequences of

*D2-D3* expansion domains of *28S*, ITS1 rRNA, *18S* rRNA, and partial *coxI* were deposited in GenBank under the accession numbers MT584052-MT584085, MT584088-MT584099, MT584086-MT584087 and MT580263-MT580274, respectively.

### 2.5.3. Description

*Female.* Body cylindrical, slightly tapering towards anterior end in a J-shape when heat relaxed. Cuticle with fine transverse striae visible in tail region, 3.2 ± 0.3 (3.0–3.5) µm thick at mid body but thicker just posterior to anus. Lateral chord 13.2 ± 2.5 (11.5–16.0) µm wide, occupying ca. 25% of corresponding body diam. Lip region hemispherical, slightly offset from body contour by a depression, 14.6 ± 0.4 (14.0–15.0) µm wide and 6.5 ± 0.6 (6.0–7.5) µm high. Odontostyle moderately long, 1.7–1.9 times longer than odontophore, the latter with well-developed flanges (13.0–15.5 µm wide). Guiding ring and guiding sheath variable in length depending on degree of protraction/retraction of stylet. Pharynx composed by a slender narrow flexible part 335–582 µm long, and a posterior muscular, cylindrical and expanded part with three nuclei. Terminal pharyngeal bulb variable in length, 112.0–149.0 µm long and 24.0–31.0 µm wide. Glandularium 110.0–129.0 µm long. Dorsal gland nucleus (DN) located at beginning of basal bulb (8.5–14.3%), ventrosublateral gland nuclei (SVN) situated *ca* halfway along bulb (52.3–67.9%) (position of gland nuclei calculated as described by Loof & Coomans [27]). Cardia conoid-rounded and variable in length, 12.0–15.0 µm long. Intestine simple, prerectum variable in size 471–516 µm long. Rectum 32.0–40.0 µm long ending in anus as a small rounded slit. Reproductive system didelphic-amphidelphic with two equally developed branches. Each branch composed of a short ovary 47–78 µm long, a reflexed oviduct 93–104 µm long with well-developed pars dilatata oviductus, a sphincter, a well-developed pars dilatata uteri, and a 254–286 µm long uterus with pseudo-Z differentiation containing numerous small granular bodies with small crystalloid bodies (6.0–12.5 µm long) and spines in low number, and presence of prominent wrinkles in the uterine wall that may be confused with spiniform structures (Figures 7 and 8); a 35.5–47.0 µm long vagina perpendicular to body axis (having 28–32% corresponding body diam. ingrowth), ovejector well-developed 36.5–50.0 µm wide, pars distalis vaginae 16.8 ± 2.4 (13.0–19.5) µm long, and pars proximalis vaginae 24.3 ± 2.4 (20.5–27.5) µm long and 26.3 ± 1.0 (21.5–29.5) µm wide, and vulva as a transverse slit. Tail short, convex-conoid on both sides, and bearing three caudal pores, ending in a rounded and broad terminus, with a very small bulge at the end in some specimens (Figure 8).

*Male.* Extremely rare, only one male individual out of 75 female specimens was found in one sample near the type locality. Morphologically similar to female except for genital system and secondary sexual features. Male genital tract diorchic with testes containing multiple rows of different stages of spermatogonia. Tail short, convex-conoid with a broadly rounded terminus and thickened outer cuticular layer. Adanal supplements paired, preceded anteriorly by a row of five irregularly spaced ventromedians supplements. Spicules paired, dorylaimoid, moderately long and slightly curved ventrally, approximately 2.5 times longer than tail length; lateral guiding pieces more or less straight or with curved proximal end.

*Juveniles.* Four developmental juvenile stages were detected and distinguished by relative body length, odontostyle and replacement odontostyle length. The 1st-stage juveniles were characterized by the replacement odontostyle inserted into odontophore base (Figure 8). In all other stages, the replacement odontostyle was posterior to the flanges of odontophore in its resting position. The correlation between body length, replacement and functional odontostyle of the type population is given in Figure 10. Lip region in all juvenile stages looks similar to that in females. Other morphological characters similar to female, except for their size and immature sexual characteristics (developing genital primordium 16.0–87.0 µm long). The first-stage juvenile was characterized by a tail elongate-conoid with characteristic subdigitate rounded terminus (c′ ratio 3.2–3.8). Tail of other developmental stages becoming progressively shorter and wider after each moult (Figure 8).

### 2.5.4. Remarks

According to the polytomous key by Loof & Luc [18] and matrix codes sorted by Archidona-Yuste et al. [6]: A (type of female genital apparatus), C (tail shape), D (c' ratio), E (vulva position), F (body length), and G (total stylet length) (in this order of main features), *X. malaka* sp. nov. is closely related to *X. subbaetense* Cai, Archidona-Yuste, Cantalapiedra-Navarrete, Palomares Rius & Castillo [11], *X. hispanum* Lamberti, Castillo, Gomez-Barcina & Agostinelli [22], *X. adenohystherum* Lamberti, Castillo, Gomez-Barcina & Agostinelli [22], *X. cohni* Lamberti, Castillo, Gomez-Barcina & Agostinelli [22], and *X. sphaerocephalum* Lamberti, Castillo, Gomez-Barcina & Agostinelli [22].

*Xiphinema malaka* sp. nov. is morphometrically almost undistinguishable from *X. subbaetense* and *X. hispanum*, from the former can only be differentiated in females by a higher a ratio (65.6–99.8 vs. 49.0–70.0), a shorter odontophore (75.0–88.0 vs. 82.0–96.5 μm), narrower lip region (14.0–15.5 vs. 15.5–18.5 μm), higher c' ratio in J1 (3.2–3.8 vs. 2.6–3.1, 2.7–3.1, respectively), and presence of male (very rare vs. absent) [11,20]. Morphologically can be differentiated from *X. subbaetense* and *X. hispanum* in pseudo-Z differentiation containing numerous small granular bodies vs. 4–5 granular bodies. It can be differentiated from *X. adenohystherum* by slightly shorter odontostyle (131.0–149.0 vs. 143.0–152.0 μm), longer tail (26.0–47.0 vs. 29.0–35.0 μm), and slightly higher a ratio (65.6–96.5 vs. 65.2–73.3). It can be differentiated from *X. sphaerocephalum* by its shorter odontostyle (131.0–149.0 vs. 143.5–168.0 μm), and shorter oral aperture-guiding ring distance (96.0–135.0 vs. 126.0–162.0 μm). Finally, *X. malaka* sp. nov. can be differentiated from *X. cohni* by its shorter odontostyle (131.0–149.0 vs. 149–174 μm), shorter oral aperture-guiding ring distance (96.0–135.0 vs. 137.0–161.0 μm), and higher c ratio (97.3–178.6 vs. 82.6–115.2). Nevertheless, it can be clearly separated by specific *28S* rRNA, ITS1 rRNA and *coxI* sequences.

## 3. Discussion

The primary objective of this study was to decipher the cryptic diversity of the *X. hispanum*-complex by applying an integrative taxonomical approaches on several new unidentified *Xiphinema* isolates from Málaga and Almería provinces (southern Spain), appearing morphologically and morphometrically indistinguishable from this species complex. Multivariate morphometric analyses proved to be useful tools for species delimitation within the genera *Longidorus* and *Xiphinema* [11,15,19,28]. These data support that *X. hispanum*-complex species comprise a model example of morphostatic speciation (genetic modifications not reflected in morphology and morphometry) [23,24], since independent approaches based on molecular analyses using ribosomal and mitochondrial sequences (haploweb and haplonet) revealed high levels of genetic diversity within these species complexes which clearly separated *X. malaka* sp. nov. from all other *X. hispanum*-complex species. These results, as well as those from previous studies, may suggest that *X. hispanum*-complex species comprises a *Xiphinema* endemic lineage, with members morphologically and morphometrically very similar, that have diversified in the Iberian peninsula, since no other records on these species have been reported outside this area [20,22,29].

Phylogenetic analyses based on three rDNA molecular markers (*D2–D3* expansion domains of *28S* rRNA gene, *ITS1* region and the partial *18S* rRNA) resulted in a general consensus of species phylogenetic positions for the majority of them, and were generally congruent with those given by previous phylogenetic analysis [6,11,19,30–33].

The results of this research support our hypothesis that biodiversity of Longidoridae in southern Spain is still not fully clarified and needs additional sampling efforts given the significant gaps in soil nematode biodiversity regarding the large number of undescribed species [34,35] and the hypothesis suggesting the Iberian Peninsula as a possible center of speciation for some groups of the family Longidoridae [6,15,36]. The recognition of this extraordinary cryptic diversity has a direct bearing on estimates of global nematode biodiversity and concepts of nematode biogeography. Regional endemicity in plant-parasitic nematodes has seldom been recognized and cosmopolitan distributions in nematodes, like other microscopic organisms, are reportedly common [37,38].

In summary, the present study confirmed the extraordinary cryptic diversity of *X. hispanum*-complex species in Andalusia and comprises a paradigmatic example of morphostatic speciation of dagger nematodes in southern Spain, which can be a potential explanation of the unusual high biodiversity within Longidoridae, considering Andalusia as a hot spot of biodiversity. However, additional similar intensive taxonomic studies are needed in other areas which can confirm this statement.

## 4. Materials and Methods

### 4.1. Nematode Isolates and Morphological Studies

No specific permits were required for the indicated fieldwork studies. The soil samples were obtained in public areas, forests and other natural areas and did not involve any endangered species or those protected in Spain, nor were the sites protected in any way.

A total of 62 individuals including 41 adults and 21 juvenile specimens from several localities in Málaga and Almería provinces (southern Spain) were used for morphological analyses (Table 1, Figure 1). Nematodes were surveyed during spring season in 2019 in natural ecosystems in Andalusia, southern Spain (Table 1). Soil samples were collected for nematode analysis with a shovel randomly selecting four to five cores at each point, and considering the upper 5–50 cm depth of soil that was close to the active plant root at each sampling spot. Nematodes were extracted from a 500-cm$^3$ sub-sample of soil by centrifugal flotation [39] and a modification of Cobb's decanting and sieving [40] methods. For morphometric studies, *Xiphinema* specimens were killed and fixed by a hot solution of 4% formalin + 1% glycerol, then processed in pure glycerin [41] as modified by De Grisse [42].

Specimens for light microscopy were killed by hot fixative using a solution of 4% formaldehyde +1% propionic acid and embedded in pure glycerine using Seinhorst's [41] method. The morphometric study of each nematode isolate included morphology-based diagnostic features in *Xiphinema* (i.e., de Man body ratios), lip region width, amphid shape, oral aperture-guiding-ring, odontostyle and odontophore length and female tail shape [7]. For line drawings of the new species, light micrographs were imported to CorelDraw ver. X7 and redrawn. The light micrographs and measurements of each nematode isolate, including important diagnostic characteristics (i.e., de Man indices, body length, odontostyle length, lip region, tail shape, amphid shape and oral aperture-guiding ring; [7]) were performed using a Leica DM6 (Wetzlar, Germany) compound microscope with a Leica DFC7000 T digital camera. For the line drawings of the new species, CorelDraw software version X7 (Corel Corporation, London, UK) was used to redraw according to the selected light micrographs.

### 4.2. DNA Extraction, Polymerase Chain Reaction (PCR) and Sequencing

For molecular analyses, in order to ensure the selected nematodes for extracting DNA were from the same species, two live nematodes from each sample were temporary mounted in a drop of 1M NaCl containing glass beads (to avoid nematode crushing/damaging specimens) to ensure specimens conformed to the unidentified isolates of *Xiphinema*. Thus, 34 individuals collected from several sampling points in Andalusia were molecularly analyzed (Table 1). All necessary morphological and morphometric data, by taking pictures and measurements using the above camera-equipped microscope, were recorded. This was followed by DNA extraction from a single specimen and polymerase chain reaction (PCR) cycle conditions as previously described [6,15]. Several sets of primers were used for PCR. A partial region of the *28S* rRNA gene including the expansion domains D2 and D3 (*D2-D3*) was amplified by using the primers D2A (5′-ACAAGTACCGTGAGGGAAAGTTG-3′) and D3B (5′-TCGGAAGGAACCAGCTACTA-3′) [43]. The Internal Transcribed Spacer region 1 (*ITS1*) separating the *18S* rRNA gene from the 5.8S rRNA gene was amplified using forward primer *18S* (5′-TTGATTACGTCCCTGCCCTTT-3′) [44] and reverse primer rDNA1 5.8S (5′-ACGAGCCGAGTGATCCACCG-3′) [45]. A partial sequence of the *18S* rRNA gene (*18S*) was amplified as previously described [46] using primers 988F (5′-CTCAAAGATTAAGCCATGC-3′), 1912R (5′-TTTACGGTCAGAACTAGGG-3′), 1813F (5′-CTGCGTGAGAGGTGAAAT-3′), and 2426R

(5′-GCTACCTTGTTACGACTTTT -3′. Finally, the portion of the cytochrome c oxidase subunit I gene (*coxI*) was amplified using the primers COIF (5′-GATTTTTTGGKCATCCWGARG-3′) and COIR (5′-CWACATAATAAGTATCATG-3′) [47]. The newly obtained sequences were deposited in the GenBank database under accession numbers indicated in Table 1 and on the phylogenetic trees.

PCR cycle conditions were one cycle of 94 °C for two min, followed by 35 cycles of 94 °C for 30 s, annealing temperature of 55 °C for 45 s, 72 °C for three min, and finally one cycle of 72 °C for 10 min. PCR products were purified after amplification using ExoSAP-IT (Affimetrix, USB products, High Wycombe, UK), quantified using a Nanodrop spectrophotometer (Nanodrop Technologies, Wilmington, DE, USA) and used for direct sequencing in both directions using the primers noted above. The resulting products were purified and run on a DNA multicapillary sequencer (Model 3130XL genetic analyser; Applied Biosystems, Foster City, CA, USA), using the BigDye Terminator Sequencing Kit v.3.1 (Applied Biosystems, Foster City, CA, USA), at the Stab Vida sequencing facilities (Caparica, Portugal). The newly obtained sequences were submitted to the GenBank database under accession numbers indicated in Table 1 and on the phylogenetic trees.

### 4.3. Species Delimitation via Multivariate Morphometric Analysis and Haplotype Networks Construction

The nine new *Xiphinema* isolates detected in this study were included in the *X. hispanum*-complex species group given the close relationships morphologically with *X. hispanum* as outlined above. An iterative analysis of morphometric and molecular data using two independent strategies of species delimitation was utilized to asses described and undescribed specimens and to determine species boundaries within this species complex.

Species delineation using morphometry was conducted with principal component analysis (PCA) in order to estimate the degree of association among species within the *X. hispanum*-complex [48]. PCA was based upon the following morphological characters: L (body length), the ratios a, c, c′, d, d′, V, odontostyle and odontophore length, lip region width and hyaline region length (Table 2) [6,7,13]. Prior to the statistical analysis, diagnostic characters were tested for collinearity [49]. We used the collinearity test based on the values of the variance inflation factor (VIF) method that iteratively excludes numeric covariates showing VIF values > 10 as suggested by Montgomery and Peck [50]. PCA was performed by a decomposition of the data matrix amongst isolates using the principal function implemented in the package psych [51]. Orthogonal varimax raw rotation was used to estimate the factor loadings. Only factors with sum of squares (SS) loadings > 1 were extracted. Finally, a minimum spanning tree (MST) based on the Euclidean distance was superimposed on the scatter plot of the *X. malaka* sp. nov.-specimens complex against the PCA axes. MST was performed using the ComputeMST function implemented in the package emstreeR [52]. All statistical analyses were performed using the R v. 3.5.1 freeware [53].

In order to detect distinct phylogenetic groups possibly representing separate species, haplotype networks (briefly, haplonet) were constructed to each of the two separate datasets, i.e., the *D2-D3* and *coxI*. Alignments were converted to the NEXUS format using DnaSP V.6 [54]; TCS networks [55] were applied in the program PopART V.1.7 [56]. Illustrations of networks were prepared using the program Adobe illustrator to add connecting curves between the haplotypes found co-occurring in heterozygous individuals [57].

### 4.4. Phylogenetic Analysis

Sequenced genetic markers in the present study (after discarding primer sequences and ambiguously aligned regions), and several *Xiphinema* spp. sequences obtained of GenBank, were used for phylogenetic reconstruction (Table 1). Outgroup taxa for each dataset were selected based on previous published studies [6,11,30,45,58]. Multiple sequence alignments of the newly obtained and published sequences were made using the FFT-NS-2 algorithm of MAFFT v. 7.450 [59]. Sequence alignments were visualized using BioEdit [60] and edited by Gblocks ver. 0.91b [61] in the Castresana Laboratory server (http://molevol.cmima.csic.es/castresana/Gblocks_server.html) using options for a less stringent selection (minimum number of sequences for a conserved or a flanking position: 50% of the number of sequences + 1; maximum number of contiguous no conserved positions: 8; minimum length of a block: 5; allowed gap positions: with half).

Phylogenetic analyses of the sequence data sets were based on Bayesian inference (BI) using MRBAYES 3.2.7a [62]. The best-fit model of DNA evolution was calculated with the Akaike information criterion (AIC) of JMODELTEST v. 2.1.7 [63]. The best-fit model, the base frequency, the proportion of invariable sites and the gamma distribution shape parameters and substitution rates in the AIC were then used in phylogenetic analyses. BI analyses were performed under a general time reversible, with a proportion of invariable sites and a rate of variation across sites (GTR + I + G) model for *D2-D3*, the partial *18S* rRNA, and the partial *coxI* gene, and under a transition model with a proportion of invariable sites and a rate of variation across sites (TIM2 +I + G). These BI analyses were run separately per dataset with four chains for $2 \times 10^6$ generations. The Markov chains were sampled at intervals of 100 generations. Two runs were conducted for each analysis. After discarding burn-in samples of 30% and evaluating convergence, the remaining samples were retained for more in-depth analyses. The topologies were used to generate a 50% majority-rule consensus tree. Posterior probabilities (PP) were given on appropriate clades. Trees from all analyses were visualized using FigTree software version v.1.42 [64].

**Author Contributions:** Conceptualization, A.A.-Y., R.C., C.C.-N., J.A.C., A.R., B.V., G.L., J.E.P.-R. and P.C., methodology, A.A.-Y., R.C., C.C.-N., J.A.C., A.R., G.L., J.E.P.-R. and P.C., software, A.A.Y., R.C., J.E.P.-R. and P.C., analysis, A.A.Y., R.C., J.E.P.-R. and P.C., resources, J.A.C., A.R., G.L. and P.C., writing, A.A.-Y., R.C., C.C.-N., J.A.C., A.R., G.L., J.E.P.-R. and P.C. All authors contributed to the final discussion data, and read and approved the final manuscript. All authors have read and agreed to the published version of the manuscript.

**Funding:** This research was supported by Spanish Ministry of Science, Innovation and Universities, grant number RTI2018-095345-B-C21, LITHOFOR ("Modulating role of LITHOlogy in the response of Mediterranean FORest ecosystems to climate change: growth, edaphological processes and future predictions") and the Humboldt Research Fellowship for Postdoctoral Researchers awarded for the first author.

**Acknowledgments:** We would like to thanks J. Martin Barbarroja and G. León Ropero (IAS-CSIC) for their excellent technical assistance in surveys and management of soil samples, and further anonymous reviewers and editors for their effort in reviewing the manuscript and helping improve this study. The first author is a recipient of Humboldt Research Fellowship for Postdoctoral Researchers at Helmholtz Centre for Environmental Research-UFZ, Leipzig, Germany. The second author acknowledges the China Scholarship Council (CSC) for financial support. The sixth author acknowledges Spanish Ministry of Economy and Competitiveness for the "Ramon y Cajal" Fellowship RYC-2017-22228.

**Conflicts of Interest:** The authors declare no conflict of interest. The funders had no role in the design of the study; in the collection, analyses, or interpretation of data; in the writing of the manuscript, or in the decision to publish the results.

## References

1. Decraemer, W.; Hunt, D. Structure and classification. In *Plant Nematology*; Perry, R.N., Moens, M., Eds.; CABI: Wallingford, UK, 2006; pp. 3–32.
2. Wyss, U. Root parasitic nematodes: An overview. In *Cellular and Molecular Aspects of Plant-Nematode Interactions. 10*; Fenoll, C., Grundler, F.M.W., Ohl, S.A., Eds.; Kluwer Academic Publisher: Dordrecht, The Netherlands, 1997; pp. 5–22.
3. Singh, S.K.; Hodda, M.; Ash, G.J. Plant-parasitic nematodes of potential phytosanitary importance, their main hosts and reported yield losses. *Bull. OEPP* **2013**, *43*, 334–374. [CrossRef]
4. Davies, L.J.; Elling, A.A. Resistance genes against plant-parasitic nematodes: A durable control strategy? *Nematology* **2015**, *17*, 249–263. [CrossRef]
5. Coomans, A.; Huys, R.; Heyns, J.; Luc, M. *Character Analysis, Phylogeny, and Biogeography of the Genus Xiphinema Cobb, 1973 (Nematoda, Longidoridae)*; Annales du Musée Royal de l'Afrique Centrale (Zoologie): Tervuren, Belgique, 2001; Volume 287.

6.  Archidona-Yuste, A.; Navas-Cortés, J.A.; Cantalapiedra-Navarrete, C.; Palomares-Rius, J.E.; Castillo, P. Remarkable diversity and prevalence of dagger nematodes of the genus *Xiphinema* Cobb, 1913 (Nematoda: Longidoridae) in olives revealed by integrative approaches. *PLoS ONE* **2016**, *11*, e0165412. [CrossRef] [PubMed]

7.  Jairajpuri, M.S.; Ahmad, W. *Dorylaimida: Free-Living, Predaceous and Plant-Parasitic Nematodes*; IBH Publishing Co: New Delhi, India, 1992.

8.  Siddiqi, M.R. *Tylenchida. Parasites of Plants and Insects*; CABI Bioscience: Egham, UK, 2000. [CrossRef]

9.  Blaxter, M.L.; De Ley, P.; Garey, J.R.; Liu, L.X.; Scheldeman, P.; Vierstraete, A.; Vanfleteren, J.R.; Mackey, L.Y.; Dorris, M.; Frisse, L.M.; et al. A molecular evolutionary framework for the Phylum Nematoda. *Nature* **1998**, *392*, 71–75. [CrossRef] [PubMed]

10.  Decraemer, W.; Robbins, R.T. The who, what and where of Longidoridae and Trichodoridae. *J. Nematol.* **2007**, *39*, 295–297.

11.  Cai, R.; Archidona-Yuste, A.; Cantalapiedra-Navarrete, C.; Palomares-Rius, J.E.; Castillo, P. New evidence of cryptic speciation in the family Longidoridae (Nematoda: Dorylaimida). *J. Zool. Syst. Evol. Res.* **2020**, *58*. [CrossRef]

12.  Gutiérrez-Gutiérrez, C.; Cantalapiedra-Navarrete, C.; Decraemer, W.; Vovlas, N.; Prior, T.; Palomares-Rius, J.E.; Castillo, P. Phylogeny, diversity, and species delimitation in some species of the *Xiphinema americanum*-group complex (Nematoda: Longidoridae), as inferred from nuclear and mitochondrial DNA sequences and morphology. *Eur. J. Plant Pathol.* **2012**, *134*, 561–597. [CrossRef]

13.  Archidona-Yuste, A.; Navas-Cortés, J.A.; Cantalapiedra-Navarrete, C.; Palomares-Rius, J.E.; Castillo, P. Unravelling the biodiversity and molecular phylogeny of needle nematodes of the genus *Longidorus* (Nematoda: Longidoridae) in olive and a description of six new species. *PLoS ONE* **2016**, *11*, e0147689. [CrossRef]

14.  Czekanski-Moir, J.; Rundell, R. The ecology of nonecological speciation and nonadaptive radiations. *Trends Ecol. Evol.* **2019**, *34*, 400–415. [CrossRef]

15.  Archidona-Yuste, A.; Navas-Cortés, J.A.; Cantalapiedra-Navarrete, C.; Palomares-Rius, J.E.; Castillo, P. Cryptic diversity and species delimitation in the *Xiphinema americanum*-group complex (Nematoda: Longidoridae) as inferred from morphometrics and molecular markers. *Zool. J. Linn. Soc.* **2016**, *176*, 231–265. [CrossRef]

16.  Taylor, C.A.; Brown, D.J.F. *Nematode Vectors of Plant Viruses*; CAB International: Wallingford, UK, 1997.

17.  Lamberti, F.; Molinari, S.; Moens, M.; Brown, D.J.F. The *Xiphinema americanum* group. I. Putative species, their geographical occurrence and distribution, and regional polytomous identification keys for the group. *Russ. J. Nematol.* **2000**, *8*, 65–84.

18.  Loof, P.A.A.; Luc, M. A revised polytomous key for the identification of species of the genus *Xiphinema* Cobb, 1913 (Nematoda: Longidoridae) with exclusion of the *X. americanum*-group. *Syst. Parasitol.* **1990**, *16*, 35–66. [CrossRef]

19.  Archidona-Yuste, A.; Navas-Cortés, J.A.; Cantalapiedra-Navarrete, C.; Palomares-Rius, J.E.; Castillo, P. Molecular phylogenetic analysis and comparative morphology resolve two new species of olive-tree soil related dagger nematodes of the genus *Xiphinema* (Dorylaimida: Longidoridae) from Spain. *Invertebr. Syst.* **2016**, *30*, 547–565. [CrossRef]

20.  Gutiérrez-Gutiérrez, C.; Palomares-Rius, J.E.; Cantalapiedra-Navarrete, C.; Landa, B.B.; Esmenjaud, D.; Castillo, P. Molecular analysis and comparative morphology to resolve a complex of cryptic *Xiphinema* species. *Zool. Scr.* **2010**, *39*, 483–498. [CrossRef]

21.  Lazarova, S.; Oliveira, C.M.G.; Prior, T.; Peneva, V.; Kumari, S. An integrative approach to the study of *Xiphinema brevicolle* Lordello and Da Costa 1961, supports its limited distribution worldwide (Nematoda: Longidoridae). *Eur. J. Plant Pathol.* **2019**, *153*, 441–464. [CrossRef]

22.  Lamberti, F.; Castillo, P.; Gomez-Barcina, A.; Agostinelli, A. Descriptions of six new species of *Xiphinema* (Nematoda, Dorylaimida) from the Mediterranean region. *Nematol. Mediterr.* **1992**, *20*, 125–139.

23.  Davis, G.M. Evolution of prosobranch snails transmitting Asian *Schistosoma*: Coevolution with *Schistosoma*: A review. *Prog. Clin. Parasitol.* **1992**, *3*, 145–204. [CrossRef]

24.  Gittenberger, E. What about non-adaptive radiation? *Biol. J. Linn. Soc.* **1991**, *3*, 263–272. [CrossRef]

25. Cai, R.; Prior, T.; Lawson, B.; Cantalapiedra-Navarrete, C.; Palomares-Rius JECastillo, P.; Archidona-Yuste, A. An integrative taxonomic study of the needle nematode complex *Longidorus goodeyi* Hooper, 1961 (Nematoda: Longidoridae) with description of a new species. *Eur. J. Plant Pathol.* **2020**, *157*, 59–81. [CrossRef]

26. Barsi, L.; Fanelli, E.; De Luca, F. A new record of *Xiphinema dentatum* Sturhan, 1978 and description of *X. paradentatum* sp. n. (Nematoda: Dorylaimida) from Serbia. *Nematology* **2017**, *19*, 925–949. [CrossRef]

27. Loof, P.A.A.; Coomans, A. On the development and location of the oesophageal gland nuclei in the Dorylaimina. *Nematologica* **1968**, *14*, 596. [CrossRef]

28. Archidona-Yuste, A.; Cantalapiedra-Navarrete, C.; Castillo, P.; Palomares-Rius, J.E. Molecular phylogenetic analysis and comparative morphology reveals the diversity and distribution of needle nematodes of the genus *Longidorus* (Dorylaimida: Longidoridae) from Spain. *Contrib. Zool.* **2019**, *88*, 1–41. [CrossRef]

29. Roca, F.; Bravo, M.A. The occurrence of *Xiphinema sphaerocephalum* Lamberti et al. and *X. hispanum* Lamberti et al. (Nematoda: Longidoridae) in Portugal with descriptions of *X. lanceolatum* sp. n. and *X. lapidosum* sp. n. *Fundam. Appl. Nematol.* **1993**, *16*, 455–465.

30. Gutiérrez-Gutiérrez, C.; Cantalapiedra-Navarrete, C.; Montes-Borrego, M.; Palomares-Rius, J.E.; Castillo, P. Molecular phylogeny of the nematode genus *Longidorus* (Nematoda: Longidoridae) with description of three new species. *Zool. J. Linn. Soc.* **2013**, *167*, 473–500. [CrossRef]

31. Tzortzakakis, E.A.; Archidona-Yuste, A.; Cantalapiedra-Navarrete, C.; Nasiou, E.; Lazanaki, M.S.; Kabourakis, E.M.; Palomares-Rius, J.E.; Castillo, P. Integrative diagnosis and molecular phylogeny of dagger and needle nematodes of olives and grapevines in the island of Crete, Greece, with description of *Xiphinema cretense* n. sp. (Nematoda, Longidoridae). *Eur. J. Plant Pathol.* **2014**, *140*, 563–590. [CrossRef]

32. Palomares-Rius, J.E.; Cantalapiedra-Navarrete, C.; Archidona-Yuste, A.; Subbotin, S.A.; Castillo, P. The utility of mtDNA and rDNA for barcoding and phylogeny of plant-parasitic nematodes from Longidoridae (Nematoda, Enoplea). *Sci. Rep.* **2017**, *7*, 10905. [CrossRef]

33. Fouladvand, Z.M.; Pourjam, E.; Castillo, P.; Pedram, M. Genetic diversity, and description of a new dagger nematode, *Xiphinema afratakhtehnsis* sp. nov., (Dorylaimida: Longidoridae) in natural forests of southeastern Gorgan, northern Iran. *PLoS ONE* **2019**, *14*, e0214147. [CrossRef]

34. Cameron, E.K.; Martins, I.S.; Lavelle, P.; Mathieu, J.; Tedersoo, L.; Gottschall, F.; Eisenhauer, N. Global gaps in soil biodiversity data. *Nat. Ecol. Evol.* **2018**, *2*, 1042–1043. [CrossRef]

35. Decaëns, T. Macroecological patterns in soil communities. *Global Ecol. Biogeogr.* **2010**, *19*, 287–302. [CrossRef]

36. Coomans, A. Phylogeny of the Longidoridae. *Russ. J. Nematol.* **1996**, *4*, 51–60.

37. Finlay, B.J. Global dispersal of free-living microbial eukaryotic species. *Science* **2002**, *296*, 1061–1063. [CrossRef] [PubMed]

38. Olson, M.; Harris, T.; Higgins, R.; Mullin, P.; Powers, K.; Olson, S.; Powers, T.O. Species delimitation and description of *Mesocriconema nebraskense* n. sp. (Nematoda: Criconematidae), a morphologically cryptic, parthenogenetic species from North American Grasslands. *J. Nematol.* **2017**, *49*, 42–66. [CrossRef] [PubMed]

39. Coolen, W.A. Methods for extraction of *Meloidogyne* spp. and other nematodes from roots and soil. In *Root-knot Nematodes (Meloidogyne species). Systematics, Biology, and Control*; Lamberti, F.F., Taylor, C.E., Eds.; Academic Press: London, UK, 1979; pp. 317–329.

40. Flegg, J.J.M. Extraction of *Xiphinema* and *Longidorus* species from soil by a modification of Cobb's decanting and sieving technique. *Ann. Appl. Biol.* **1967**, *60*, 429–437. [CrossRef]

41. Seinhorst, J.W. Killing nematodes for taxonomic study with hot f.a. 4:1. *Nematologica* **1966**, *12*, 178. [CrossRef]

42. De Grisse, A.T. Redescription ou modification de quelques techniques utilisées dans l'étude des nematodes phytoparasitaires. *Meded. Rijksfak. Landbouwwet. Gent* **1969**, *34*, 351–369.

43. De Ley, P.; Felix, M.A.; Frisse, L.A.; Nadler, S.; Sternberg, P.; Thomas, W. Molecular and morphological characterisation of two reproductively isolated species with mirror-image anatomy (Nematoda: Cephalobidae). *Nematology* **1999**, *1*, 591–612. [CrossRef]

44. Vrain, T.C.; Wakarchuk, D.A.; Levesque, A.C.; Hamilton, R.I. Intraspecific rDNA Restriction Fragment Length Polymorphism in the *Xiphinema americanum* group. *Fund. Appl. Nematol.* **1992**, *15*, 563–573.

45. Cherry, T.; Szalanski, A.L.; Todd, T.C.; Powers, T.O. The internal transcribed spacer region of *Belonolaimus* (Nemata: Belonolaimidae). *J. Nematol.* **1997**, *29*, 23–29.

46. Holterman, M.; Van Der Wurff, A.; Van Den Elsen, S.; Van Megen, H.; Bongers, T.; Holovachov, O.; Bakker, J.; Helder, J. Phylum-wide analysis of SSU rDNA reveals deep phylogenetic relationships among nematodes and accelerated evolution toward crown clades. *Mol. Phylogenet. Evol.* **2006**, *23*, 1792–1800. [CrossRef]

47. Lazarova, S.S.; Malloch, G.; Oliveira, C.M.G.; Hübschen, J.; Neilson, R. Ribosomal and mitochondrial DNA analyses of *Xiphinema americanum*-group populations. *J. Nematol.* **2006**, *38*, 404–410.

48. Legendre, P.; Legendre, L. *Numerical Ecology*; Elsevier: Oxford, UK, 2012; 1006.

49. Zuur, A.F.; Ieno, E.N.; Elphick, C.S. A protocol for data exploration to avoid common statistical problems. *Methods Ecol. Evol.* **2010**, *1*, 3–14. [CrossRef]

50. Montgomery, D.C.; Peck, E.A. *Introduction to Linear Regression Analysis*; Wiley: New York, NY, USA, 1992; 688p.

51. Revelle, W. *psych: Procedures for Personality and Psychological Research*; Delay Discounting of Different Outcomes 21; North-western University: Evanston, IL, USA, 2020. Available online: https://CRAN.R-project.org/package=psychVersion=1.9.12.31 (accessed on 14 May 2020).

52. Quadros, A. emstreeR: Tools for Fast Computing and Plotting Euclidean Minimum Spanning Trees. R Package Version 2.2.0. 2019. Available online: https://CRAN.R-project.org/package=emstreeR (accessed on 14 May 2020).

53. R Core Team. *R: A language and Environment for Statistical Computing*; R Foundation for Statistical Computing: Vienna, Austria, 2019.

54. Rozas, J.; Ferrer-Mata, A.; Sánchez-DelBarrio, J.C.; Guirao-Rico, S.; Librado, P.; Ramos-Onsins, S.E.; Sánchez-Gracia, A. DnaSP 6: DNA sequence polymorphism analysis of large data sets. *Mol. Biol. Evol.* **2017**, *34*, 3299–3302. [CrossRef] [PubMed]

55. Clement, M.; Snell, Q.; Walker, P.; Posada, D.; Crandall, K. TCS: Estimating Gene Genealogies. In Proceedings of the 6th International Parallel and Distributed Processing Symposium, Fort Lauderdale, FL, USA, 15–19 April 2002; IEEE Computer Society: Fort Lauderdale, FL, USA, 2002; p. 184.

56. Leigh, J.W.; Bryant, D. PopART: Full-feature software for haplotype network construction. *Methods Ecol. Evol.* **2015**, *6*, 1110–1116. [CrossRef]

57. Flot, J.F.; Couloux, A.; Tillier, S. Haplowebs as a graphical tool for delimiting species: A revival of Doyle's "field for recombination" approach and its application to the coral genus *Pocillopora* in Clipperton. *BMC Evol. Biol.* **2010**, *10*, 372. [CrossRef] [PubMed]

58. He, Y.; Subbotin, S.A.; Rubtsova, T.V.; Lamberti, F.; Brown, D.J.F.; Moens, M. A molecular phylogenetic approach to Longidoridae (Nematoda: Dorylaimida). *Nematology* **2005**, *7*, 111–124. [CrossRef]

59. Katoh, K.; Rozewicki, J.; Yamada, K.D. MAFFT online service: Multiple sequence alignment, interactive sequence choice and visualization. *Brief. Bioinform.* **2019**, *20*, 1160–1166. [CrossRef] [PubMed]

60. Hall, T.A. BioEdit: A user-friendly biological sequence alignment editor and analysis program for windows 95/98/NT. *Nucleic Acids Symp. Ser.* **1999**, *41*, 95–98.

61. Castresana, J. Selection of conserved blocks from multiple alignments for their use in phylogenetic analysis. *Mol. Biol. Evol.* **2000**, *17*, 540–552. [CrossRef]

62. Ronquist, F.; Huelsenbeck, J.P. MRBAYES 3: Bayesian phylogenetic inference under mixed models. *Bioinformatics* **2003**, *19*, 1572–1574. [CrossRef]

63. Darriba, D.; Taboada, G.L.; Doallo, R.; Posada, D. jModelTest 2: More models, new heuristics and parallel computing. *Nat. Methods* **2012**, *9*, 772. [CrossRef]

64. Rambaut, A. FigTree v1.4.2, A Graphical Viewer of Phylogenetic Trees. 2014. Available online: http://tree.bio.ed.ac.uk/software/figtree/ (accessed on 15 September 2020).

**Publisher's Note:** MDPI stays neutral with regard to jurisdictional claims in published maps and institutional affiliations.

*plants*

*Article*

# Global Distribution of the Reniform Nematode Genus *Rotylenchulus* with the Synonymy of *Rotylenchulus macrosoma* with *Rotylenchulus borealis*


Juan E. Palomares-Rius [1], Ilenia Clavero-Camacho [1], Antonio Archidona-Yuste [2], Carolina Cantalapiedra-Navarrete [1], Guillermo León-Ropero [1], Sigal Braun Miyara [3], Gerrit Karssen [4,5] and Pablo Castillo [1,*]

1   Instituto de Agricultura Sostenible (IAS), Consejo Superior de Investigaciones Científicas (CSIC), Avda, Menéndez Pidal s/n, 14004 Córdoba, Spain; palomaresje@ias.csic.es (J.E.P.-R.); iclavero@ias.csic.es (I.C.-C.); ccantalapiedra@ias.csic.es (C.C.-N.); gleon@ias.csic.es (G.L.-R.)
2   Helmholtz Centre for Environmental Research—UFZ, Department of Ecological Modelling, Permoserstrasse 15, 04318 Leipzig, Germany; antonio.archidona-yuste@ufz.de
3   Nematology and Chemistry Units, Agricultural Research Organization (ARO), The Volcani Center, Department of Entomology, Bet Dagan 50250, Israel; sigalhor@volcani.agri.gov.il
4   Nematology Research Unit, Department of Biology, Ghent University, K.L. Ledeganckstraat 35, 9000 Ghent, Belgium; gerrit.karssen@ugent.be
5   National Plant Protection Organization, Wageningen Nematode Collection, P.O. Box 9102, 6700 HC Wageningen, The Netherlands
*   Correspondence: p.castillo@csic.es



Citation: Palomares-Rius, J.E.; Clavero-Camacho, I.; Archidona-Yuste, A.; Cantalapiedra-Navarrete, C.; León-Ropero, G.; Braun Miyara, S.; Karssen, G.; Castillo, P. Global Distribution of the Reniform Nematode Genus *Rotylenchulus* with the Synonymy of *Rotylenchulus macrosoma* with *Rotylenchulus borealis*. Plants **2021**, *10*, 7. https://dx.doi.org/10.3390/plants10010007

Received: 24 November 2020
Accepted: 21 December 2020
Published: 23 December 2020



**Publisher's Note:** MDPI stays neutral with regard to jurisdictional claims in published maps and institutional affiliations.





**Abstract:** Reniform nematodes of the genus *Rotylenchulus* are semi-endoparasites of numerous herbaceous and woody plant roots that occur largely in regions with temperate, subtropical, and tropical climates. In this study, we compared 12 populations of *Rotylenchulus borealis* and 16 populations of *Rotylenchulus macrosoma*, including paratypes deposited in nematode collections, confirming that morphological characters between both nematode species do not support their separation. In addition, analysis of molecular markers using nuclear ribosomal DNA (*28S*, *ITS1*) and mitochondrial DNA (*coxI*) genes, as well as phylogenetic approaches, confirmed the synonymy of *R. macrosoma* with *R. borealis*. This study also demonstrated that *R. borealis* (= *macrosoma*) from Israel has two distinct rRNA gene types in the genome, specifically the two types of *D2-D3* (A and B). We provide a global geographical distribution of the genus *Rotylenchulus*. The two major pathogenic species (*Rotylenchulus reniformis* and *Rotylenchulus parvus*) showed their close relationship with warmer areas with high annual mean temperature, maximum temperature of the warmest month, and minimum temperature of the coldest month. The present study confirms the extraordinary morphological and molecular diversity of *R. borealis* in Europe, Africa, and the Middle East and comprises a paradigmatic example of remarkable flexibility of ecological requirements within reniform nematodes.

**Keywords:** Bayesian inference; cytochrome c oxidase subunit 1; distribution; *D2-D3* expansion domains of *28S* rRNA gene; *ITS1*; phylogeny


## 1. Introduction

Reniform nematodes of the genus *Rotylenchulus* are an economically important polyphagous group of highly adapted obligate plant parasites that parasitize numerous plants and crops usually associated with temperate, subtropical, and tropical climates [1]. The genus *Rotylenchulus* Linford and Oliveira [2] comprise 11 valid species; some of them are distributed worldwide, whereas others have shown a limited distribution [1,3,4]. This genus has been reported in 77 countries of Africa, Asia, Europe, North and South America, and Australia [1,3,4]. The influence of future global climate change could shorten the life cycle of these nematodes and may expand the distribution of well-adapted species to drought conditions [5,6]. However, other factors such as the low population density in soil, no apparent harvest losses in some crops, or the difficulties for an accurate identification for some *Rotylenchulus* species could thwart their precise geographical distribution. For these reasons, *Rotylenchulus* spp. could be regarded as a "neglected pathogen", but also as a potentially dangerous pathogen in the future because of new ecological conditions predicted in global climate change scenarios [7]. Consequently, an updating of the global distribution of this group of nematodes allowed us to know the climatic conditions adapted to each species, which are essential to predict the response of this genus to climate change [8,9].

*Rotylenchulus* spp. show high intraspecific variability of some morphological diagnostic features in immature females (the developmental stage usually employed for species identification) [3], and for this reason, it is necessary to use molecular markers for precise species identification. In this regard, the use of rRNA markers is challenging due to the previously noted presence of several gene copies that are not well homogenized in the genome, and for this reason, several different amplicon sizes and associated sequences can be observed [4]. A prominent example of this high intraspecific variability was established in the study on several populations of *Rotylenchulus macrosoma* by Dasgupta et al. [10] and *R. borealis* by Loof and Oostenbrink [11].

In 1952, Oostenbrink found a population of reniform-shaped nematode in a soil sample from Arnhem (The Netherlands). Subsequent examination and comparison with published descriptions showed that the new nematode represented an undescribed species, proposed as *Rotylenchulus borealis* Loof & Oostenbrink [11], referring to its occurrence in northern countries, since the other species of the genus were mainly known from the tropical and subtropical regions [11]. Some years later, Dasgupta et al. [10] revised the genus *Rotylenchulus* and described a new species from olive in Hulda, Israel, closely related to *R. borealis*, named *Rotylenchulus macrosoma* (original spelling *macrosomus*). *R. macrosoma* differed from the former by its larger body length of immature females and males (0.52–0.64 mm, 0.50–0.68 mm vs. 0.37–0.46 mm, 0.40–0.49 mm in *R. borealis*, respectively), larger female stylet (18–22 vs. 13–16 μm in *R. borealis*), and longer hyaline portion of immature female tail (h = 13–18 vs. 9–13 μm in *R. borealis*). These limited differences between both species have been confirmed by posterior morphometrics of several African populations studied by Germani [12], as well as the recent *R. macrosoma* populations studied from Europe [3,9].

In 2003, Castillo et al. [13] detected a population of reniform nematodes infecting the roots of wild olive trees (*Olea europea* L. ssp. *sylvestris*) on a sandy soil in Cádiz province, southern Spain, which was identified as *R. macrosoma*. Morphometric of the Spanish population agreed with the original description of *R. macrosoma*, except for a shorter stylet length (15–18 *vs.* 18–22 μm), which was considered as an intraspecific variability. Later on, in 2016, Van den Berg et al. [3] provided morphological and molecular characterization of 6 out of 11 presently known species of *Rotylenchulus*, including three Spanish populations (two and one from Cádiz and Seville provinces, respectively) of *R. macrosoma*, which formed a separate and well-supported clade within phylogenetic trees of *D2-D3* expansion segments of *28S* rRNA, *ITS*, and *hsp90* genes [3]. This study also reported high levels of intraspecific and intra-individual variations of rRNA with two or more distinct types of rRNA genes, namely, type A and B [3]. These phylogenetic relationships were confirmed by posterior studies on additional new reports of *R. macrosoma* populations from several European countries including the Czech Republic, France, Germany, Greece, Hungary, Italy, Portugal, Romania, Serbia, and Spain [4,9]. In a recent study on the integrative characterization of plant-parasitic nematodes of potato in Rwanda, Niragire [14] provided morphological and molecular data of a population of *R. macrosoma* from Burera (North Rwanda), but no sequences were deposited in the National Center for Biotechnology Information (NCBI) database. Molecular data available for *R. borealis* is a *28S* rRNA sequence obtained from a Belgian population (MK558206) and the mentioned sequence for Burera clustered together with the Spanish *R. macrosoma* populations [14]. However, this Belgian population (Oudenaarde, Belgium) of *R. borealis* was not mentioned in the associated paper with the NCBI sequence and no morphological data were available alongside it [15]. This sequence has a 99.45% identity with *R. macrosoma*-KT003748 from Spain. Recently, Qing et al. [16] studied the rRNA variation (intragenomic polymorphism) across 30 terrestrial nematode species and sequenced *28S* and *ITS1* from a population of *R. macrosoma* in Israel, which clustered together in the same clade with *R. macrosoma* populations from Spain and Crete (Greece) and clearly separated from other *Rotylenchulus* spp. Finally, in the last months, one new *28S* rRNA sequence of *R. borealis* from New Delhi, India, was deposited on the NCBI database, MT775429 (95% identity with *R. macrosoma* KT003748 from Spain and 94% identity with *R. borealis* MK558206 from Belgium). All these concerns prompted us to carry out an integrative taxonomic analysis of *R. borealis* from the Netherlands in order to confirm the validity of these species or their synonymization with *R. macrosoma*.

The objectives of this study were (1) to morphometrically and molecularly characterize several populations of *R. macrosoma* from Europe and a population of *R. borealis* from the Netherlands, as well as paratypes of both species deposited in Nematode Collections, and to compare them with previous records; (2) to study the phylogenetic relationships of the European and Dutch populations of *R. macrosoma* and *R. borealis* and compare them with available sequenced populations of these species to establish their validity; and (3) to provide a clear view of the global distribution and the current climatic conditions that affect the distribution of species within the genus *Rotylenchulus*.

## 2. Results

*2.1. Morphometric Comparison of Paratypes and Several Populations of Rotylenchulus Borealis and Rotylenchulus Macrosoma*

We detected similar morphological traits in the comparison of 12 populations of *R. borealis* and 16 populations of *R. macrosoma* (Figure 1, Tables 1–8), but ordinary morphometric differences among both species grouped within the three main diagnostic characters of immature females originally used for separating both species (namely, body length, stylet length, and hyaline tail region length) (Figure 2), being the major differences in the original species descriptions. Our data indicated that mean body length of all 12 populations of *R. borealis* was 401.7 μm, whereas the mean for 16 populations of *R. macrosoma* was 483.0 μm. Similarly, stylet and hyaline tail region lengths were 14.25 μm, 7.8 μm vs. 17.28 μm, 12.5 μm, respectively (Tables 2–8). No differences were detected between the paratype immature females and males of *R. borealis* and the original description, as well as the new studied population from Huissen, Betuwe region (close to the type locality), the Netherlands (Table 2). However, of the two paratype immature females of *R. macrosoma* examined from Wageningen Nematode Collection (WANECO) and United States Department of Agriculture (USDA) nematode collections, both specimens showed a stylet length slightly lower than 18.0 μm (Table 5), and representing a lower measure to that provided in the type population from olive at Hulda, Israel, and quite close to several European populations, such as Spanish populations from Jerez and Huévar del Aljarafe, Cretan populations from Petrokefali and Limnes, or the Rwandan population from Burera. Nevertheless, immature female body and hyaline tail region lengths were similar to those provided in the original description.

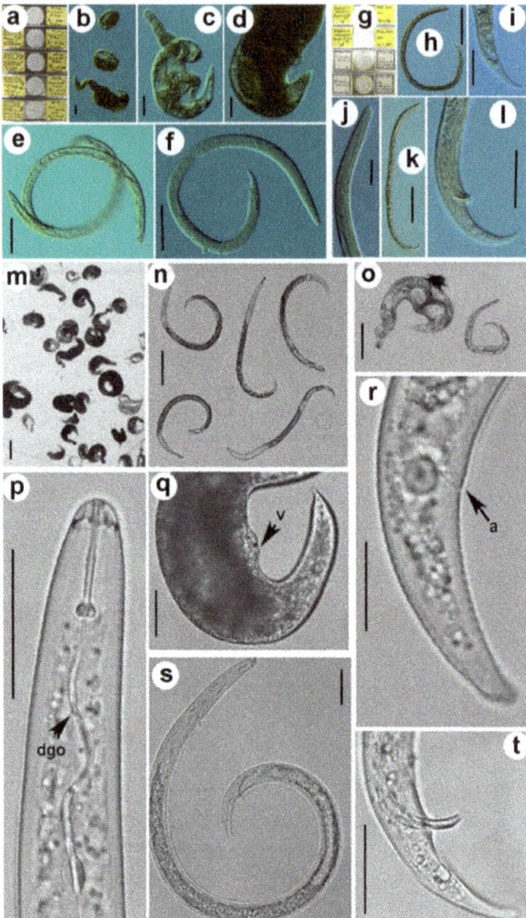

**Figure 1.** Comparative morphology among paratype specimens of *Rotylenchulus borealis* from the Netherlands (**a–f**), paratype specimens of *Rotylenchulus macrosoma* from Israel (**g–l**), and a population of *Rotylenchulus macrosoma* from Hungary (**m–t**). (**a,g**) slides deposited in Wageningen Nematode Collection (WANECO) and United States Department of Agriculture (USDA) nematode collections; (**b–d,m,o,q**) mature females; (**e, h–j,n,p,r**) immature females; (**f,k,l,n,s,t**) = males. Abbreviations: a = anus; dgo = dorsal gland opening; V = vulva. Scale bars: (**b–d,h,k,m–o**) 100 µm; (**e,f**) 50 µm; (**i,j,l,p,q,s,t**) 20 µm; (**r**) 10 µm.

**Table 1.** Populations sampled for *Rotylenchulus* spp. from two localities in the Netherlands and Israel used in this study.

| Locality, Country | Nematode Code | *D2-D3* | *ITS1* | *coxI* |
|---|---|---|---|---|
| *Rotylenchulus borealis* | | | | |
| Huissen, Betuwe region (the Netherlands) | AV23 | MW173970 | MW1742399 | MW182432 |
| Huissen, Betuwe region (the Netherlands) | AV25 | MW173971 | MW174240 | - |
| Huissen, Betuwe region (the Netherlands) | AV26 | MW173972 | - | MW182433 |
| Huissen, Betuwe region (the Netherlands) | AV27 | MW173973 | MW174241 | MW182434 |
| Huissen, Betuwe region (the Netherlands) | AV28 | MW173974 | MW174242 | MW182435 |
| Huissen, Betuwe region (the Netherlands) | AV29 | MW173975 | - | |
| Huissen, Betuwe region (the Netherlands) | AV30 | MW173976 | - | |
| *Rotylenchulus macrosoma* | | | | |
| Beit She'an (Israel) | C26 | MW173977 | - | - |
| Beit She'an (Israel) | C27 | MW173978 | - | - |
| Beit She'an (Israel) | C29 | MW173979 | - | - |
| Beit She'an (Israel) | C30 | MW173980 | - | - |
| Beit She'an (Israel) | C31 | MW173981 | MW174243 | - |
| Beit She'an (Israel) | C32 | MW173982 | MW174244 | - |
| Beit She'an (Israel) | C45 | MW173983 | MW174245 | - |
| Beit She'an (Israel) | C46 | MW173984 | MW174246 | - |
| Beit She'an (Israel) | C47 | MW173985 | - | - |

(-) Not obtained or not performed.

**Figure 2.** Range (minimum and maximum) comparative key diagnostic measures of immature females (body, stylet, and hyaline female tail lengths) for separating among *R. borealis* and *R. macrosoma* populations in decreasing chronological order of publication.

**Table 2.** Measurements of immature females and males of *Rotylenchulus borealis* from type locality in the Netherlands and several localities in Africa. All measurements are in micrometers and in the form mean ± SD (range).

| Character/Ratio | Grasses, Arnhem, the Netherlands, Paratypes | | Paratypes from WANECO Nematode Collection, Re-Measured in This Study | | (R. variabilis = R. borealis) Rumex sp., Inyanga Orchard Area, Southern Rhodesia (Dasgupta et al. [10] | | Upland Rice, Ferkessedougou, Ivory Coast [12] | | Sweet Potato, Ferkessedougou, Ivory Coast [12] | |
|---|---|---|---|---|---|---|---|---|---|---|
| | Immature Females | Males | Immature Females | Males | Immature Females | Males | Immature Females | Males | Immature Females | Males |
| n | 20 | 40 | 3 | 5 | 22 | 21 | 18 | 17 | 5 | 8 |
| L [a] | (370–460) | (400–490) | 408 ± 24 (381–424) | 418 ± 12 (404–435) | (300–370) | (340–410) | 420 (360–550) | 480 (400–540) | 470 (430–500) | 520 (480–570) |
| a | (22.5–32.5) | (30.3–40.2) | 27.9 ± 2.3 (25.4–29.9) | 24.1 ± 2.8 (21.6–28.9) | (22–26) | (22–33) | 27 (23–31) | 29.2 (25–42) | 29 (26–31) | 28 (24–36) |
| b | (2.5–3.4) | (3.2–4.0) | 4.1 ± 0.2 (3.9–4.2) | 3.8 ± 0.2 (3.4–3.9) | (3.3–3.9) | (3.1–4.1) | 2.9 (2.0–3.3) | 4.4 (3.8–5.2) | 3.0 (2.8–3.6) | 4.5 (4.5–5.0) |
| c | (11.3–14.8) | (12.1–15.8) | 14.9 ± 0.5 (14.6–15.5) | 14.7 ± 1.0 (13.6–16.1) | (13.0–16.0) | (14–20) | 14.7 (12.5–17.3) | 16 (14–18) | 14.3 (12.0–16.0) | 15 (11–18) |
| c′ | - | - | 2.8 ± 0.2 (2.6–2.9) | 2.8 ± 0.1 (2.6–2.9) | (2.6–3.2) | - | 3.4 (3.0–4.0) | - | 3.5 (3.1–4.4) | - |
| V or T | (57.6–64.8) | (25.0–54.0) | 62.2 ± 1.0 (61.0–63.0) | 39.7 ± 13.5 (23.4–56.4) | (59–66) | (29–51) | 61 (57–67) | - | 62 (59–66) | - |
| o | - | - | 124.4 ± 3.7 (121.4–128.6) | 136.7 ± 8.1 (130.8–150.0) | (120–138) | - | 131 (113–160) | - | 124 (113–147) | - |
| DGO | - | - | 17.0 ± 1.0 (16.0–18.0) | 17.2 ± 0.8 (16.0–18.0) | - | - | - | - | - | - |
| Stylet length | (13.0–16.0) | (12.0–14.0) | 13.7 ± 0.6 (13.0–14.0) | 12.6 ± 0.5 (12.0–13.0) | (13.0–15.0) | (10.0–12.0) | 14.0 (12.0–15.0) | 13.0 (10.0–14.0) | 14.5 (13.0–15.0) | 14.0 (13.0–14.0) |
| Lip region width | - | - | 6.3 ± 0.6 (6.0–7.0) | 6.5 ± 0.5 (6.0–7.0) | - | - | - | - | - | - |
| Tail length | - | - | 27.3 ± 1.5 (26.0–29.0) | 28.6 ± 2.3 (26.0–32.0) | - | - | - | - | - | - |
| h | - | - | 10.0 ± 1.0 (9.0–11.0) | 9.9 ± 0.7 (9.0–11.0) | (3.0–6.0) | (3.0–7.0) | 9.0 (7.0–12.0) | 8.0 (6.0–10.0) | 8.2 (6.0–11.0) | 9.0 (6.0–11.0) |
| Spicule length | - | (20.0–21.0) | - | 21.2 ± 0.8 (20.0–22.0) | - | (19.0–23.0) | - | 22.7 (18.0–24.0) | - | 23.0 (20.0–26.0) |
| Gubernaculum length | - | (7.0–8.0) | - | 7.0 ± 0.7 (6.0–8.0) | - | (7.0–9.0) | - | 9.0 (6.0–10.0) | - | 10.0 (8.0–12.0) |

[a] a = body length/maximum body width; b = body length/pharyngeal length; c = body length/tail length; c′ = tail length/body width at anus; DGO = distance from stylet base to dorsal gland opening; h = hyaline tail region length; o = (DGO/stylet length) × 100; V = (distance from anterior end to vulva/body length) × 100.

**Table 3.** Measurements of immature females and males of *Rotylenchulus borealis* from several localities in Africa. All measurements are in micrometers and in the form mean ± SD (range).

| Character/Ratio | Sweet Potato, Bouaké, Ivory Coast [12] | | Upland Rice, Bambari, Central African Republic [12] | | Cotton, North Republic of Benin [12] | |
| --- | --- | --- | --- | --- | --- | --- |
| | Immature Females | Males | Immature Females | Males | Immature Females | Males |
| n | 6 | 2 | 9 | 16 | 6 | 9 |
| L [a] | 420 (370–500) | 500, 520 | 390 (370–420) | 380 (360–410) | 330 (300–360) | 360 (320–370) |
| a | 25 (24–26) | 26, 27 | 25.2 (23–27) | 27 (21–30) | 23 (21–26) | 24 (21–28) |
| b | 2.9 (2.6–3.4) | 5.0, 5.7 | 2.8 (2.6–3.2) | 3.4 (2.7–3.9) | 2.7 (2.6–2.9) | 3.3 (2.6–3.7) |
| c | 16.7 (13.4–25.5) | 14.6, 17.2 | 15.5 (14.0–17.2) | 17 (14–21) | 14.2 (12.7–15.4) | 15.0 (8.7–16.7) |
| c′ | 3.2 (2.9–3.3) | - | 2.7 (2.3–3.1) | - | 2.7 | - |
| V or T | 62 (60–64) | - | 63.7 (61–65) | - | 63.6 (64–65) | - |
| o | 119 (100–127) | - | 112 (100–128) | - | 134 (130–143) | - |
| DGO | | | 6.8 (5.0–8.0) | 6.0 (3.0–10.0) | 6.0 (5.0–8.0) | - |
| Stylet length | 14.0 (13.0–15.0) | - | 15.0 (14.0–15.0) | 12.0 (11.0–13.0) | 13.0 (13.0–14.0) | 6.3 (4.0–8.0) |
| Lip region width | - | - | - | 22.0 (17.0–25.0) | - | 19.0 (17.0–22.0) |
| Tail length | - | - | - | 8.0 (7.0–11.0) | - | 8.0 (7.0–10.0) |
| h | 7.5 (5.0–11.0) | 8.0, 9.0 | (3.0–6.0) | (3.0–7.0) | 9.0 (7.0–12.0) | 8.0 (6.0–10.0) |
| Spicule length | - | 22.0, 24.0 | - | (19.0–23.0) | - | 22.7 (18.0–24.0) |
| Gubernaculum length | - | - | - | (7.0–9.0) | - | 9.0 (6.0–10.0) |

[a] a = body length/maximum body width; b = body length/pharyngeal length; c = body length/tail length; c′ = tail length/body width at anus; DGO = distance from stylet base to dorsal gland opening; h = hyaline tail region length; o = (DGO/stylet length) × 100; V = (distance from anterior end to vulva/body length) × 100.

**Table 4.** Measurements of immature females and males of *Rotylenchulus borealis* from several localities in Africa and the Slovak Republic. All measurements are in micrometers and in the form mean ± SD (range).

| Character/Ratio | Cotton, North Cameroon [12] | | Peanut, Burkina Faso [12] | | Corn, Somotor, Slovak Republic [17] | | Grasses, the Netherlands, This Study | |
|---|---|---|---|---|---|---|---|---|
| | Immature Females | Males | Immature Females | Males | Immature Females | Males | Immature Females | Males |
| n | 10 | 10 | 15 | 2 | 8 | 5 | 5 | 10 |
| L [a] | 370 (320–400) | 390 (360–420) | 400 (360–450) | 480, 490 | 428 ± 18 (410–457) | 445 ± 17 (416–459) | 427 ± 361 (381–465) | 432 ± 28 (400–490) |
| a | 22.6 (18–29) | 25.6 (23–28) | 24.5 (21–31) | 21, 31 | 29 ± 2.3 (27.0–34.5) | 31 ± 1.5 (28.7–32.1) | 27.5 ± 3.9 (22.5–32.5) | 29.2 ± 4.3 (25.1–40.2) |
| b | 2.6 (2.3–3.0) | 3.1 (2.4–3.6) | 2.7 (2.4–3.2) | 2.4, 3.2 | 4.2 ± 0.3 (3.9–4.6) | 4.0 ± 0.1 (3.6–3.9) | 3.2 ± 0.5 (2.5–3.8) | 3.6 ± 0.2 (3.2–4.0) |
| c | 15.5 (12–21) | 14.9 (13.5–16.0) | 15.8 (13.8–18.1) | 17.4, 21.3 | 13.9 ± 1.4 (12.7–16.7) | 14.0 ± 0.1 (12.8–15.3) | 13.8 ± 1.6 (11.3–15.0) | 14.6 ± 0.9 (13.3–16.0) |
| c′ | 3.2 (2.0–4.3) | - | 2.6 (2.2–3.2) | - | 3.4 ± 0.3 (2.9–3.8) | 3.0 ± 0.3 (2.6–3.5) | 3.0 ± 0.1 (2.9–3.1) | 2.5 ± 0.2 (2.3–2.9) |
| V or T | 62 (59–64) | - | 62 (56–65) | - | 63 ± 1.3 (62–65) | - | 61.9 ± 3.2 (57.0–65.0) | 39.2 ± 10.0 (25.0–54.0) |
| o | 139 (121–167) | - | 131 (108–171) | - | 145 ± 12.8 (122–163) | 145 (130–160) | 119.3 ± 11.9 (106.7–133.3) | 120.7 ± 5.0 (114.3–129.0) |
| DGO | - | - | - | - | - | - | 17.6 ± 1.5 (16.0–20.0) | 16.8 ± 0.6 (16.0–18.0) |
| Stylet length | 13.6 (12.0–14.0) | 12.5 (11.0–14.0) | 14.4 (13.0–15.0) | 13.0, 14.0 | 15.5 ± (15.0–16.5) | 13.0 ± 0.8 (11.5–13.5) | 14.8 ± 1.1 (13.0–16.0) | 13.7 ± 0.6 (12.0–14.0) |
| Lip region width | - | - | - | - | - | - | 6.7 ± 0.6 (6.0–7.0) | 6.4 ± 0.4 (6.0–7.0) |
| Tail length | - | - | - | - | - | - | 30.7 ± 1.2 (30.0–33.0) | 30.5 ± 1.5 (28.5–33.0) |
| h | 4.9 (3.0–7.0) | 7.0 (4.0–9.0) | 6.5 (5.0–8.0) | 7.0, 10.0 | 10.0 ± 2.3 (7.0–14.5) | 10.5 ± 1.0 (9.5–11.5) | 10.0 ± 0.5 (9.5–10.5) | 9.8 ± 0.8 (9.0–10.5) |
| Spicule length | - | 21.0 (18.0–25.0) | - | 19.0, 21.0 | - | 19.0 ± 0.7 (18.5–20.0) | - | 20.4 ± 0.5 (20.0–21.0) |
| Gubernaculum length | - | 7.0 (4.0–9.0) | - | 8.0, 9.0 | - | 7.0 ± 0.4 (6.5–7.5) | - | 7.4 ± 0.5 (7.0–8.0) |

[a] a = body length/maximum body width; b = body length/pharyngeal length; c = body length/tail length; c′ = tail length/body width at anus; DGO = distance from stylet base to dorsal gland opening; h = hyaline tail region length; o = (DGO/stylet length) × 100; V = (distance from anterior end to vulva/body length) × 100.

**Table 5.** Measurements of immature females and males of *Rotylenchulus borealis* (= *R. macrosoma*) from type locality in Israel and several localities in Spain. All measurements are in micrometers and in the form mean ± SD (range).

| Character/Ratio | Olive, Hulda, Israel, Type Population [10] | | Paratypes from WANECO and USDA NEMATODE Collections, Re-Measured in This Study | | Olive, Growth Chamber Built Population [18] | Wild Olive, Vejer, Cádiz, Spain [13] | | Cultivated Olive, Jerez de la Frontera, Cádiz, Spain [3] | |
|---|---|---|---|---|---|---|---|---|---|
| | Immature Females | Males | Immature females | Males | Immature Females | Immature Females | Males | Immature Females | Male |
| n | 21 | 21 | 2 | 2 | 11 | 12 | 11 | 6 | 3 |
| L [a] | 520–640 | 500–568 | 612, 634 | 493, 517 | 490 (470–510) | 453 ± 28 (408–510) | 467 ± 13 (449–495) | 476 ± 26 (438–502) | 475.0 ± 28 (446–501) |
| a | 30–38 | 30–41 | 27.6, 37.1 | 32.9, 39.5 | 26.8 (24.5–29.5) | 29.8 ± 2.1 (26.3–34.2) | 31.5 ± 2.1 (27.5–34.0) | 30.1 ± 1.2 (28.7–31.4) | 31.7 ± 0.3 (31.3–31.9) |
| b | 3.8–5.7 | 3.7–5.7 | 3.6, 4.1 | 3.8, 5.2 | 3.5 (3.0–3.8) | 3.9 ± 0.3 (3.5–4.4) | 4.7 ± 0.7 (3.5–5.2) | 3.7 ± 0.3 (3.3–4.3) | 3.4 ± 0.3 (3.1–3.6) |
| c | 12–16 | 12–16 | 14.6, 15.3 | 12.0, 12.6 | 12.7 (11.8–14.7) | 13.5 ± 1.3 (11.7–16.8) | 13.9 ± 0.6 (13.1–15.0) | 15.6 ± 1.3 (13.8–17.6) | 15.2 ± 0.3 (14.9–15.4) |
| c′ | 3.7–5.0 | - | 3.5, 4.2 | 3.5, 3.9 | - | 3.7 ± 0.5 (2.8–4.4) | 3.2 ± 0.4 (2.6–3.9) | 3.3 ± 0.4 (2.8–4.0) | 3.1 ± 0.2 (3.0–3.3) |
| V or T | 63.0–68.0 | 20–33 | 62.3, 65.5 | 21.9, 28.6 | 62.1 (58.9–63.3) | 62 ± 2 (59–64) | 33 ± 5 (25–42) | 61.5 ± 1.9 (59.0–64.0) | 25.0 ± 4.1 (20.9–29.1) |
| o | 139.0–188.0 | - | 126.5, 142.9 | 138.5, 141.0 | 134.4 (122.0–140.1) | 152 ± 15 (126–183) | 171 ± 15 (142–188) | 135.8 ± 17.4 (116.0–156.0) | 147.1 ± 8.5 (137.5–154.0) |
| DGO | - | - | 21.5, 25.0 | 18.0, 19.0 | - | 25 ± 2 (22–27) | 23 ± 2 (19.0–26.0) | 24.0 ± 1.4 (22.0–26.0) | 21.0 ± 1.0 (20.0–22.0) |
| Stylet length | 18.0–22.0 | 13.0–16.0 | 17.0, 17.5 | 13.0, 13.5 | 19.7 (18.2–21.1) | 16 ± 1 (15–18) | 14 ± 1 (12–15) | 17.8 ± 1.7 (16.0–20.0) | 15.0 ± 1.0 (14.0–16.0) |
| Lip region width | - | - | - | 5.0, 6.5 | - | - | - | 6.6 ± 0.7 (6.0–7.0) | 6.3 ± 0.6 (6.0–7.0) |
| Tail length | - | - | 41.5, 42.0 | 39.0, 43.0 | - | 34 ± 4 (26–40) | 34 ± 2 (30–36) | 30.8 ± 3.5 (27.0–36.0) | 31.3 ± 1.5 (30.0–33.0) |
| h | 13.0–18.0 | 15.0–23.0 | 18.5 | 10.0, 12.0 | 14.6 11.5–18.2) | 10 ± 1 (9–12) | 11 ± 1 (10–12) | 10.5 ± 1.4 (9.0–12.0) | 10.0 ± 1.0 (9.0–11.0) |
| Spicule length | - | 21.0–24.0 | - | 21.0, 21.5 | | - | 22 ± 2 (19–25) | - | 21.3 ± 1.5 (20.0–23.0) |
| Gubernaculum length | - | 8.0–10.0 | - | 6.5, 7.0 | | - | 9 ± 1 (8–10) | - | 9.0 ± 1.0 (8.0–10.0) |

[a] a = body length/maximum body width; b = body length/pharyngeal length; c = body length/tail length; c′ = tail length/body width at anus; DGO = distance from stylet base to dorsal gland opening; h = hyaline tail region length; o = (DGO/stylet length) × 100; V = (distance from anterior end to vulva/body length) × 100.

**Table 6.** Measurements of immature females and males of *Rotylenchulus borealis* (= *R. macrosoma*) from cultivated olive in Spain and Crete, Greece. All measurements are in micrometers and in the form mean ± SD (range).

| Character/Ratio | Cultivated Olive, Huévar del Aljarafe, Seville Province, Spain [3] | | Cultivated Olive, Petrokefali, Crete, Greece [9] | | Cultivated Olive, Limnes, Crete, Greece [9] | |
|---|---|---|---|---|---|---|
| | Immature Females | Males | Immature Females | Males | Immature Females | Male |
| n | 10 | 10 | 10 | 5 | 10 | 10 |
| $L^a$ | 484 ± 30 | 478 ± 31 | 467 ± 27 | 468 ± 31 | 488 ± 31 | 463 ± 33 |
| | (432–520) | (432–514) | (432–506) | (433–503) | (428–526) | (418–516) |
| a | 29.7 ± 1.5 | 29.9 ± 1.4 | 28.9 ± 1.9 | 30.8 ± 2.1 | 29.7 ± 1.0 | 28.2 ± 1.8 |
| | (27.6–32.1) | (27.6–31.9) | (26.1–31.6) | (27.1–32.0) | (28.5–31.4) | (26.1–31.3) |
| b | 3.7 ± 0.3 | 3.6 ± 0.3 | 3.6 ± 0.3 | 3.4 ± 0.3 | 3.7 ± 0.2 | 3.5 ± 0.3 |
| | (3.3–4.3) | (3.0–4.1) | (3.3–4.3) | (3.1–3.6) | (3.4–4.1) | (3.2–4.0) |
| c | 15.4 ± 1.2 | 14.8 ± 0.7 | 15.3 ± 0.9 | 14.7 ± 0.9 | 15.4 ± 1.0 | 13.8 ± 1.3 |
| | (14.1–17.6) | (13.4–15.4) | (13.8–17.0) | (13.1–15.4) | (14.1–17.1) | (11.6–15.4) |
| c′ | 3.1 ± 0.3 | 3.0 ± 0.1 | 3.3 ± 0.3 | 3.0 ± 0.2 | 3.1 ± 0.4 | 2.9 ± 0.1 |
| | (2.6–4.0) | (2.8–3.1) | (2.8–4.0) | (2.8–3.3) | (2.6–3.8) | (2.8–3.1) |
| V or T | 62.6 ± 2.2 | 27.1 ± 3.8 | 61.4 ± 2.0 | 31.8 ± 1.3 | 62.4 ± 1.9 | 31.3 ± 5.3 |
| | (59.0–66.0) | (21.1–32.2) | (58.0–64.0) | (30.0–33.0) | (60.0–65.0) | (21.5–37.1) |
| o | 138.0 ± 10.4 | 135.1 ± 3.8 | 129.8 ± 16.6 | 142.5 ± 7.2 | 128.8 ± 8.4 | 140.3 ± 5.2 |
| | (126.3–156.3) | (129.4–140.0) | (105.0–156.3) | (133.3–153.0) | (116.7–138.0) | (133.3–147.0) |
| DGO | 23.8 ± 1.3 | 21.6 ± 1.1 | 23.8 ± 1.5 | 20.8 ± 1.3 | 23.6 ± 1.3 | 21.6 ± 1.7 |
| | (22.0–26.0) | (20.0–23.0) | (21.0–26.0) | (20.0–23.0) | (21.0–25.0) | (19.0–24.0) |
| Stylet length | 17.3 ± 1.2 | 16.0 ± 0.9 | 18.5 ± 1.7 | 14.6 ± 0.5 | 17.4 ± 1.4 | 15.4 ± 0.5 |
| | (16.0–19.0) | (15.0–17.0) | (16.0–21.0) | (14.0–15.0) | (15.5–20.0) | (15.0–16.0) |
| Lip region width | 6.6 ± 0.6 | 6.5 ± 0.5 | 6.7 ± 0.7 | 6.6 ± 0.5 | 6.7 ± 0.8 | 6.5 ± 0.6 |
| | (6.0–7.5) | (6.0–7.0) | (6.0–7.5) | (6.0–7.0) | (6.0–8.0) | (6.0–7.5) |
| Tail length | 31.5 ± 3.2 | 32.4 ± 2.3 | 30.6 ± 2.8 | 31.4 ± 0.9 | 31.8 ± 3.0 | 33.6 ± 2.4 |
| | (27.0–37.0) | (29.0–35.0) | (27.0–36.0) | (30.0–32.0) | (28.0–37.0) | (29.0–37.0) |
| h | 10.8 ± 1.6 | 10.6 ± 0.5 | 10.4 ± 1.1 | 10.2 ± 1.3 | 10.6 ± 1.3 | 10.5 ± 0.8 |
| | (9.0–13.0) | (10.0–11.0) | (9.0–12.0) | (9.0–11.0) | (9.0–12.0) | (9.0–12.0) |
| Spicule length | - | 22.6 ± 1.4 | - | 21.8 ± 1.3 | - | 22.4 ± 1.4 |
| | | (21.0–24.0) | | (20.0–23.0) | | (20.0–24.0) |
| Gubernaculum length | - | 10.0 ± 0.9 | - | 9.2 ± 0.8 | - | 10.2 ± 0.8 |
| | | (9.0–11.0) | | (8.0–10.0) | | (9.0–11.0) |

[a] a = body length/maximum body width; b = body length/pharyngeal length; c = body length/tail length; c′ = tail length/body width at anus; DGO = distance from stylet base to dorsal gland opening; h = hyaline tail region length; o = (DGO/stylet length) × 100; V = (distance from anterior end to vulva/body length) × 100.

**Table 7.** Measurements of immature females and males of *Rotylenchulus borealis* (= *R. macrosoma*) from potato in Rwanda and almond-peach rootstock and corn from several localities in Europe. All measurements are in micrometers and in the form mean ± SD (range).

| Character/Ratio | Potato, Burera, North Rwanda [14] | | Almond-Peach Rootstock, Montañana, Zaragoza, Spain, This Study | | Corn, Bečej, Vojvodina, Serbia, This Study | | Corn, Moretta, Cuneo, Italy, This Study | |
|---|---|---|---|---|---|---|---|---|
| | Immature Females | Males | Immature Females | Males | Immature Females | Males | Immature Females | Males |
| n | 6 | 1 | 10 | 10 | 10 | 10 | 10 | 10 |
| L [a] | 403 ± 8 (395–416) | 462 | 461 ± 36 (401–517) | 482 ± 33 (410–533) | 425 ± 13 (401–440) | 461 ± 39 (400–505) | 427 ± 11 (411–441) | 436.7 ± 28 (405–483) |
| a | 24.3 ± 0.7 (23.5–25.5) | 29.2 | 27.6 ± 2.4 (25.1–31.9) | 28.1 ± 1.2 (26.5–30.4) | 26.2 ± 1.4 (24.3–28.2) | 25.5 ± 1.6 (22.1–27.3) | 25.9 ± 1.7 (23.0–28.0) | 27.6 ± 1.8 (25.3–31.2) |
| b | 4.1 ± 0.1 (4.0–4.3) | 4.1 | 3.1 ± 0.3 (2.6–3.6) | 4.1 ± 0.2 (3.8–4.6) | 3.5 ± 0.4 (2.9–4.2) | 3.9 ± 0.4 (3.4–4.6) | 3.6 ± 0.4 (2.9–4.2) | 3.9 ± 0.2 (3.0–3.3) |
| c | 13.1 ± 0.7 (12.4–14.2) | 13.6 | 12.6 ± 1.3 (11.0–14.9) | 13.7 ± 1.7 (11.0–17.6) | 12.6 ± 0.7 (12.0–14.0) | 15.3 ± 1.4 (13.7–17.9) | 12.6 ± 0.8 (11.8–14.5) | 13.2 ± 0.9 (12.3–14.6) |
| c′ | 2.9 ± 0.3 (2.5–3.3) | 2.5 | 3.5 ± 0.3 (2.8–4.0) | 2.9 ± 0.3 (2.4–3.6) | 3.7 ± 0.2 (3.4–3.9) | 2.7 ± 0.2 (2.4–3.0) | 3.6 ± 0.2 (3.2–3.8) | 3.1 ± 0.1 (3.0–3.3) |
| V or T | 63.3 ± 1.2 (62.1–64.9) | - | 61.1 ± 0.9 (59.6–62.7) | 27.1 ± 5.3 (18.7–35.1) | 59.9 ± 0.6 (59–61) | 30.5 ± 3 (26–35) | 60.1 ± 1.0 (58.0–61.8) | 23.4 ± 2.1 (21.5–26.0) |
| o | - | - | 127.4 ± 9.6 (112.5–141.2) | 137.4 ± 15.4 (117.9–163.0) | 120 ± 7 (113.3–135.5) | 131 ± 10 (102–133) | 118.0 ± 7.9 (106.3–129.0) | 122.7 ± 7.9 (114.3–143.0) |
| DGO | 20.2 ± 1.6 (18.0–22.0) | - | 20.8 ± 2.4 (17.5–24.0) | 19.1 ± 2.1 (16.0–22.0) | 17.9 ± 1.6 (16.0–21.0) | 18.6 ± 1.1 (17.0–21.0) | 17.7 ± 1.5 (16.0–20.0) | 17.4 ± 1.0 (16.0–20.0) |
| Stylet length | 17.3 ± 0.3 (17.0–18.0) | 16.8 | 16.3 ± 0.8 (15.0–17.0) | 13.9 ± 0.3 (13.5–14.5) | 14.9 ± 0.6 (14.0–16.0) | 14.2 ± 0.3 (13.5–14.5) | 15.0 ± 0.7 (14.0–16.0) | 14.2 ± 0.4 (13.5–15.0) |
| Lip region width | - | - | 6.6 ± 0.4 (6.0–7.0) | 6.1 ± 0.4 (5.5–6.50) | 6.5 ± 0.3 (6.0–7.0) | - | 6.5 ± 0.3 (6.0–7.0) | 6.5 ± 0.4 (6.0–7.0) |
| Tail length | 30.9 ± 1.3 (29.0–32.0) | 34.0 | 37.1 ± 5.1 (27.0–41.5) | 35.6 ± 5.0 (25.5–43.0) | 33.8 ± 1.5 (31.0–35.5) | 30.3 ± 4.1 (23.5–35.0) | 33.9 ± 2.0 (30.0–36.5) | 33.6 ± 1.3 (32.0–36.0) |
| h | - | - | 12.3 ± 1.6 (9.5–14.0) | 11.6 ± 2.0 (7.5–14.5) | 12.2 ± 1.4 (10.5–15.0) | 8.7 ± 1.2 (7.0–10.0) | 12.5 ± 1.7 (10.5–15.0) | 8.5 ± 0.9 (7.0–10.0) |
| Spicule length | - | - | - | 21.9 ± 1.2 (20.0–23.5) | - | 22.3 ± 1.3 (20.0–24.5) | - | 22.9 ± 0.7 (22.0–24.0) |
| Gubernaculum length | - | - | - | 9.2 ± 1.0 (8.0–11.0) | - | 8.0 ± 0.7 (7.0–9.0) | - | 7.7 ± 0.3 (7.0–8.0) |

[a] a = body length/maximum body width; b = body length/pharyngeal length; c = body length/tail length; c′ = tail length/body width at anus; DGO = distance from stylet base to dorsal gland opening; h = hyaline tail region length; o = (DGO/stylet length) × 100; V = (distance from anterior end to vulva/body length) × 100.

**Table 8.** Measurements of immature females and males of *Rotylenchulus borealis* (= *R. macrosoma*) from corn and wheat from several localities in Europe. All measurements are in micrometers and in the form mean ± SD (range).

| Character/Ratio | Corn, Le Sen, Landes, France, This Study | | Wheat, Mihail Kogalniceau, Romania, This Study | | Corn, Létavertes, Hajdú-Bihar, Hungary, This Study | | Corn, Möckmühl, Heilbronn, Germany, This Study | |
|---|---|---|---|---|---|---|---|---|
| | Immature Females | Males | Immature Females | Males | Immature Females | Males | Immature Females | Males |
| n | 10 | 10 | 10 | 10 | 10 | 10 | 10 | 10 |
| L [a] | 428 ± 22 (396–473) | 444 ± 21 (411–477) | 435 ± 29 (387–468) | 435 ± 25 (409–485) | 430 ± 14 (412–458) | 432 ± 19.3 (411–474) | 422 ± 16 (409–429) | 417 ± 7 (409–429) |
| a | 26.3 ± 1.3 (24.2–28.3) | 28.3 ± 1.8 (25.7–31.2) | 27.3 ± 1.4 (24.9–29.1) | 28.1 ± 1.8 (25.6–30.3) | 27.0 ± 0.8 (25.8–28.3) | 28.9 ± 2.5 (26.3–32.2) | 27.3 ± 0.7 (26.2–28.3) | 27.6 ± 1.6 (25.6–29.8) |
| b | 3.6 ± 0.4 (3.2–4.3) | 3.9 ± 0.2 (3.6–4.1) | 3.7 ± 0.3 (3.3–4.2) | 3.8 ± 0.2 (3.5–4.0) | 3.7 ± 0.1 (3.6–4.0) | 3.8 ± 0.1 (3.5–4.0) | 3.6 ± 0.1 (3.4–3.8) | 3.7 ± 0.2 (3.4–4.1) |
| c | 12.8 ± 0.8 (11.4–14.3) | 13.5 ± 0.5 (12.6–14.0) | 14.0 ± 0.9 (12.6–15.5) | 13.0 ± 0.5 (12.6–14.0) | 14.1 ± 0.4 (13.5–14.8) | 14.1 ± 1.0 (12.4–15.4) | 14.0 ± 0.2 (13.7–14.3) | 13.1 ± 1.0 (11.6–15.3) |
| c′ | 3.2 ± 0.2 (2.9–3.4) | 3.0 ± 0.2 (2.7–3.2) | 3.0 ± 0.2 (2.8–3.3) | 3.1 ± 0.2 (2.8–3.3) | 3.0 ± 0.1 (2.9–3.3) | 2.9 ± 0.2 (2.7–3.3) | 3.0 ± 0.1 (2.9–3.3) | 3.1 ± 0.2 (2.8–3.3) |
| V or T | 60.5 ± 0.8 (59.5–62.0) | 31.0 ± 6.8 (26.5–43.0) | 60.3 ± 0.9 (58.1–61.5) | 32.2 ± 3.1 (29.5–37.1) | 60.6 ± 0.5 (60.0–61.5) | 33.9 ± 7.4 (27.0–48.1) | 60.6 ± 0.5 (60.0–61.5) | 33.9 ± 7.4 (27.0–48.1) |
| o | 120.1 ± 7.3 (106.7–130.0) | 117.6 ± 6.9 (107.1–129.0) | 124.0 ± 9.7 (106.7–135.5) | 124.5 ± 8.5 (113.3–138.0) | 121.3 ± 8.8 (106.7–137.9) | 125.5 ± 7.6 (115.4–138.0) | 125.0 ± 7.5 (113.3–135.7) | 121.5 ± 6.6 (107.1–129.0) |
| DGO | 17.9 ± 1.5 (16.0–20.0) | 16.8 ± 0.91 (15.0–18.0) | 19.0 ± 1.9 (16.0–21.0) | 17.5 ± 0.6 (17.0–18.5) | 18.3 ± 1.3 (16.0–20.0) | 17.3 ± 1.0 (15.0–18.5) | 18.6 ± 1.1 (17.0–20.0) | 17.0 ± 0.8 (15.0–18.0) |
| Stylet length | 14.9 ± 0.6 (14.0–16.0) | 14.3 ± 0.9 (14.0–15.0) | 15.3 ± 0.5 (14.5–16.0) | 14.1 ± 0.6 (13.0–15.0) | 15.1 ± 0.4 (14.5–15.5) | 13.8 ± 0.5 (13.0–14.5) | 14.9 ± 0.6 (14.0–15.5) | 14.0 ± 0.4 (13.5–15.0) |
| Lip region width | 6.6 ± 0.3 (6.0–7.0) | 6.5 ± 0.3 (6.0–7.0) | 6.6 ± 0.2 (6.5–7.0) | 6.5 ± 0.4 (6.0–7.0) | 6.7 ± 0.3 (6.5–7.0) | 6.6 ± 0.4 (6.0–7.0) | 6.6 ± 0.3 (6.0–7.0) | 6.4 ± 0.5 (6.0–7.0) |
| Tail length | 33.5 ± 1.4 (31.0–35.5) | 33.3 ± 0.7 (32.0–34.0) | 31.0 ± 1.5 (29.0–34.0) | 31.8 ± 1.9 (32.0–38.0) | 30.6 ± 1.3 (29.0–33.0) | 31.2 ± 2.3 (28.0–36.0) | 30.1 ± 1.0 (28.0–31.0) | 32.5 ± 2.0 (28.0–36.0) |
| h | 11.4 ± 0.8 (10.5–12.5) | 10.0 ± 1.5 (8.0–15.5) | 10.4 ± 0.6 (9.5–11.5) | 10.3 ± 1.8 (8.0–12.5) | 10.1 ± 0.4 (9.5–10.5) | 9.9 ± 0.6 (9.0–11.0) | 10.0 ± 0.4 (9.5–10.5) | 9.7 ± 1.3 (8.0–12.5) |
| Spicule length | - | 22.5 ± 0.6 (22.0–23.5) | - | 22.2 ± 0.3 (22.0–23.0) | - | 21.8 ± 2.5 (21.0–22.5) | - | 21.6 ± 0.8 (20.0–22.5) |
| Gubernaculum length | - | 7.8 ± 0.3 (7.5–8.0) | - | 7.7 ± 0.4 (7.0–8.0) | - | 7.5 ± 0.4 (7.0–8.0) | - | 7.4 ± 0.4 (7.0–8.0) |

[a] a = body length/maximum body width; b = body length/pharyngeal length; c = body length/tail length; c′ = tail length/body width at anus; DGO = distance from stylet base to dorsal gland opening; h = hyaline tail region length; o = (DGO/stylet length) × 100; V = (distance from anterior end to vulva/body length) × 100.

*2.2. Molecular Characterisation and Phylogenetic Analysis of Rotylenchulus Borealis and Rotylenchulus Macrosoma Populations*

The amplification of *D2-D3* expansion domains of *28S* rRNA, *ITS1* rRNA, and *coxI* genes of *R. borealis* and *R. macrosoma* populations yielded single fragments of ≈900 bp, 1100 bp, and 450 bp, respectively, on the basis of gel electrophoresis and, in the case of the Israel population, from cloning of the PCR product. Sixteen new sequences from the *D2-D3* of *28S* rRNA gene and eight new sequences from ITS1 rRNA gene were obtained in this study (7 and 9, and 4 and 4, from the Netherlands and Israel, respectively). Four new *coxI* sequences from the Netherlands were deposited in GenBank; however, due to lack of material, it was not possible to obtain *coxI* sequences from Israel. Type B-D2D3 sequence of *R. macrosoma* from Israel was obtained for the first time in this study (MW173975). *D2-D3* for *R. borealis* (MW173970-MW173976) showed a low intraspecific variability with 1–5 different nucleotides and 0 indels (99% similarity). Similarly, intraspecific variability for *D2-D3* in *R. macrosoma* from Israel was slightly higher, with 6–17 different nucleotides and 0–2 indels (97–99% similarity). The molecular diversity of this marker between *R. borealis* (MW173970-MW173976) from the Netherlands and *R. macrosoma* (MW173977-MW173985) from Israel populations was also low, with 5–22 different nucleotides and 0–2 indels (96–99% similarity). *D2-D3* sequences of *R. macrosoma* from Israel (MW173977-MW173985) differed in 0–10 nucleotides and 0 indels (99% similarity) when compared with sequences of *R. macrosoma* deposited in the NCBI database from Spain, Belgium, Serbia, Romania, Hungary, and Portugal, and with *Rotylenchulus* sp. 191_7 (MK558208) and *R. borealis* (MT775429) from Ethiopia and New Delhi in 32, 44 bp, 0, 1 indels (95%, 94% similarity), respectively. Similarly, *D2-D3* sequences of *R. borealis* from the Netherlands (MW173970-MW173976) differed in 14–21 nucleotides and 0 indels (97–98% similarity) when compared with sequences of *R. macrosoma* deposited in the NCBI database from Spain, Belgium, Serbia, Romania, Hungary, and Portugal, and with *Rotylenchulus* sp. 191_7 (MK558208) and *R. borealis* (MT775429) from Ethiopia and New Delhi in 41, 39 bp, 0 indels (94%, 94% similarity), respectively.

The *ITS1* region showed a low intraspecific variability for *R. borealis* (MW174239-MW174242) from the Netherlands, with 0–6 different nucleotides and 0–1 indels (98–100% similarity). Similarly, intraspecific variability for *ITS1* in *R. macrosoma* from Israel (MW174243-MW174246) was low, with 0–11 different nucleotides and 0–4 indels (98–100% similarity). The molecular diversity of this marker between *R. borealis* from the Netherlands (MW174239-MW174242) and *R. macrosoma* from Israel (MW174243-MW174246) populations was also low, with 0–24 different nucleotides and 0–12 indels (95–100% similarity). *ITS1* sequences of *R. macrosoma* from Israel (MW174243-MW174246) differed in 19–32 nucleotides and 1–8 indels (94–96% similarity) when compared with sequences of *R. macrosoma* deposited in the NCBI database from Spain and Greece, and with *Rotylenchulus reniformis* (KF999979) from Japan in 92 bp, 26 indels (86% similarity). Similarly, *ITS1* sequences of *R. borealis* from the Netherlands (MW174239-MW174242) differed in 13–42 nucleotides and 1–11 indels (94–98% similarity) when compared with sequences of *R. macrosoma* deposited in the NCBI database from Spain and Greece, and with *R. reniformis* (KP018567) from China in 137 bp, 54 indels (83% similarity).

The partial *coxI* gene for *R. borealis* from the Netherlands (MW182432-MW182435) showed a low intraspecific variability with 0–8 different nucleotides and 0 indels (98–100% similarity). These sequences differed in 0–47 nucleotides and 0 indels (89–100% similarity) with sequences of *R. macrosoma* deposited in the NCBI database from Spain, Serbia, Romania, Hungary, and Greece, and with *Rotylenchulus parvus* (MK558211) from Tanzania in 64 bp, 4 indels (85% similarity). All molecular markers suggest that populations of *R. borealis* from the Netherlands and *R. macrosoma* from Israel are conspecific.

Phylogenetic relationships among *Rotylenchulus* species inferred from analyses of *D2-D3* expansion domains of *28S* rRNA, *ITS1*, and partial *coxI* gene sequences using Bayesian inference (BI) are shown in Figures 3–5, respectively. The phylogenetic trees generated with the two nuclear and the mitochondrial markers included 123, 77, and 38 sequences, with 704, 888, and 355 positions in length, respectively (Figures 3–5). *D2-D3* tree of *Rotylenchulus* spp. showed two moderately supported clades including *R. borealis* type A and type B sequences (posterior probabilities (PP) = 0.87, 0.93, respectively), including *R. reniformis*, *Rotylenchulus macrodoratus*, and *Rotylenchulus macrosomoides* (Figure 3). All sequences of *R. borealis* from the Netherlands (MW173970-MW173976) and Belgium (MK558206), as well as those of *R. borealis* (= *R. macrosoma*) from Israel and all the sequences from Spain, Serbia, Romania, Hungary, and Greece deposited in the NCBI database clustered together in a highly supported clade (PP = 1.00) and were well separated (PP = 1.00) from *28S* of *R. borealis* (MT775429) from New Delhi (Figure 3).

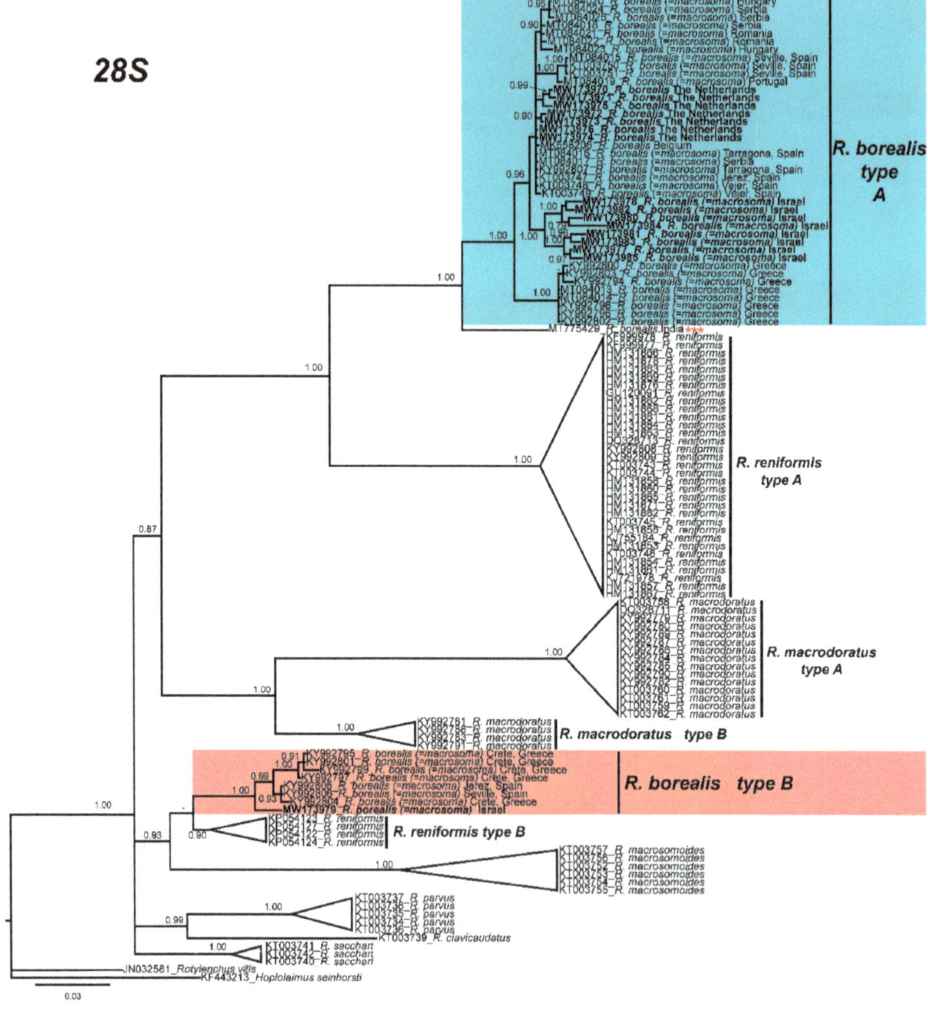


**Figure 3.** Phylogenetic relationships within the genus *Rotylenchulus*. Bayesian 50% majority rule consensus tree as inferred from D2 and D3 expansion domains of 28S rRNA sequence alignment under the general time-reversible model of sequence evolution with correction for invariable sites and a gamma-shaped distribution (GTR + I + G). Posterior probabilities of more than 0.70 are given for appropriate clades. Newly obtained sequences in this study are shown in bold. Scale bar = expected changes per site. Some branches were collapsed for improving readability of *Rotylenchulus* species. *** Sequence that needs to be revised under integrative taxonomical approaches.

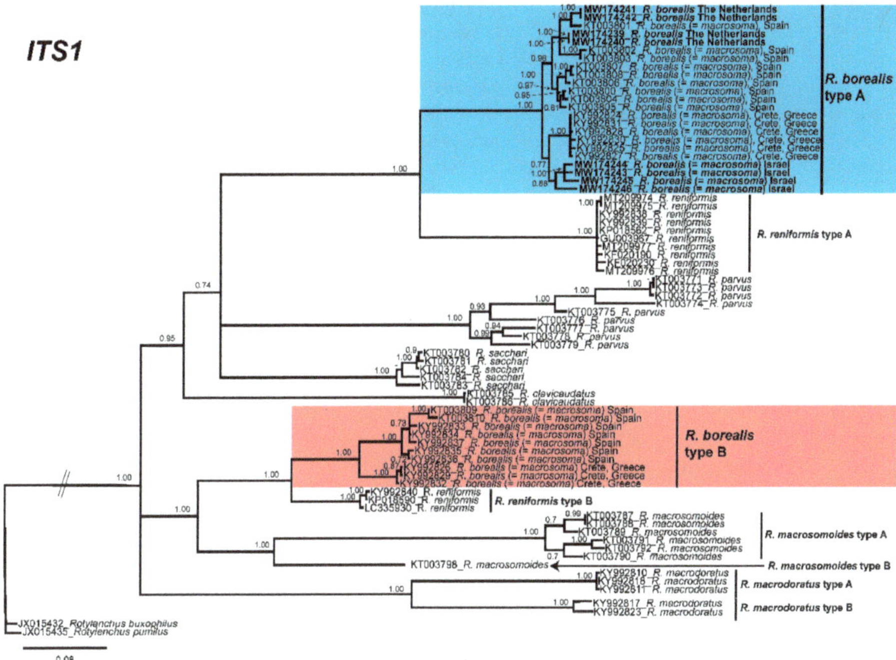

**Figure 4.** Phylogenetic relationships within the genus *Rotylenchulus*. Bayesian 50% majority rule consensus tree as inferred from *ITS1* rRNA sequence alignment under the general time-reversible model of sequence evolution with correction for invariable sites and a gamma-shaped distribution (GTR + I + G). Posterior probabilities of more than 0.70 are given for appropriate clades. Newly obtained sequences in this study are shown in bold. Scale bar = expected changes per site.

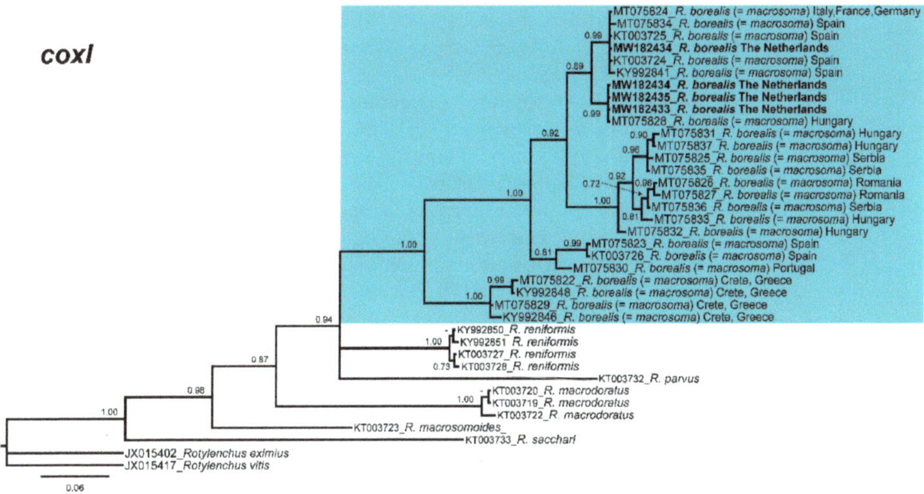

**Figure 5.** Phylogenetic relationships within the genus *Rotylenchulus*. Bayesian 50% majority rule consensus tree as inferred from *coxI* mitochondrial DNA (mtDNA) sequence alignment under the general time-reversible model of sequence evolution with correction for invariable sites and a gamma-shaped distribution (GTR + I + G). Posterior probabilities of more than 0.70 are given for appropriate clades. Newly obtained sequences in this study are shown in bold. Scale bar = expected changes per site.

The 50% majority rule consensus *ITS1* BI tree also showed two clades, one moderately and the other well supported including *R. borealis* type A and type B sequences (PP = 0.95, 1.00, respectively), including *R. reniformis*, *R. parvus*, *Rotylenchulus sacchari*, and *Rotylenchulus clavicaudatus* (Figure 4). All sequences of *R. borealis* from the Netherlands (MW174239-MW174242) and those of *R. borealis* (= *macrosoma*) from Israel and all the sequences from Spain and Greece deposited in the NCBI database clustered together in a highly supported clade (PP = 1.00).

Finally, the phylogenetic relationships of *Rotylenchulus* species inferred from analysis of partial *coxI* gene sequences showed several clades that were well defined (Figure 5). All sequences of *R. borealis* from the Netherlands (MW182432-MW182435) and sequences from several European countries (Germany, Greece, Hungary, Italy, Portugal, Romania, Serbia, and Spain) deposited in the NCBI database clustered together in a highly supported clade (PP = 1.00).

### 2.3. Global Distribution Rotylenchulus spp.

We detected that the genus *Rotylenchulus* exhibited a worldwide distribution across a wide variety of environments and climatic zones (Figure 6). We found that *Rotylenchulus* spp. are widely distributed in warm temperature ($-3\,°C$ < annual mean temperature < $+18\,°C$) and arid (annual precipitation < 300 mm) climate zones, with seven different species for both types, and to a lesser extent in equatorial (annual mean temperature $\geq +18\,°C$) and snow (mean temperature of the coldest month $\leq -3\,°C$) climate zones, with four and one species, respectively (Figure 6). We did not detect species in the polar (mean temperature of the warmest month < $+10\,°C$) climate zone (Figure 6). It should be noted that highest diversity of species, although less frequently found, seems to be in the southern part of Africa with mainly warm temperate and arid climatic zones (Figure 6). The species distribution observed in this study revealed that the genus *Rotylenchulus* is adapted to heterogeneous climatic conditions, with an annual mean temperature of 19.14 $°C$, but ranging from 8.36 to 28.58 $°C$, and a mean annual precipitation of 1026.97 mm, but ranging from 1 to 3583.00 mm. This suggests that the occurrence of *Rotylenchulus* species in areas with extremely low values in annual precipitation (i.e., desert lands in Egypt and Iraq; Figure 6) could be due to the establishment of an irrigation regime in agricultural ecosystems. Only four species were reported more than three times in literature review, i.e., *R. borealis* (= *R. macrosoma*), *R. macrodoratus*, *R. parvus*, and *R. reniformis* (Figure 6). The most widely distributed species was *R. reniformis*, followed by *R. parvus*, both reported in all continents except Antarctica (Africa, North and South America, Asia, Australia, and Europe), and *R. borealis* in Africa, Europe, and Middle East Asia (Figure 6). Bioclimatic variables (BIOCLIM) based on temperature (annual mean temperature (BIO1), maximum temperature of warmest month (BIO5), and minimum temperature of coldest month (BIO6)) showed significantly different temperature conditions on the distribution of these most common species (Figure 7). The two major pathogenic species (*R. reniformis* and *R. parvus*) were mainly distributed in tropical, temperate, and arid climates, showing their close relationship with warmer areas with high annual mean temperature, max temperature of the warmest month, and minimum temperature of the coldest month, ranging from 9.55 to 21.11 $°C$, 24.00 to 3583.00 mm and 14.79 to 26.99 $°C$, 1.00 to 1773.00 mm, respectively (Figures 6 and 7). *Rotylenchulus macrodoratus* showed a distribution in temperate climate with annual mean temperature and precipitation ranging from 12.32 to 19.23 $°C$ and 526.00 to 1013.00 mm, respectively (Figure 7). The climatic plasticity of *R. borealis* is remarkable in relationship with annual mean temperature and precipitation, ranging from 8.36 to 28.58 $°C$ and 160.00 to 1998.00 mm, respectively (Figure 7). *Rotylenchulus borealis* (= *R. macrosoma*) showed statistically significant differences in lower annual mean temperature, max temperature of the warmest month, and min temperature of the coldest month in comparison to *R. parvus* and *R. reniformis* (Figure 7). However, only *R. reniformis* showed statistically significant differences in higher annual precipitation in comparison to the other studied species (Figure 7). Other bioclimatic variables are shown in Figure S1.

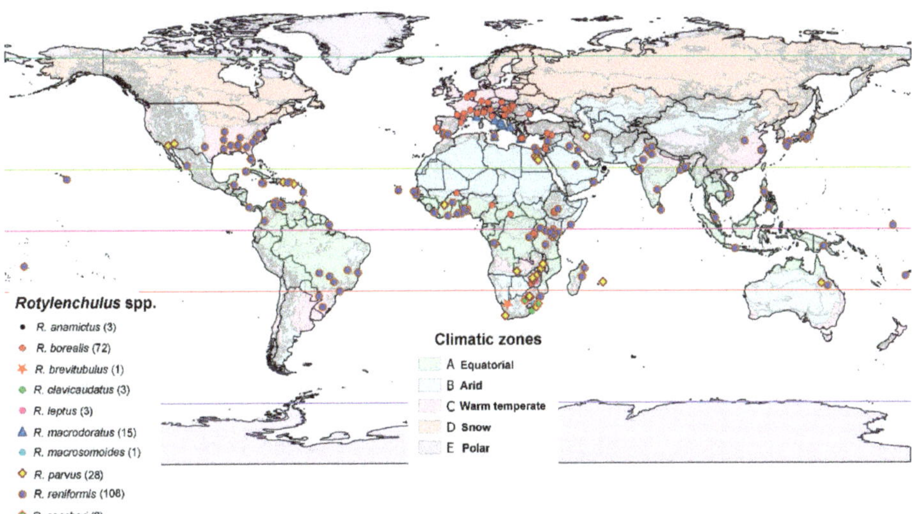


**Figure 6.** World map distribution of *Rotylenchulus* species across different climate conditions. Climatic zones based on type of vegetation [19]. In the species list, the number in brackets indicates the locations cited for each species.

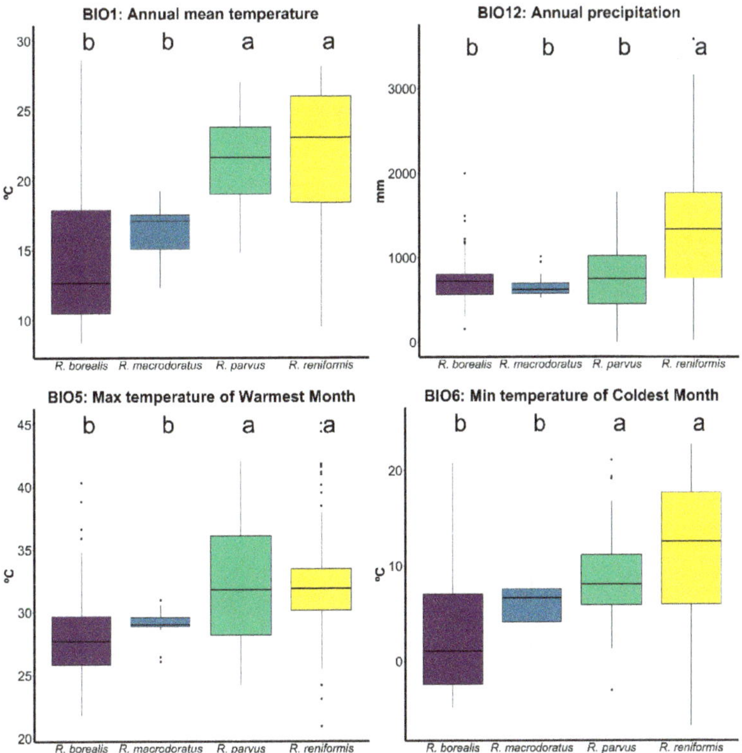


**Figure 7.** Annual mean temperature, annual precipitation, maximum temperature of warmest month, and minimum temperature of coldest month for *Rotylenchulus* species with $\geq 3$ reports (each single dot correspond to a species report). The different lowercase letters indicate the differences in each bioclimatic variable between species. They were tested using ANOVA with a level of significance of $p < 0.05$.

## 3. Discussion

The primary objective of this study was to decipher the intraspecific diversity of *R. borealis* and *R. macrosoma* by applying integrative taxonomical approaches on several new unidentified *Rotylenchulus* populations from Europe, appearing morphological and morphometrically undistinguishable. Additionally, we aimed to provide new insights into the global distribution and climatic requirements of the genus *Rotylenchulus*.

The resemblance between the mature females of *R. borealis* and *R. macrosoma*, as well as the general similarity between these two species also in their male and immature female forms, host preferences, and host tissue reactions was emphasized by Cohn and Mordechai [18] studying a topotype population of *R. macrosoma* from olive under growth chamber conditions. Our morphometric studies in this research support that both species do not have major differences in basic morphology or in morphometric informative characters such as immature female body length, stylet length, tail hyaline region, and spicules morphology and morphometry, showing a remarkable example of a close phylogenetic relationship of both species. The results on our new measurements on *R. macrosoma* immature female paratype specimens from WANECO and USDA nematode collections suggest that the range in stylet length could probably be shorter than that provided in the original description [10], but unfortunately no other paratypes could be studied. The morphometric comparison of an important number of populations from *R. borealis* and *R. macrosoma* exhibited morphometric variation normally expected among populations of the same *Rotylenchulus* species. The higher values in all of the three main distinguishing morphometric characters between both species were detected in Israel, Crete, and a Spanish population from Huévar del Aljarafe (southern Spain), but these differences do not justify the separation in two different species [3,4,9,10].

In the present study, in which sequence data obtained from *28S* and *ITS1* rRNA genes and *coxI* mitochondrial DNA (mtDNA) gene was analyzed, specimens from populations identified as representing *R. borealis* and *R. macrosoma* from the Netherlands and several European countries, including Israel, respectively, clustered together as a single group. This grouping was well supported by the high bootstrap values in the phylogenetic analysis, thereby supporting the synonymization of *R. macrosoma* with *R. borealis*, as already emphasized by Cohn and Mordechai [18].

Phylogenetic analyses based on three molecular markers (*D2-D3* expansion domains of *28S* rRNA gene, *ITS1* region, and the partial *coxI* mtDNA) resulted in a general consensus of species phylogenetic positions clustering *R. borealis* population from the Netherlands with *R. macrosoma* from Israel, together with all other *R. macrosoma* populations previously reported in several European countries. These phylogenetic analyses were congruent with those given by previous studies [3,4,9,16,20], and phylogeny of the *28S* rRNA and ITS regions confirm that *R. borealis* population from the Netherlands is conspecific with *R. macrosoma* from Israel and all other populations from Europe. Our results on *28S* rRNA phylogeny also suggest that *R. borealis* (MT775429) from New Delhi could not be considered conspecific with *R. borealis* and needs to be revised under integrative taxonomical approaches for confirming its specific status. The genus *Rotylenchulus* has rRNA genes that exhibit high levels of intraspecific and intra-individual variation [3,9,16]. However, they seem functional through the reconstruction of secondary structure models and mutation mapping using *R. reniformis* sequences [3]. Qing et al. [16] suggested that these different sequences are paralogs located in different rRNA clusters or chromosomes and that these tandem arrays may still be expanding in number.

Longer stylet specimens do not seem to be associated with differences in molecular markers (as some Andalusian populations with longer stylet were molecularly associated with other species with shorter stylets) (Figures 3–5). Other characters (body length and hyaline tail region length), as shown in Figure 2, seem to be very variable for African populations of *R. borealis*. Palomares-Rius et al. [4], in a broad molecular study of *R. borealis* (= *R. macrosoma*), also studied the molecular species separation, with the results showing incongruent results for species separation between Cretan and other European populations for *R. borealis* (= *R. macrosoma*), even with the relatively high molecular differences between both population groups. In our case, the new population of *R. borealis* found in the Netherlands in this study, and the sequence deposited in GenBank from Belgium (MK558206), had an even lower molecular similarity with other former *R. macrosoma* populations from Crete, Greece, fully supporting our idea of conspecificity.

Thus, the morphological and morphometric results of both species groups, together with the high molecular similarity among ribosomal and mitochondrial genes of both species groups, do not support the validity of *R. macrosoma* as a separate species and give sufficient basis for the synonymization of *R. macrosoma* n. syn. with *R. borealis*. Since the description of *R. borealis* was in 1962 and that of *R. macrosoma* in 1968, the name *R. borealis* has priority over *R. macrosoma*; thus, *R. macrosoma* is proposed here as a junior synonym of the former.

Climate is a critical environmental determinant of the distribution of plant-parasitic nematodes and a key driver of their reproduction and survival [21]. Temperature, moisture, and availability of host plants are three of the most important factors governing the distribution, spread, and symptom development in plants from plant-parasitic nematodes [21,22], including reniform nematodes. The wide distribution of *Rotylenchulus* species likely resulted from an exceptionally wide host range, as well as their ability to survive extended periods in a dehydrated state [1]. Anhydrobiotic *Rotylenchulus* forms have been documented dispersing long distances in dust storms [23]; however, human dispersion through agriculture activities need also to be considered [4]. The influence of annual precipitation on *Rotylenchulus* spp. distribution suggests that this factor may be not as important as expected. However, the majority of the recorded points have crops with irrigation, and this could change the natural precipitation conditions and importance for these species. In particular, the widespread presence of *R. borealis* in localities at higher latitude in Northern Europe and lower latitude in several central African countries indicated and adaptation to heterogeneous climatic conditions and probably survival strategies for colder and warmer winters and humid to dry soil conditions. Similarly, the cosmopolitan distribution of *R. parvus* can be related to the wide range of temperature reproduction (20 to 35 °C) and survival (4 to 35 °C), as suggested by Dasgupta and Raski [24]. Climate change could expand *R. borealis* to upper latitudes as climate will warm and this will fulfil the ecological requirements of this species, one of the most adapted to lower temperatures among the four most distributed species (*R. borealis*, *R. macrodoratus*, *R. parvus*, and *R. reniformis*). Interestingly, the major diversity of the genus is from sub-Saharan Africa, with the exception of *R. macrodoratus* (Mediterranean distribution) and *R. leptus* (Arabian Peninsula). Siddiqi [25] proposed the idea about the origin of this genus in the Afrotropical (Ethiopian) zoogeographical region, comprising Africa (south of the Sahara); the southern part of the Arabian Peninsula; and various islands, including Madagascar. This idea was reinforced with phylogenetic analysis [3]. However, only three species (*R. borealis*, *R. parvus*, and *R. reniformis*) have been able to colonize different continents with wide ecological requirements, as was shown in this research. Additionally, to these ecological requirements for species distribution, other factors such as survival in anhydrobiotic stage or resting eggs could help with the dispersal of this species to other agricultural areas in the world.

In summary, the present study confirmed the synonymy of *R. macrosoma* with *R. borealis*, and thus the genus comprises 10 valid species. Our data also demonstrate the extraordinary morphological and molecular diversity of *R. borealis* in Europe, Africa, and the Middle East and comprise a paradigmatic example of remarkable flexibility of climatic requirements within reniform nematodes. Nevertheless, despite frequent surveys in different continents of the world, the number of sites studied is still low. Therefore, further surveys are still needed in unsampled geographical areas and climatic conditions, both in plantations and indigenous forests with the aim to identify additional *Rotylenchulus* species.

## 4. Materials and Methods

### 4.1. Nematode Populations and Morphometric Studies

One of the authors (G. Karssen) visited the type locality of *R. borealis*, and the place reported in the original description was lost, i.e., was filled up by new building of houses. Nevertheless, this author detected a new population of *R. borealis* in another location close near the type locality, at Huissen, Betuwe region, the Netherlands. This new population, together with mounted paratypes from of *R. macrosoma* and *R. borealis* from the nematode collections Wageningen Nematode Collection (WANECO; slides WT106, WT107, WT110, WT111, and #1025 NT and #1026 NT) and USDA Nematode Collection kindly provided by Dr. Z. A. Handoo (slides T-594p and T-595p), were used for morphological studies.

In addition, some new European reports recently detected and associated with corn and wheat [4] were measured in order to carry out a morphometric comparison with all the measured populations of both species (Tables 2–8). All these populations were compared with the morphometry of all previously studied populations of both species, including a total of 12 populations of *R. borealis* and 16 populations of *R. macrosoma*.

Nematodes were extracted from 500 cm$^3$ of soil by centrifugal flotation [26] method. For morphometric studies, *Rotylenchulus* specimens were killed and fixed by a hot solution of 4% formalin + 1% glycerol, then processed in pure glycerin [27], as modified by De Grisse [28]. The light micrographs and measurements of each nematode population including important diagnostic characteristics (i.e., de Man indices, body length, stylet length, lip region, tail length, etc.) were performed using a Leica DM6 compound microscope with a Leica DFC7000 T digital camera. Nematodes were identified at the species level using an integrative approach combining molecular and morphological techniques to achieve efficient and accurate identification [4,9]. For each nematode population, key diagnostic characters were determined, including body length, stylet length, a ratio (body length/maximum body width), c′ ratio (tail length/body width at anus), V ratio ((distance from anterior end to vulva/body length) × 100), and o ratio ((distance from stylet base to dorsal pharyngeal opening/body length) × 100) [9], and the sequencing of specific DNA fragments (described below) confirmed the identity of the nematode species for each population.

### 4.2. DNA Extraction, PCR, and Sequencing

For molecular analyses, in order to ensure that the selected nematodes for extracting DNA are from the same species, we temporary mounted 2 live nematodes from each sample in a drop of 1M NaCl containing glass beads (to avoid nematode crushing/damaging specimens) to ensure specimens conformed to the unidentified populations of *Rotylenchulus*. All necessary morphological and morphometric data by taking pictures and measurements using the above camera-equipped microscope were recorded. This was followed by DNA extraction from a single specimen and polymerase chain reaction (PCR) cycle conditions, as previously described [4,9]. PCR and sequencing of the Dutch population was performed at the Institute for Sustainable Agriculture, Spanish National Research Council (IAS-CSIC) facility, whereas for the Israeli population at Agricultural Research organization (ARO)-Volcani Center, Israel. Several sets of primers were used for PCR. A partial region of the *28S* rRNA gene including the expansion domains D2 and D3 (*D2-D3*) was amplified by using the primers D2A (5′-ACAAGTACCGTGAGGGAAAGTTG-3′) and D3B (5′-TCGGAAGGAACCAGCTACTA-3′) [29]. The internal transcribed spacer region (*ITS*) was amplified using forward primer TW81 (5′-GTTTCCGTAGGTGAACCTGC-3′) and reverse primer AB28 (5′-ATATGCTTAAGTTCAGCGGGT -3′) [30]. The *coxI* gene was amplified using the primers JB3 (5′-TTTTTTGGGCATCCTGAGGTTTAT-3′) and JB5 (5′-AGCACCTAAACTTAAAACATAATGAAAATG-3′) [31]. The PCR cycling conditions for the *28S* rRNA primers were as follows: 94 °C for 2 min, followed by 35 cycles of 94 °C for 30 s, an annealing temperature of 55 °C for 45 s, and 72 °C for 1 min, and 1 final cycle of 72 °C for 10 min. The PCR cycling for *coxI* primers was as follows: 95 °C for 15 min, 39 cycles at 94 °C for 30 s, 53 °C for 30 s, and 68 °C for 1 min, followed by a final extension at 72 °C for 7 min. PCR volumes were adapted to 25 μL for each reaction, and primer concentrations were as described in De Ley et al. [29] and Bowles et al. [31]. We used 5x HOT FIREpol Blend Master Mix (Solis Biodyne, Tartu, Estonia) in all PCR reactions. The PCR products were purified after amplification using ExoSAP-IT (Affimetrix, USB products, Kandel, Germany) and used for direct sequencing in both directions with the corresponding primers. Israeli amplification products were cloned before sequencing using pGEM-T easy vector systems (Promega). The resulting products were purified and run in a DNA multicapillary sequencer (Model 3130XL Genetic Analyzer; Applied Biosystems, Foster City, CA, USA), using the BigDye Terminator Sequencing Kit v.3.1 (Applied Biosystems) at the Stab Vida sequencing facility (Caparica, Portugal). The sequence chromatograms of the 2 markers (*coxI* and *D2-D3* expansion segments of *28S* rRNA) were analyzed using DNASTAR LASERGENE SeqMan v. 7.1.0. Basic local alignment search tool (BLAST) at the National Center for Biotechnology Information (NCBI) was used to confirm the species identity of the DNA sequences obtained in this study [32]. The newly obtained sequences were deposited in the GenBank database under accession numbers indicated on the phylogenetic trees and in Table 1.

### 4.3. Phylogenetic Analysis

Sequenced genetic markers in the present study (after discarding primer sequences and ambiguously aligned regions) and several *Rotylenchulus* spp. sequences obtained from GenBank were used for phylogenetic reconstruction (Table 1). Outgroup taxa for each dataset were selected on the basis of previous published studies [3,4,9]. Multiple sequence alignments of the newly obtained and published sequences were made using the Fast Fourier transform-normalized similarity matrix (FFT-NS-2) algorithm of MAFFT v. 7.450 [33]. Sequence alignments were visualized using BioEdit [34] and edited by Gblocks ver. 0.91b [35] in Castresana Laboratory server (http://molevol.cmima.csic.es/castresana/Gblocks_server.html) using options for a less stringent selection (minimum number of sequences for a conserved or a flanking position: 50% of the number of sequences + 1; maximum number of contiguous no conserved positions: 8; minimum length of a block: 5; allowed gap positions: with half).

Phylogenetic analyses of the sequence datasets were based on Bayesian inference (BI) using MRBAYES 3.2.7a [36]. The best-fit model of DNA evolution was calculated with the Akaike information criterion (AIC) of JMODELTEST v. 2.1.7 [37]. The best-fit model, the base frequency, the proportion of invariable sites, and the gamma distribution shape parameters and substitution rates in the AIC were then used in phylogenetic analyses. BI analyses were performed under a general time reversible, with a proportion of invariable sites and a rate of variation across sites (GTR + I + G) model for *D2-D3*, *ITS1* rRNA, and the partial *coxI* gene. These BI analyses were run separately per dataset with 4 chains for $2 \times 10^6$ generations. The Markov chains were sampled at intervals of 100 generations. Two runs were conducted for each analysis. After discarding burn-in samples of 30% and evaluating convergence, we retained the remaining samples for more in-depth analyses. The topologies were used to generate a 50% majority-rule consensus tree. Posterior probabilities (PP) were given on appropriate clades. Trees from all analyses were visualized using FigTree software version v.1.42 [38].

### 4.4. Data Collection of Global Distribution of Rotylenchulus spp. and Statistical Analysis

The species distribution data of *Rotylenchulus* spp. were exhaustively compiled from the national and regional nematofauna records worldwide from databases (Google Scholar, Web of Sciences, Scopus, and PubMed) and specialized literature (nematological and phytopathological journals) during the period 2020–1940. We selected only those articles satisfying one the following criteria for this review: (1) contained geographical information about the presence and/or abundance of reniform nematodes (*Rotylenchulus* spp.); (2) contained data on their taxonomy, morphology, molecular identification, ecology, pathogenicity, and provided localities of each population. Articles lacking information about geographic coordinates were cross-checked using Quantum GIS v. 3.12.0 [39]. Nevertheless, since *R. reniformis* has been associated with hundreds of crops and native plants in many regions of the world (on the four aforementioned databases we found 9640, 1377, 446, and 189 studies, respectively), only selected reports concerning geographical information were selected, and duplicity of reported localities were not included.

We used bioclimatic predictors (BIOCLIM) based on temperature and precipitation [40] to detect environmental conditions associated with the global distribution of *Rotylenchulus* spp. and to compare the climate spaces for the different species. Additionally, we plotted the global distribution *Rotylenchulus* spp. across climate zones on the basis of the type of vegetation [19]. Only species with more than 3 reported populations were plotted in order to assess the range of climatic variables for each species. Species with type locality only or occasional records were omitted.

The analysis on the bioclimatic variables for *Rotylenchulus* spp. with more than 3 reported populations was concentrated in 18 variables: BIO1 (Annual mean temperature), BIO2 [Mean Diurnal Range (Mean of monthly (max temp-min temp)], BIO3 [Isothermality, (BIO2/BIO7) * 100], BIO4 [Temperature seasonality, (standard deviation * 100)], BIO5 (maximum temperature of the warmest month), BIO6 (minimum temperature of the coldest month), BIO7 [temperature annual range (BIO5-BIO6)], BIO9 (mean temperature of driest quarter), BIO10 (mean temperature of the warmest quarter), BIO 15 (precipitation seasonality, coefficient of variation), and BIO18 (precipitation of the warmest quarter). To detect the influence on *Rotylenchulus* spp. of the different bioclimatic variables, we used one-way ANOVA among species conducted using the R v. 3.5.1 freeware [41]

**Supplementary Materials:** The following material is available online at https://www.mdpi.com/22 23-7747/10/1/7/s1. Figure S1. BIOCLIM variables for *Rotylenchulus* species with $\geq$ 3 reports. BIO3 [Isothermality, (BIO2/BIO7) * 100], BIO4 [Temperature seasonality, (SD * 100)], BIO7 [Temperature annual range (BIO5-BIO6)], BIO9 (mean temperature of driest quarter), BIO10 (mean temperature of the warmest quarter), BIO 15 [precipitation seasonality (CV)], BIO17 (precipitation of driest quarter), and BIO18 (precipitation of the warmest quarter).

**Author Contributions:** Conceptualization, J.E.P.-R., I.C.-C., A.A.-Y., G.L.-R., C.C.-N., S.B.M., G.K., and P.C.; methodology, J.E.P.-R., I.C.-C., A.A.-Y., G.L.-R., C.C.-N., S.B.M., G.K., and P.C.; software, J.E.P.-R., I.C.-C., A.A.-Y., G.L.-R., C.C.-N., and P.C.; analysis, J.E.P.-R., I.C.-C., A.A.-Y., G.L.-R., C.C.-N., and P.C.; resources, J.E.P.-R. and P.C.; writing, J.E.P.-R., I.C.-C., A.A.-Y., C.C.-N., S.B.M., G.K., and P.C. All authors contributed to the final discussion data, and read and approved the final manuscript. All authors have read and agreed to the published version of the manuscript.

**Funding:** This research was supported by grant RTI2018-095925-A-100 from Ministerio de Ciencia, Innovación y Universidades, Spain, grant 201740E042 "Análisis de diversidad molecular, barcoding, y relaciones filogenéticas de nematodos fitoparásitos en cultivos mediterráneos" from the Spanish National Research Council (C.S.I.C.), and by the Humboldt Research Fellowship for Postdoctoral Researchers awarded to the third author (A.A.-Y.).

**Data Availability Statement:** The datasets generated during and/or analysed during the current study are available from the corresponding author on reasonable request.

**Acknowledgments:** We would like to thank J. Martin Barbarroja (IAS-CSIC) for their excellent technical assistance in surveys and management of soil samples, as well as further anonymous reviewers and editors for their effort in reviewing the manuscript and helping improve this study. This research is part of the PhD project of the second author. The second author is a recipient of a contract from Ministry of Science and Innovation for Predoctoral Researchers in Spain, PRE2019-090206. The third author is a recipient of the Humboldt Research Fellowship for Postdoctoral Researchers at the Helmholtz Centre for Environmental Research-UFZ, Leipzig, Germany. We acknowledge support of the publication fee by the CSIC Open Access Publication Support Initiative through its Unit of Information Resources for Research (URICI).

**Conflicts of Interest:** The authors declare no conflict of interest. The funders had no role in the design of the study; in the collection, analyses, or interpretation of data; in the writing of the manuscript; or in the decision to publish the results.

## References


1. Robinson, A.F.; Inserra, R.N.; Caswell-Chen, E.P.; Vovlas, N.; Troccoli, A. *Rotylenchulus* species: Identification distribution, host ranges, and crop plant resistance. *Nematropica* **1997**, *27*, 127–180.
2. Linford, M.B.; Oliveira, J.M. *Rotylenchulus reniformis* nov. gen., n. sp., a nematode parasite of roots. *Proc. Helm. Soc. Wash.* **1940**, *7*, 35–42.
3. Van den Berg, E.; Palomares-Rius, J.E.; Vovlas, N.; Tiedt, L.R.; Castillo, P.; Subbotin, S.A. Morphological and molecular characterisation of one new and several known species of the reniform nematode, *Rotylenchulus* Linford & Oliveira, 1940 (Hoplolaimidae: Rotylenchulinae), and a phylogeny of the genus. *Nematology* **2016**, *18*, 67–107.
4. Palomares-Rius, J.E.; Archidona-Yuste, A.; Cantalapiedra-Navarrete, C.; Azpilicueta, A.; Saborido, A.; Tzortzakakis, E.A.; Cai, R.; Castillo, P. New distribution and molecular diversity of the reniform nematode *Rotylenchulus macrosoma* Dasgupta, Raski and Sher, 1968 (Nematoda: Rotylenchulinae) in Europe. *Phytopathology* **2020**, *110*. in press. [CrossRef]
5. Colagiero, M.; Ciancio, A. Climate change and nematodes: Expected effects and perspectives for plant protection. *Redia* **2011**, *94*, 113–118.
6. Jones, L.M.; Koehler, A.K.; Trnka, M.; Balek, J.; Challinor, A.J.; Atkinson, H.J.; Urwin, P.E. Climate change is predicted to alter the current pest status of *Globodera pallida* and *G. rostochiensis* in the United Kingdom. *Glob. Chang. Biol.* **2017**, *23*, 4497–4507. [CrossRef]
7. IPCC. Intergovernmental Panel on Climate Change. 2019. Available online: https://archive.ipcc.ch/index.htm (accessed on 17 September 2020).
8. Bellard, C.; Bertelsmeier, C.; Leadley, P.; Thuiller, W.; Courchamp, F. Impacts of climate change on the future of biodiversity. *Ecol. Lett.* **2012**, *15*, 365–377. [CrossRef]
9. Palomares-Rius, J.E.; Cantalapiedra-Navarrete, C.; Archidona-Yuste, A.; Tzortzakakis, E.A.; Birmpilis, I.G.; Vovlas, N.; Subbotin, S.A.; Castillo, P. Prevalence and molecular diversity of reniform nematodes of the genus *Rotylenchulus* (Nematoda: Rotylenchulinae) in the Mediterranean Basin. *Eur. J. Plant Pathol.* **2018**, *150*, 439–455. [CrossRef]
10. Dasgupta, D.R.; Raski, D.J.; Sher, S.A. A revision of the genus *Rotylenchulus* Linford and Oliveira, 1940 (Nematoda: Tylenchidae). *Proc. Helminthol. Soc. Wash.* **1968**, *35*, 169–192.
11. Loof, P.A.A.; Oostenbrink, M. *Rotylenchulus borealis* n. sp. with a key to the species of *Rotylenchulus*. *Nematologica* **1962**, *7*, 83–90.
12. Germani, G. Morphological and biometrical characters of three West-African species of *Rotylenchulus* Linford & Oliveira, 1940 (Nematoda: Tylenchida). *Rev. Nématol.* **1978**, *1*, 241–250.
13. Castillo, P.; Vovlas, N.; Troccoli, A. The reniform nematode, *Rotylenchulus macrosoma*, infecting olive in southern Spain. *Nematology* **2003**, *5*, 23–29. [CrossRef]
14. Niragire, I. Morphological and Molecular Characterization of Plant-Parasitic Nematodes of Potato (*Solanum tuberosum* I.) in Rwanda. Master's Thesis, University of Ghent, Ghent, Belgium, 2018; p. 44.
15. Singh, P.R.; Kashando, B.E.; Couvreur, M.; Karssen, G.; Bert, W. Plant-parasitic nematodes associated with sugarcane in Kilimanjaro, Tanzania. *J. Nematol.* **2020**, *52*, e2020-59. [CrossRef]
16. Qing, X.; Bik, H.; Yergaliyev, T.M.; Gu, J.; Fonderie, P.; Brown-Miyara, S.; Szitenberg, A.; Bert, W. Widespread prevalence but contrasting patterns of intragenomic rRNA polymorphisms in nematodes: Implications for phylogeny, species delimitation and life history inference. *Mol. Ecol. Resour.* **2020**, *20*, 318–332. [CrossRef]
17. Lišková, M.; Troccoli, A.; Vovlas, N.; Sasanelli, N. On the occurrence of *Rotylenchulus borealis* in the Slovak Republic. *Helminthologia* **2002**, *39*, 165–167.
18. Cohn, E.; Mordechai, M.M. Morphology and parasitism of the mature female of *Rotylenchulus macrosomus*. *Rev. Nématol.* **1988**, *11*, 385–389.
19. Kottek, M.; Grieser, J.; Beck, C.; Rudolph, B.; Rubel, F. World map of the Köppen-Geiger climate classification updated. *Meteorol. Z.* **2006**, *15*, 259–263. [CrossRef]
20. Riascos-Ortiz, D.; Mosquera-Espinosa, A.T.; De Agudelo, F.V.; de Oliveira, C.M.G.; Muñoz-Flórez, J.E. Morpho-molecular characterization of Colombian and Brazilian populations of *Rotylenchulus* associated with *Musa* spp. *J. Nematol.* **2019**, *51*, e2019-47. [CrossRef]
21. Thakur, M.P.; Tilman, D.; Purschke, O.; Ciobanu, M.; Cowles, J.; Isbell, F.; Wragg, P.D.; Eisenhauer, N. Climate warming promotes species diversity, but with greater taxonomic redundancy, in complex environments. *Sci. Adv.* **2017**, *3*. [CrossRef]
22. Neher, D.A. Ecology of plant and free-living nematodes in natural and agricultural soil. *Annu. Rev. Phytopathol.* **2010**, *48*, 371–394. [CrossRef]
23. Gaur, H.S. Dissemination and mode of survival of nematodes in dust storms. *Ind. J. Nematol.* **1988**, *18*, 84–98.
24. Dasgupta, D.R.; Raski, D.J. The biology of *Rotylenchulus parvus*. *Nematologica* **1968**, *14*, 429–440.
25. Siddiqi, M.R. *Tylenchida Parasites of Plants and Insects*, 2nd ed.; CABI Publishing: Wallingford, UK, 2000.
26. Coolen, W.A. Methods for extraction of *Meloidogyne* spp. and other nematodes from roots and soil. In *Root-Knot Nematodes (Meloidogyne species). Systematics, Biology, and Control*; Lamberti, F.F., Taylor, C.E., Eds.; Academic Press: London, UK, 1979; pp. 317–329.
27. Seinhorst, J.W. Killing nematodes for taxonomic study with hot f.a. 4:1. *Nematologica* **1966**, *12*, 178. [CrossRef]
28. De Grisse, A.T. Redescription ou modification de quelques techniques utilisées dans l'étude des nematodes phytoparasitaires. *Meded. Rijksfak. Landbouwet.* **1969**, *34*, 351.

29. De Ley, P.; Felix, M.A.; Frisse, L.A.; Nadler, S.; Sternberg, P.; Thomas, W. Molecular and morphological characterisation of two reproductively isolated species with mirror-image anatomy (Nematoda: Cephalobidae). *Nematology* **1999**, *1*, 591–612. [CrossRef]

30. Subbotin, S.A.; Vierstraete, A.; De Ley, P.; Rowe, J.; Waeyenberge, L.; Moens, M.; Vanfleteren, J.R. Phylogenetic relationships within the cyst-forming nematodes (Nematoda, Heteroderidae) based on analysis of sequences from the ITS regions of ribosomal DNA. *Mol. Phylogenet. Evol.* **2001**, *21*, 1–16. [CrossRef]

31. Bowles, J.; Blair, D.; McManus, D.P. Genetic variants within the genus *Echinococcus* identified by mitochondrial DNA sequencing. *Mol. Biochem. Parasitol.* **1992**, *54*, 165–174. [CrossRef]

32. Altschul, S.; Gish, W.; Miller, W.; Myers, E.W.; Lipman, D.J. Basic local alignment search tool. *J. Mol. Biol.* **1990**, *215*, 403–410. [CrossRef]

33. Katoh, K.; Rozewicki, J.; Yamada, K.D. MAFFT online service: Multiple sequence alignment, interactive sequence choice and visualization. *Brief. Bioinform.* **2019**, *20*, 1160–1166. [CrossRef]

34. Hall, T.A. BioEdit: A user-friendly biological sequence alignment editor and analysis program for windows 95/98/NT. *Nucleic Acids Symp. Ser.* **1999**, *41*, 95–98.

35. Castresana, J. Selection of conserved blocks from multiple alignments for their use in phylogenetic analysis. *Mol. Biol. Evol.* **2000**, *17*, 540–552. [CrossRef] [PubMed]

36. Ronquist, F.; Huelsenbeck, J.P. MRBAYES 3: Bayesian phylogenetic inference under mixed models. *Bioinformatics* **2003**, *19*, 1572–1574. [CrossRef] [PubMed]

37. Darriba, D.; Taboada, G.L.; Doallo, R.; Posada, D. jModelTest 2: More models, new heuristics and parallel computing. *Nat. Methods* **2012**, *9*, 772. [CrossRef] [PubMed]

38. Rambaut, A. FigTreev1.4.2. A Graphical Viewer of Phylogenetic Trees. 2014. Available online: http://tree.bio.ed.ac.uk/software/figtree/ (accessed on 20 October 2020).

39. QGIS Development Team. QGIS Geographic Information System. Open Source Geospatial Foundation Project. 2020. Available online: http://qgis.osgeo.org (accessed on 12 November 2020).

40. Fick, S.E.; Hijmans, R.J. WorldClim 2: New 1km spatial resolution climate surfaces for global land areas. *Int. J. Climatol.* **2017**, *37*, 4302–4315. [CrossRef]

41. R_Core_Team. R: A language and environment for statistical computing. In *R Foundation for Statistical Computing*; R Foundation: Vienna, Austria, 2018; Available online: https://www.R-project.org/ (accessed on 6 October 2020).

*Review*

# Optimizing Sampling and Extraction Methods for Plant-Parasitic and Entomopathogenic Nematodes

Mahfouz M. M. Abd-Elgawad 

National Research Center, Plant Pathology Department, Agricultural and Biological Research Division, El-behooth St., Dokki 12622, Egypt; mahfouzian2000@yahoo.com

**Abstract:** Plant-parasitic and entomopathogenic nematodes (PPNs and EPNs) are key groups in crop production systems. This study aims at optimizing nematode sampling and extraction methods to benefit integrated pest management (IPM) through (a) management of PPNs and (b) use of EPNs. The impacts of these methods on PPNs and EPNs to achieve cost-effective and efficient IPM programs are presented. The common misuses of sampling and extraction methods are discussed. Professionals engaged in IPM should consider sampling the reliability level in the light of the intended goal, location, crop value, susceptibility, nematode species, and available funds. Logical sampling methodology should be expanded to integrate various factors that can recover extra EPN isolates with differential pathogenicity. It should seek for the best EPN-host matching. Merits of repeated baiting for EPN extraction from soil and sieving for PPN recovery from suspensions are presented. Their extraction values may be modelled to quantify the efficiency of nematode separation. The use of proper indices of dispersion to enhance the biocontrol potential of EPNs or save costs in nematicidal applications is ideally compatible with IPM programs. Selecting an extraction method may sometimes require further tests to find the best extraction method of the existing fauna and/or flora. Cons and pros of modern sampling and extraction techniques are highlighted.

**Keywords:** index of dispersion; IPM; modelling; molecular approaches; sampling and extraction

**Citation:** Abd-Elgawad, M.M.M. Optimizing Sampling and Extraction Methods for Plant-Parasitic and Entomopathogenic Nematodes. *Plants* **2021**, *10*, 629. https://doi.org/10.3390/plants10040629

Academic Editors: Zafar Handoo and Mihail Kantor

Received: 3 February 2021
Accepted: 1 March 2021
Published: 26 March 2021

**Publisher's Note:** MDPI stays neutral with regard to jurisdictional claims in published maps and institutional affiliations.



## 1. Introduction

Plant-parasitic and entomopathogenic nematodes (PPNs and EPNs) represent two key groups as damaging [1] and beneficial [2] organisms, respectively in crop production systems. Their sampling and extraction methods should be optimized to benefit integrated pest management (IPM) through (a) management of PPNs and (b) use of EPNs. Addressing both groups might be a little tricky but the soil is their original habitat. They have a few sampling and extraction issues related to assessing their populations, distribution patterns, and interactions with many other factors within the context of IPM.

## 2. Sampling Goal and Conceived Scenario

As sampling pertains importantly to every aspect of nematode study and management, its significance and drawbacks will cover all related scopes. Sampling of PPNs is basically intended to detect, identify, and estimate their population densities in soil or plant tissues. Its timing, pattern, intensity, tool, and the associated material sampled, all depend on the desired goal, carefully conceived scenario to avoid problems and allocated funds. For example, heavily nematode-infected plants may consequently possess too small a root system to support many PPNs, whereas samples from nearby less infected plants may harbor more nematodes for relatively large root system. Soil samples preferably obtained from the rhizosphere are often used to count PPN number per unit (either volume in $cm^3$ or weight in g, but it is quite better in this case to express nematode number per g of feeder roots in the same volume of soil. This is especially important to avoid discrepancy of PPN population densities relative to plant damage. Clearly, this issue will result in a false correlation between nematode population levels and plant growth parameters/yields based on using either volume or weight unit, not both.

Sampling may also be utilized in a survey, advisory service, research, or relating population level to specific biological/ecological factor(s) or production practices. Plant root, instead of soil samples, may sometimes be good alternatives. Even one individual of any root-knot nematode (RKN) *Meloidogyne* species in a root sample of a highly susceptible crop, may call for PPN control measure, be it (regulatory, cultural, and sanitary methods, nematicide, rotation, resistant variety); e.g., according to the number of RKN-galls, more than one of these control methods may be properly used in IPM [1,3]. This applies for most species/varieties of solanaceous (tomato, eggplant, potato, pepper) and cucurbitaceous (melon, watermelon, squash, pumpkin, zucchini, cucumber) crops. For advisory service, soil and roots should be sampled for PPNs at planting or pre-planting of annual crops.

Sampling to relate the nematode population level to specific biological/ecological factor(s) or production practices can effectively contribute in IPM. It can monitor population level and impact of any biological control agent (BCA) for its further development. It can also detect harmful organisms to prevent their suppression of any beneficial invertebrate [4] in IPM. In such cases, sampling may be done just at the planting and harvest times. However, more informative sampling times may be better. It would preferably fit different growth stages of the plant. This enables pest control operators to know whether prevalence of natural or introduced BCA gradually or rapidly decline with each stage. This approach clearly addresses IPM for both PPNs and EPNs. It may monitor BCAs (e.g., EPNs against insect pests or fungi/bacteria against PPNs) for different IPM programs.

### 3. Ecological Considerations and Concepts

It is well known that there is inherited sampling error, but the most accurate samples should be obtained from locations and at times when population size is greatest in general [5,6]. Inaccuracy of assessing nematode population level is known as sampling error. Precision/reliability is the probability of getting a specified degree of sampling accuracy. Both should be considered for sound IPM. Sampling reliability is used either in terms of the standard error to mean ratio ($E$) or the ratio of the half-width of the confidence interval to the mean ($D$) of the samples [7,8]. The reliability level acceptable as a basis for PPN control in IPM decisions may vary due to the location, crop, nematode species, and available fund or personnel.

Sampling should be done to get accurate data on the pest's ecology for effective IPM. Its design and timing should also enable us to grasp BCA ecology and biology as well as host-BCA interactions. Edaphic and crop factors (e.g., soil properties, cultivar susceptibility, nematode-economic threshold, planting, harvest times, and previous crops), and climatic factors (rainfall, temperature, humidity, solar efficacy) may add better perception for the used IPM strategy. These variables can reveal the positive or negative role of a specific production practice in IPM. Generally, pesticide usage, tillage, crop rotation, and fallow periods can adversely disrupt BCA populations [9]. Biological control of PPNs using an introduced BCA may not be as effective in various settings as that of indigenous BCA due to ecological validity. Soil moisture and texture [10], salinity [11], mulching [12], and pH [13] were also found to modulate EPN populations directly or indirectly by influencing their hosts or enemies [4].

Though often used, random soil samples suffer from the possibility that samples may chance to target an unimportant range of biotic and edaphic factors. So, most soil nematodes and related organisms remain unsampled. In contrast, stratified random sampling can upgrade efficiency to assess population densities and related factors if variations in a stratum are obviously less than that among strata. So, dividing the strata should be based on factors known to the farmer; e.g., difference in soil characteristics, productivity of previous crops, or susceptibility of these crop varieties to PPN-infection. Stratified random sampling may not only offer better estimate of PPN population levels but can also lower pest-control cost via IPM of individual or uniform strata. Regular zigzag patterns with dimensions smaller than the nematode foci can adequately sample PPNs and offer proper weight to the larger, non-infested area as nematodes mostly have clumped distribution [7]. Such patterns can more accurately assess the population density than random sampling especially when more sampling points are taken across plant rows than within rows [5].

Nematode extraction should consider the related settings and nematode genera. For example, extraction of nematode cysts (genera Heterodera and Globodera) may differ from that used for *Meloidogyne* spp. [6,14]. Proper extraction techniques should best fit the existing organisms (e.g., protozoa, fungi, bacteria, invertebrate predators, omnivores, and microarthropods) as the extraction efficiency varies among species. Sucrose centrifugation is the most efficient method for microarthropod extraction. It is) used [15] as a model to study nematodes and their natural enemies such as collembolan and acari mites. Such techniques as species-specific primers and probes in quantitative polymerase chain reaction (qPCR) assessment, colony culture to count colony forming units per unit of root weight, sucrose centrifugation, Baermann funnel, sieving, or baiting with EPN-susceptible host may vary in the extraction efficiency [6,14,16].

Various factors can share in cost-effective and efficient IPM programs. Cost can be reduced via more efficient sampling procedures. A common mistake is to assume that there is always a linear relation between sampling cost and sample size. The variation due to laboratory procedures in the sampling and extraction methodology are mostly unknown and may even exceed field variation that requires more samples. One would rather improve methods instead of reducing samples. A big gap in the accuracy and precision of nematode counts resulting from inter-laboratory proficiency tests was reported [17]. The reasons may be the different custom-made equipment, laboratory-specific adaptations, and/or relative operator's experience. Manufacturers of sampling and extraction tools should continuously contact the related stakeholder for tools' fine-tuning and upgrading. The tests should further be expanded to evaluate and fix the quality of the laboratories' own methods especially in developing countries. This will help to gain insights into possible trends and potential refinements. Mechanized sampling could improve the accuracy and precision, but it requires well-qualified operator on the mechanical sampling equipment (e.g., operate the tractor in a sound and safe manner, sound review for the map of sampled area and handling of the samples, bags, and bag holder).

### 4. Sampling Tools

Conventional soil samplers [14] such as augers to obtain cores are often used in developed countries while an ordinary spade, bladed shovel, or hand trowel is frequently used in developing countries. The use of these variable samplers for similar IPM programs may lead to erratic results. The difference in volume/area of the sampling units may influence the obtained distribution patterns of the pest or BCA [18]. Though acceptable, they lack in the standardization of the used sampler which may falsely contribute to the value of the same index of nematode dispersion used (Table 1). Even sampling for similar objectives is taken with cores that may differ in diameters (e.g., 17, 18, 20, and 25 mm) from one trial to another. This may lead to inconsistent results and misinterpretation of data. For instance, sampling the same site with two concentric circles (as core or unit area) might unexpectedly reveal different spatial patterns of the same population. These patterns (Figure 1) are so different that the nematode counts would require log (for aggregated distribution) or square root (for random distribution) data transformations to equalize experimental treatment variances; a pre-requisite to use parametric statistical methods such as analysis of variance, regression, and correlation [19]. Moreover, adopting a standardized sampler can grant sound comparison between different trials and expand analysis of individual trials for perfection of the conclusions. For vertical distribution, deep-rooted crops require deeper sampling (e.g., grape; 60 cm depth) than shallow-rooted ones (e.g., squash; 20 cm depth) but generally a depth of 30 cm can target the nematodes in the zone of their highest density. A standard core of 2 cm diameter with adjustable depths may be suggested unless it stifles an innovation or experimental goal. Notably, this suggestion avoids other drawbacks because characteristics of a distribution pattern are often dependent on the "standard" scale over which it is processed. Manufacturers and suppliers of such tools would preferably consult pest control operators to standardize their products for better IPM.

**Table 1.** Comparison of index of aggregation ($I_a$)* values of five studies on entomopathogenic nematode [EPN] distributions using different sampling approaches in various regions.

| EPN Studied Population | Form of the Measured EPN | $I_a$ Value | Comments (Location) | Reference |
|---|---|---|---|---|
| *Heterorhabditis bacteriophora*-infective juveniles (IJ) applied uniformly, in one, or nine patches on Kentucky bluegrass | EPN-infected *Galleria mellonella* larvae over time | All mean values were less than one but differed ($p \leq 0.05$) until 20 weeks, no more, after EPN application | The values suggest a more even distribution than a random one (NJ/USA) | [20] |
| Natural populations of *Steinernema feltiae* and *S. affine* in grassland plots | IJ assigned to one of 4 groups of increasing physiological age | The values ranged 1.27–1.45 | All values indicate aggregated distribution (Merelbeke/Belgium) | [21] |
| *H. bacteriophora* or *S. carpocapsae*-infected *G. mellonella* larvae applied within 24 h of initial IJ emergence to cultivated fields and adjoining grassy border plots | *H. bacteriophora* and *S. carpocapsae*-IJ recovered from *G. mellonella* larvae baits applied several times after the cadavers | Range <1 to >2. Mean values differed between EPN species in bait traps and between soil management regimes at 48 h and 16 days after placing the cadavers, respectively | Spatial distributions dispersed from a grassy border to the adjacent cultivated field plots were more aggregated for *H. bacteriophora* than for *S. carpocapsae* (OH/USA) | [22] |
| *Steinernema diaprepesi*, *Heterorhabditis indica*, and *Heterorhabditis zealandica* | EPN were measured using quantitative qPCR during a 6-month citrus orchard survey | The values ranged 0.8–1.3 over 6 months and could be compared with those of the fungus and Diaprepes root weevil | Highly significant spatial associations between *Fusarium solani* and EPN communities of up to three EPN species (FL, USA) | [23] |
| Natural populations of *H. indica* in citrus groves | EPN-infected *G. mellonella* larvae | 0.913 | $I_a$ refers to uniform distribution pattern (Giza, Egypt) | [24] |

$I_a$ = the observed value of distance to regularity/the mean randomized value [25]; qPCR: quantitative polymerase chain reaction.

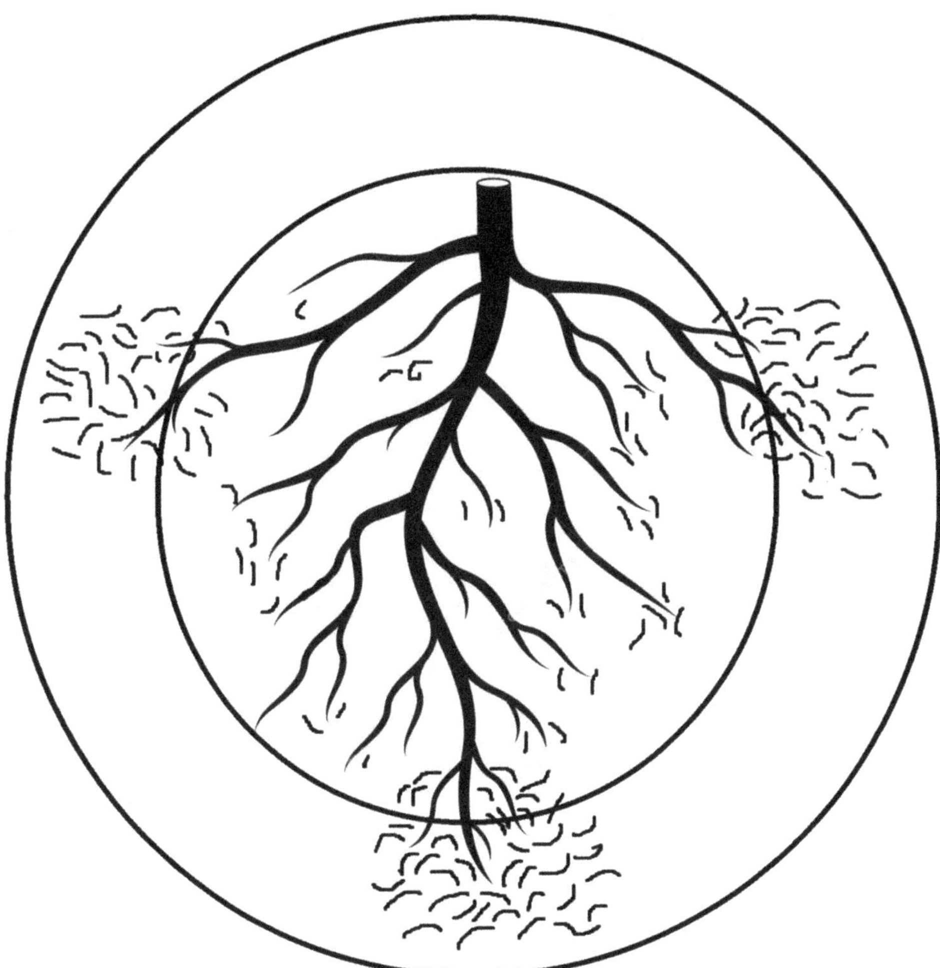

**Figure 1.** Two quadrat sizes are represented by concentric circles. The inner circle represents random nematode distribution around plant main root and the outer circle represents clumped nematode distribution around lateral fibrous roots as well.

### 5. Addressing Nematode Distribution Patterns

These patterns, revealed by sound sampling, can enable pest control operators to: (1) choose plant material that best fit to specific locations, (2) leverage variable rate methods for the used nematicides, and (3) characterize relationships between organisms in space and time for careful IPM. Stuart and Gaugler [25] stressed that nematode clumped distribution can have great ramifications at the community levels by changing the dynamics of parasitism, predation, and competition. Such spatial (horizontal or vertical) and temporal distributions may be compared with one or more of the relevant biotic and physical forces for better development of IPM. Moreover, definite models [26–28] (4) can serve in the nematode-count transformation to fulfill accurate treatment comparisons.

However, samples often become costly to offer these merits of distribution patterns. So, a trade-off between objectivity and cost is necessary. For convenience, recent trends offer different accuracy levels for the same sample size and tactic to meet affordability. Iteration was also used to further improve optimum sample size [29]. Increasing the cost by increasing the number of cores or samples, or both, should be weighed against the benefit it provides via accuracy and reliability [30]. Collecting more small cores, though costly, offers a more accurate mean estimate than an equivalent amount of soil collected as fewer large cores [5]. Moreover, collecting numbers of cores/sub-samples from a targeted area before mixing into a composite sample to homogenize variance of nematode counts may reduce costs though it introduces another potential source of statistical variability. A modification is to take subsample(s) from the composite sample to estimate the population density and/or the numbers of BCA/endospores associated with the nematodes per sample to reduce costs.

## 6. Indices of Nematode Dispersion

We should cautiously suggest type and number of these indices to fit the goal of the work. For instance, contrary to Taylor's Power law (TPL) [31], Spatial Analysis by Distance Indices (SADIE) has geographic coordinates. [32] reviewed geostatistical models as another group that can apply sample values and locations simultaneously to depict spatial patterns and estimate values at unsampled locations. Yet, this group does not offer tests to assess the statistical significance of the patterns but SADIE software can determine the statistical probability level of spatial association between organisms or the same organism at different times [33]. So, these indices can complement each other to show more aspects of the distribution patterns. Gorny et al. [34] manipulated two indices to set sound sampling protocols and determined specific sites for nematicide usage. Moreover, Wu et al. [23] used SADIE to prove regulation of EPNs by a natural enemy. Therefore, the use of such indices to enhance biocontrol potential or to save costs in nematicidal applications is ideally compatible with IPM.

Conceivably, nematode spatial patterns are more representative in samples taken far apart which will be more impacted by various microhabitats than samples taken close to each other. This concept could be backed by using both semivariogram and SADIE analyses together to better grasp PPN and/or EPN spatial patterns and spatiotemporal dynamics [34,35]. To facilitate its use, SADIE program in terms of its major indices and graphical displays were recently reviewed [36]. It was integrated with other methods to study soil food webs in citrus orchards in order to develop new biocontrol approach that can serve in IPM [32,37].

Complementary methods [38,39] can optimally detect spatial heterogeneity when clusters are situated on elongated or square domains and near to the edges of the surveyed sites. They can reveal clusters with small radius and in sample size smaller than that of SADIE as well as adjust for the absolute location or the magnitude of the counts.

## 7. Other Examples to Optimize EPN Sampling and Extraction

A main challenge facing the use of EPN is to broaden the EPN species/strain library in order to provide suitable matches of nematodes to target pests. This will certainly optimize their benefits as biocontrol agents. The wide variation of EPN sampling makes results from a definite case-study difficult to generalize. Nevertheless, it is quite evident that the percentage of samples positive for EPN in many typical surveys worldwide are relatively low; <35% [12,40]. There is a dire need to increase it to likely offer new strains and upgrade EPN-host matching. So, novel sampling concepts to get EPN with high recovery frequency value and differential pathogenicity should be further sought. One such recent concept relies on combining four factors. The factors are favorable sampling method, time, site targeting, and use of multiple extraction technique. This combination could recover EPN from the seven surveyed groves and from 61.7% soil samples [24]. On the contrary, only one EPN-positive out of 593 soil samples was detected also in Egypt [41]. However, they used random sampling and single baiting cycle. Moreover, the EPN isolates recovered via rational sampling showed so variable pathogenicity to the strawberry white grub, *Temnorhynchus baal* [42]. Using such criteria or other new concepts to optimize EPN sampling and recovery frequency value should be further tested and expanded.

Another example is the invasive mole crickets as major pests of pastures, turf, and vegetables in the Caribbean Basin where *Steinernema scapterisci* is the only EPN species utilized efficiently as classical biocontrol strategy. It is used against the mole crickets [43]. Classical biocontrol should be expanded via directing sampling and extraction techniques to isolate BCA from environments where the organisms will presumably have had to develop the desired trait [44].

Specifically, the extraction technique using multiple *Galleria*-baiting cycles proved more effective than a single cycle in several studies [25,28,45]. Moreover, stressing by crowding, abiotic/biotic factors in soil, or presence under other suboptimal conditions may prevent or delay the nematode activity for infection [46,47]. Optimizing conditions for infection may gradually revert the EPN activity to infect the baiting insect in a consecutive cycle. Repeated extraction via baiting cycles can usually provide optimal conditions and longer time, for such a revision. Hence, it allows for differential pathogenicity of EPN too. A common technique is to keep the soil samples at about 23 °C in suitable cups with 4 last instar *Galleria mellonella* larvae as baits per cup in each cycle. Soil is sometimes watered to remain almost at field capacity during the extraction cycles. Each cup is inspected twice weekly in the first 3–4 weeks but once thereafter. Each cycle ends by inspecting the cups to: *i)* isolate insect cadavers with symptoms of EPN infection. These cadavers are transferred to White traps [48] to fulfill Koch's postulates, and/or *ii)* discard the other dead insects. A new following cycle begins with replacing the infected cadavers/dead insects by new living *G. mellonella* larvae. Suspect cadavers that failed to produce EPN-infective juveniles (IJs) are considered negative. The first cycle may be repeated 5–10 times [25,28] depending on the magnitude of EPN-positive samples. Other modifications to improve the baiting method are possible. They may include screening for EPN by using the target insect pest species; e.g., citrus root weevil, *Diaprepes abbreviatus* [49] or pecan weevil, *Curculio caryae* [50], as baits to achieve adequate EPN-host matching. Moreover, two model insect species/baiting at different temperatures to increase and diversify the recovery of EPN were tried [50,51]. These trends to find ecologically adaptable and effective BCAs should not be limited to a specific region or pest. Biocontrol components can strengthen IPM programs by using indigenous, or to a less degree introduced, EPN against the target 'baiting' pest or via setting the best EPN-host matching.

## 8. Other Sampling and Extraction Methods

EPNs in soil may be detected directly under binocular microscope via dissecting or enzymatic hydrolysis of the EPN-infected-cadavers or indirectly by scoring the cadavers per sample. Other methods of extracting EPNs from soil or their host insects [16,52] and PPNs from soil or plant tissues [5–7,14] were reviewed. Pest control operators must consider their relative merits and demerits for perfection of IPM. For instance, sieving and centrifugation using a sucrose gradient may directly extract and quantify dead and live EPN-IJs and PPNs from soil samples. The method may recover a larger proportion of EPNs in soil than insect baiting. It is less biased due to differential pathogenicity among EPNs extracted via the baiting method. However, it is rarely used to recover EPNs as it is more labor-intensive and require taxonomic expertise for the recovered nematodes [16]. Baermann funnel method and its modifications can extract only live nematodes. Selecting a method may sometimes require further tests to find the most efficient extraction method of the existing fauna and flora related to IPM [15].

## 9. Quantifying Extraction Efficiency of EPNs with a Model Used for PPNs

Nematode extraction via sieving, mostly favored for PPNs. or insect baiting, often used for EPNs, is based on physical (aperture sizes of the sieves) or biological (susceptibility of the baiting insects) background, respectively. So, it is exciting to find out their extraction efficiency herein via modeling. To test efficiency of sieving processes, the PPN suspension is poured through a stack of like sieves, and the recovery on each sieve is assessed. So, the cumulative recovery is related to number of times sieved. [53] related the number of sieving to percentage recovery of PPNs in the formula: Percentage recovery = $100\,(1 - a^x)$ where *a* is proportion of total number of PPNs present of a given length which pass through the sieve, and *x* is number of times sieved. The equation is used herein for EPN extraction too where *a* is the proportional loss at each *Galleria*-baiting cycle, and *x* is the number of repetitions (baiting cycles). The raw data of two EPN surveys were applied to the formula where 6 [28] or 10 [24] *Galleria*-baiting cycles produced positive samples (Figure 2). Herein, the practical % recovery of EPNs vs. theoretical corresponding values were 79% vs. 99% and 74.3% vs. 98.3% for % recovery of EPNs from mango [28] and citrus [24] orchards, respectively (Figure 2). This formula may offer approximate quantification of separation efficiency during the extraction processes. It allows consideration of the benefit to be gained by devoting more time and resources into the used EPN separation techniques [14,54] for IPM.

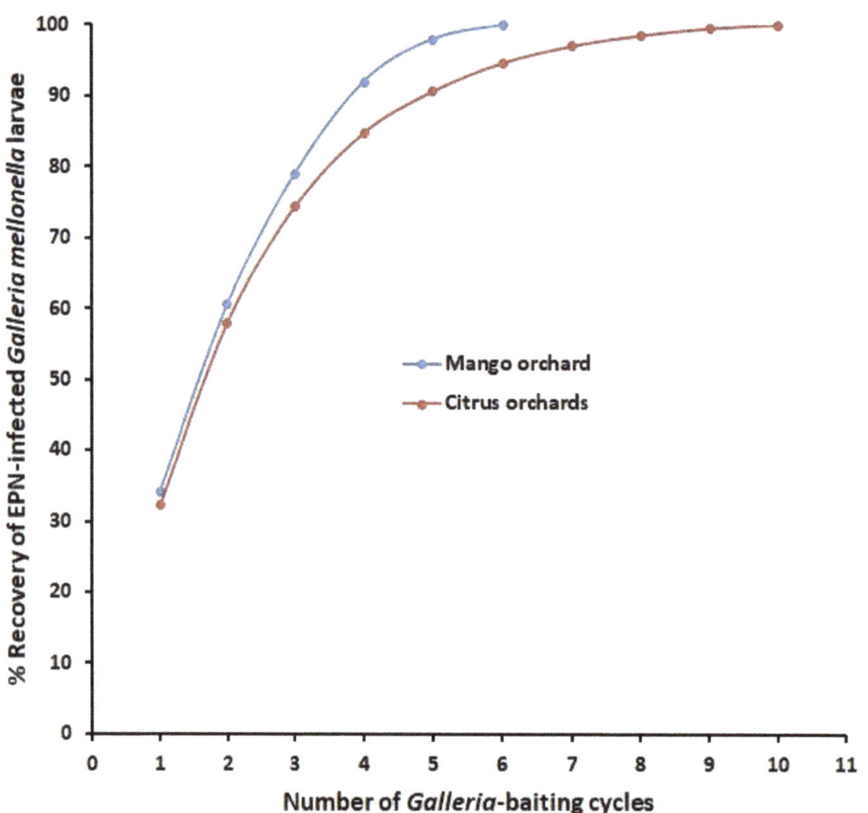


**Figure 2.** Calculated relationships between number of *Galleria*-baiting cycles and percentage recovery of entomopathogenic nematodes-infected insects for surveys of mango and citrus orchards.

## 10. Molecular vs. Traditional Sampling and Extraction Technology

Limitations of traditional sampling and extraction methods are apparent. Notwithstanding the utility of a series of extractions using the above-mentioned methods to significantly enhance the PPN- and EPN-separation efficiency, they do not provide a full recovery rate [6,45]. Moreover, not all EPN species can be isolated using just one insect bait species [55]. The most common *Galleria*-baiting method can hamper the laboratory maintenance of certain EPN species (e.g., *Steinernema kraussei*).

Sampling and extraction of biochemicals are relatively newer approaches. Relevant assays [6,16] may extract proteins or isozymes from the nematodes (e.g., for identification) or from their hosts (e.g., for measuring enzyme activity of a host species/cultivar related to its compatible or incompatible reaction to nematode infection). These accurate assays may designate PPN susceptible or resistant plant cultivars and assess the contribution of BCAs in priming the plant against PPNs [56]. Extraction of isozymes has enabled the study of species diversity, frequency, and abundance to study the nature conservancy and biodiversity. Moreover, new isozyme phenotypes may be detected particularly in conserved areas that may thrive our grasping of biogeography and ecology of key species such as RKNs [57]. Moreover, sampling methods to detect and measure volatiles in the soil atmosphere in situ can enable the study of chemical cues that are critical to communicate across various trophic levels of different organisms. Hence, they can assist in grasping the IPM scenario in the soil [32]. However, reliable results can often be obtained with nematodes at a specific developmental stage.

In contrast, DNA-based diagnostics do not rely on the express products of the genome and are independent of environmental influence or developmental stage [6]. Significant gains in sampling and extraction of nematodes and their related organisms are in progress due to introducing the polymerase chain reaction (PCR) and relevant techniques [6,19,58]. The relationship between the EPN numbers in soil samples extracted by conventional techniques and the numbers recovered via qPCR approaches could be established by [59] as a base to count EPN via the molecular technique. The novel set of primers and probes integrated with the qPCR systems could then optimize a protocol for extracting nematodes and DNA from soil samples. The protocol can detect even one EPN added to a nematode community [60]. This method could detect and quantify soil-inhabiting organisms (EPNs and their related nematophagous fungi, ectoparasitic bacteria, and competitor free-living nematodes) in Florida citrus groves and examine the EPN soil food web in various ecological settings [16]. Campos-Herrera et al. [61] used qPCR to reveal sympatric distributions of EPN species and detected their low numbers in samples where the insect baiting method failed.

These molecular tools were integrated with appropriate models, e.g., indices of dispersion, in order to: (1) clarify soil food webs that modulate the rates of a herbivore-disease complex [37], (2) prove regulation of EPNs by a natural enemy where manipulating a soil property (pH) can enhance biocontrol of an insect pest [4], and (3) examine geospatial relationships between native EPN and the fungus *Fusarium solani* in citrus habitats [23]. Such gains can enable us to better conceptualize biological control potential of pests and pathogens within sound IPM context.

New molecular methods are still in the pipeline or are of limited geographic scale. Using species-specific primer-probe combinations and the high throughput sequencing [62] to characterize nematode communities and their natural enemies in soil are often used in developed countries. These methods are generally costly and require a variety of reagents and equipment of medium-high technology levels that are rarely produced in developing countries. Their cost issue will exacerbate if the local currency has gone a drastic exchange rate. A current limitation is that qPCR will identify and quantify only those organisms for which the molecular toolkits are employed. It does not reveal the presence of those species not screened for, or species for which the qPCR was not developed. Therefore, in areas where EPN diversity is not well known, the insect-bait method is done to isolate new and/or unexpected species [16]. The insect-baiting can detect new species and provide their activity (ability to kill) data.

The primer/probe combination is designed to be specific for a single species, but discovery of closely related species in the sampled area might increase the likelihood of cross-amplification. So, optimizing the approach in a new system is recommended. It requires great skill in molecular biology. If not, contradictory results may be due to imperfectly carried out tests. Moreover, contamination of the used reagents may indicate false positives for some EPN species. In this case, re-sampling and repeating the tests will be required and increase the costs. Finally, qPCR and insect-baiting may or may not agree.

The qPCR method indicated high numbers of IJs, but no insects were infected when the same soil was baited [63]. So, more studies are needed to trust the merits and demerits of each technique (qPCR and insect-bait).

In parallel to EPN, adequate methods for DNA extraction from the PPNs and related BCA are going ahead. For instance, techniques using beacon probe qPCR to detect, quantify, and surveil PPN antagonists in samples are applied. Regaieg et al. [64] used this technique to evaluate capability of the fungus *Verticillium leptobactrum* to colonize RKN-egg masses. Its accurate quantification of the *V. leptobactrum* DNA over the egg masses can help in unraveling the complexity of the soil ecology that has many biological and physical factors. These methods can identify pathogens such as PPNs and discriminate resident microbial populations and cells or propagules which form the released BCA [65]. Isolation of BCAs and genomic DNA extraction from the organisms are described elsewhere [66]. The methods are ideally used collectively; combining morphology, biochemical, and molecular attributes of the organism. This strategy is necessary to strengthen diagnose, define species boundaries, and offer a comprehensive database for BCA and PPN species that can serve IPM programs [57]. Multiplex PCR can detect one or several species in a nematode mixture by a single PCR test, thus decreasing diagnostic time and costs. Cautious must be exercised in this technique for identifying several nematode targets in one assay. It is limited by the available primer pairs that can be used in a reaction and the number of bands that can be identified without giving false-positive results [6]. It requires precise optimization of the reaction conditions for the primer sets used simultaneously in the test.

In conclusion, advances in IPM programs related to nematology can be achieved via optimizing sampling and extraction methods. Solving their related issues via perseverance will lead to gain more experience and refine current methods. The price of related devices on which new technologies are based usually drops rapidly after a short marketing time. So, it is expected that decreasing costs for sequence analyses will allow its wider application for diagnostics and quantification of nematodes and related organisms. This optimism will serve IPM programs concerning nematodes in many ways such as unravelling the complexity of nematode interactions in soil and characterizing their food webs, taxonomy, and best EPN-host matching.

**Funding:** This research was funded by STDF, US-Egyptian project cycle 17 grant number 172 and The NRC in-house project No. 12050105 entitled Pesticide alternatives against soil-borne pathogens and pests attacking economically important solanaceous crops. funded by The National Research Centre.

**Acknowledgments:** This article is supported in part by the US-Egypt Project cycle 17 (no. 172) entitled "Preparing and evaluating IPM tactics for increasing strawberry and citrus production." The study was also supported in part by the NRC In-house project No. 12050105 entitled "Pesticide alternatives against soil-borne pathogens and pests attacking economically important solanaceous crops". The author thanks Larry Duncan and Zafar Handoo for their valuable comments on the manuscript.



## References

1. Sikora, D.C.; Hallmann, J.; Timper, P. (Eds.) *Plant Parasitic Nematodes in Subtropical and Tropical Agriculture*; CABI: Boston, MA, USA, 2018.
2. Abd-Elgawad, M.M.; Askary, T.H.; Coupland, J. *Biocontrol Agents: Entomopathogenic and Slug Parasitic Nematodes*; CAB International: Wallingford, UK, 2017.
3. Abd-Elgawad, M.M.; Askary, T.H. Impact of phytonematodes on agriculture economy. In *Biocontrol Agents of Phytonematodes*; Askary, T.H., Martinelli, P.R.P., Eds.; CABI: Wallingford, UK, 2015; pp. 3–49.
4. Campos-Herrera, R.; Stuart, R.J.; Pathak, E.; El-Borai, F.E.; Duncan, L.W. Temporal patterns of entomopathogenic nematodes in Florida citrus orchards: Evidence of natural regulation by microorganisms and nematode competitors. *Soil Biol. Biochem.* **2019**, *128*, 193–204. [CrossRef]
5. Been, T.H.; Schomaker, C.H. Distribution patterns and sampling. In *Plant Nematology*, 2nd ed.; Perry, R.N., Moens, M., Eds.; CABI: Wallingford, UK, 2013; pp. 331–358.

6.  Hallmann, J.; Subbotin, S.A. Methods for extraction, processing and detection of plant and soil nematodes. In *Plant Parasitic Nematodes in Subtropical and Tropical Agriculture*; Sikora, R.A., Coyne, D., Hallmann, J., Timper, P., Eds.; CABI: Boston, MA, USA, 2018; pp. 87–119.

7.  Duncan, L.W.; Phillips, M.S. Sampling root-knot nematodes. In *Root-Knot Nematodes*; Perry, R.N., Moens, M., Starr, J.L., Eds.; CABI: St. Albans, UK, 2009; pp. 275–300.

8.  Abd-Elgawad, M.M. Accuracy and Precision of Phytonematode Sampling Plans. *Agric. Eng. Int. CIGR J.* **2017**, *19*, 6–15.

9.  Timper, P. Conserving and Enhancing Biological Control of Nematodes. *J. Nematol.* **2014**, *46*, 75–89.

10. Duncan, L.W.; Stuart, R.J.; El-Borai, F.E.; Campos-Herrera, R.; Pathak, E.; Giurcanu, M.; Graham, J.H. Modifying orchard planting sites conserves entomopathogenic nematodes, reduces weevil herbivory and increases citrus tree growth, survival and fruit yield. *Biol. Control* **2013**, *64*, 26–36. [CrossRef]

11. Nielsen, A.L.; Spence, K.O.; Nakatani, J.; Lewis, E.E. Effect of soil salinity on entomopathogenic nematode survival and behaviour. *Nematology* **2011**, *3*, 859–867. [CrossRef]

12. Hussaini, S.S. Entomopathogenic nematodes: Ecology, diversity and geographical distribution. In *Biocontrol Agents: Entomopathogenic and Slug Parasitic Nematodes*; Abd-Elgawad, M.M.M., Askary, T.H., Coupland, J., Eds.; CABI: Wallingford, UK, 2017; pp. 88–142.

13. Campos-Herrera, R.; Pathak, E.; El-Borai, F.E.; Schumann, A.; Abd-Elgawad, M.M.; Duncan, L.W. New citriculture system suppresses native and augmented entomopathogenic nematodes. *Biol. Control* **2013**, *66*, 183–194. [CrossRef]

14. Van Bezooijen, J. *Methods and Techniques for Nematology (A Manual)*; Wageningen University: Wageningen, The Netherlands, 2006.

15. Dritsoulas, A.; Duncan, L.W. Optimizing for taxonomic coverage: A comparison of methods to recover mesofauna from soil. *J. Nematol.* **2020**, *52*, 1–9. [CrossRef]

16. Campos-Herrera, R.; Lacey, L.A. Methods for studying the ecology of invertebrate diseases and pathogens. In *Ecology of Invertebrate Diseases*; Hajek, A.E., Shapiro-Ilan, D.I., Eds.; John Wiley and Sons: Hoboken, NJ, USA, 2018; pp. 19–48.

17. Berg, W.V.D.; Hartsema, O.; Nijs, L.J.D. Statistical analysis of nematode counts from interlaboratory proficiency tests. *Nematology* **2014**, *16*, 229–243. [CrossRef]

18. Abd-Elgawad, M.M. Towards optimization of entomopathogenic nematodes for more service in the biological control of insect pests. *Egypt. J. Biol. Pest Control* **2019**, *29*, 77. [CrossRef]

19. Elliott, J.M. *Some Methods for the Statistical Analysis of Samples of Benthic Invertebrates*; Freshwater Biological Association, Science Publication No. 15; The Ferry House: Ambleside, UK, 1971.

20. Wilson, M.J.; Lewis, E.E.; Yoder, F.; Gaugler, R. Application pattern and persistence of the entomopathogenic nematode Heterorhabditis bacteriophora. *Biol. Control* **2003**, *26*, 180–188. [CrossRef]

21. Spiridonov, S.E.; Moens, M.; Wilson, M.J. Fine scale spatial distributions of two entomopathogenic nematodes in a grassland soil. *Appl. Soil Ecol.* **2007**, *37*, 192–201. [CrossRef]

22. Bal, H.K.; Acosta, N.; Cheng, Z.; Grewal, P.S.; Hoy, C.W. Effect of habitat and soil management on dispersal and distribution patterns of entomopathogenic nematodes. *Appl. Soil Ecol.* **2017**, *121*, 48–59. [CrossRef]

23. Wu, S.-Y.; El-Borai, F.E.; Graham, J.H.; Duncan, L.W. Geospatial relationships between native entomopathogenic nematodes and Fusarium solani in a Florida citrus orchard. *Appl. Soil Ecol.* **2019**, *140*, 108–114. [CrossRef]

24. Abd-Elgawad, M.M. Can rational sampling maximise isolation and fix distribution measure of entomopathogenic nematodes? *Nematology* **2020**, *22*, 907–916. [CrossRef]

25. Stuart, R.; Gaugler, R. Patchiness in Populations of Entomopathogenic Nematodes. *J. Invertebr. Pathol.* **1994**, *64*, 39–45. [CrossRef]

26. Duncan, L.W.; Ferguson, J.J.; Dunn, R.A.; Noling, J.W. Application of Taylor's Power Law to Sample Statistics of *Tylenchulus semipenetrans* in Florida Citrus. *J. Nematol.* **1989**, *21*, 707–711. [PubMed]

27. Abd-Elgawad, M. Spatial distribution of the phytonematode community in Egyptian citrus groves. *Fundam. Appl. Nematol.* **1992**, *14*, 367–373.

28. Abd-Elgawad, M.M. Spatial patterns of *Tuta absoluta* and heterorhabditid nematodes. *Russ. J. Nematol.* **2014**, *22*, 89–100.

29. Abd-Elgawad, M.M. Managing nematodes in Egyptian citrus orchards. *Bull. Natl. Res. Centre* **2020**, *44*, 41. [CrossRef]

30. Ferris, H.; Goodell, P.B.; McKenry, M.V. Sampling for nematodes. *Calif. Agric.* **1981**, *35*, 13–15.

31. Taylor, L.R. Aggregation, Variance and the Mean. *Nat. Cell Biol.* **1961**, *189*, 732–735. [CrossRef]

32. Campos-Herrera, R.; Ali, J.G.; Diaz, B.M.; Duncan, L.W. Analyzing spatial patterns linked to the ecology of herbivores and their natural enemies in the soil. *Front. Plant Sci.* **2013**, *4*, 378. [CrossRef]

33. Perry, J.N. Spatial Analysis by Distance Indices. *J. Anim. Ecol.* **1995**, *64*, 303. [CrossRef]

34. Gorny, A.M.; Hay, F.S.; Esker, P.; Pethybridge, S.J. Spatial and spatiotemporal analysis of Meloidogyne hapla and Pratylenchus penetrans populations in commercial potato fields in New York, USA. *Nematology* **2020**, *23*, 139–151. [CrossRef]

35. Stuart, R.J.; Barbercheck, M.E.; Grewal, P.S. Entomopathogenic nematodes in the soil environment: Distributions, interactions and the influence of biotic and abiotic factors. In *Nematode Pathogenesis of Insects and Other Pests*; Campos-Herrera, R., Ed.; Springer: Cham, Switzerland, 2015; pp. 97–138.

36. Winder, L.; Alexander, C.; Griffiths, G.; Holland, J.; Wooley, C.; Perry, J. Twenty years and counting with SADIE: Spatial Analysis by Distance Indices software and review of its adoption and use. *Rethink. Ecol.* **2019**, *4*, 1–16. [CrossRef]

37. Campos-Herrera, R.; El-Borai, F.E.; Duncan, L.W. Wide interguild relationships among entomopathogenic and free-living nematodes in soil as measured by real time qPCR. *J. Invertebr. Pathol.* **2012**, *111*, 126–135. [CrossRef] [PubMed]

38. Lavigne, C.; Ricci, B.; Franck, P.; Senoussi, R. Spatial analyses of ecological count data: A density map comparison approach. *Basic Appl. Ecol.* **2010**, *11*, 734–742. [CrossRef]

39. Li, B.; Madden, L.V.; Xu, X. Spatial analysis by distance indices: An alternative local clustering index for studying spatial patterns. *Methods Ecol. Evol.* **2012**, *3*, 368–377. [CrossRef]

40. Stock, S.P. Diversity, biology and evolutionary relationships. In *Nematode Pathogenesis of Insects and Other Pests*; Campos-Herrera, R., Ed.; Springer: Cham, Switzerland, 2015; pp. 3–27.

41. Abdel-Razek, S.A.; Hussein, M.; Shehata, I. Isolation and identification of indigenous entomopathogenic nematode (EPN) isolate from Egyptian fauna. *Arch. Phytopathol. Plant Prot.* **2018**, *51*, 197–206. [CrossRef]

42. Shehata, I.E.; Hammam, M.M.A.; El-Borai, F.E.; Duncan, L.W.; Abd-Elgawad, M.M.M. Comparison of virulence, reproductive potential, and persistence among local Heterorhabditis indica populations for the control of Temnorhynchus baal (Reiche & Saulcy) (Coleoptera: Scarabaeidae) in Egypt. *Egypt. J. Biol. Pest Control.* **2019**, *29*, 32. [CrossRef]

43. Nguyen, K.B.; Smart, G.C., Jr. *Steinernema scapterisci* n. sp. (Steinernematidae: Nematoda). *J. Nematol.* **1990**, *22*, 187–199. [PubMed]

44. Gaugler, R.; Han, R. Production technology. In *Entomopathogenic Nematology*; Gaugler, R., Ed.; CABI: Wallingford, UK, 2002; pp. 289–310.

45. Fan, X.; Hominick, W.M. Efficiency of the *Galleria* (wax moth) baiting technique for recovering infective stages of entomopathogenic rhabditids (Steinernematidae and Heterorhabditidae) from sand and soil. *Rev. Nematol.* **1991**, *14*, 381–387.

46. Griffin, C.T. Behaviour and population dynamics of entomopathogenic nematodes following application. In *Nematode Pathogenesis of Insects and Other Pests*; Campos-Herrera, R., Ed.; Springer: Cham, Switzerland, 2015; pp. 139–163.

47. Lewis, E.E.; Hazir, S.; Hodson, A.; Gulcu, B. Trophic relationships of entomopathogenic nematodes in agricultural habitats. In *Nematode Pathogenesis of Insects and Other Pests*; Campos-Herrera, R., Ed.; Springer: Cham, Switzerland, 2015; pp. 139–163.

48. White, G.F. A method for obtaining infective nematode larvae from cultures. *Science* **1927**, *66*, 302–303. [CrossRef]

49. Duncan, L.W.; Graham, J.H.; Dunn, D.C.; Zellers, J.; McCoy, C.W.; Nguyen, K. Incidence of endemic entomopathogenic nematodes following application of *Steinernema riobrave* for control of *Diaprepes abbreviatus*. *J. Nematol.* **2003**, *35*, 178–186.

50. Shapiro-Ilan, D.I.; Gardner, W.A.; Fuxa, J.R.; Wood, B.W.; Nguyen, K.B.; Adams, B.J.; Humber, R.A.; Hall, M.J. Survey of Entomopathogenic Nematodes and Fungi Endemic to Pecan Orchards of the Southeastern United States and Their Virulence to the Pecan Weevil (Coleoptera: Curculionidae). *Environ. Entomol.* **2003**, *32*, 187–195. [CrossRef]

51. Campos-Herrera, R.; Gutierrez, C. Comparative study of entomopathogenic nematode isolation usingGalleria mellonella(Pyralidae) andSpodoptera littoralis(Noctuidae) as baits. *Biocontrol Sci. Technol.* **2008**, *18*, 621–626. [CrossRef]

52. Woodring, J.L.; Kaya, H.K. *Steinernematid and Heterorhabditid Nematodes: A Handbook of Techniques*; Southern Cooperative Series Bulletin No. 331; Arkansas Agricultural Experiment Station: Fayetteville, AR, USA, 1988.

53. Seinhorst, J.W. Extraction methods for nematodes inhabiting soil. In *Progress in Soil Zoology*; Murphy, P.W., Ed.; Butterworths: London, UK, 1962; pp. 243–256.

54. Ferris, H. Extraction efficiencies and population estimation. In *Vistas on Nematology*; Veech, J.A., Dickson, D.W., Eds.; Society of Nematologists: Hyattsville, MD, USA, 1987; pp. 59–63.

55. Adams, J.B.; Nguyen, K.B. Taxonomy and systematics. In *Entomopathogenic Nematology*; Gaugler, R., Ed.; CABI: Wallingford, UK, 2002; pp. 1–33.

56. Molinari, S.; Leonetti, P. Bio-control agents activate plant immune response and prime susceptible tomato against root-knot nematodes. *PLoS ONE* **2019**, *14*, e0213230. [CrossRef]

57. Carneiro, R.M.D.G.; Silva, F.S.O.; Correia, V.R. Methods and tools currently used for the identification of plant parasitic nematodes. In *Nematology-Concepts, Diagnosis and Control*; Shah, M.M., Mahamood, M., Eds.; InTech Open: Rijeka, Croatia, 2017; pp. 19–35. [CrossRef]

58. Dritsoulas, A.; Campos-Herrera, R.; Blanco-Pérez, R.; Duncan, L.W. Comparing high throughput sequencing and real time qPCR for characterizing entomopathogenic nematode biogeography. *Soil Biol. Biochem.* **2020**, *145*, 107793. [CrossRef]

59. Torr, P.; Spiridonov, S.E.; Heritage, S.; Wilson, M.J. Habitat associations of two entomopathogenic nematodes: A quantitative study using real-time quantitative polymerase chain reactions. *J. Anim. Ecol.* **2007**, *76*, 238–245. [CrossRef]

60. Campos-Herrera, R.; Johnson, E.; El-Borai, F.; Stuart, R.; Graham, J.; Duncan, L. Long-term stability of entomopathogenic nematode spatial patterns in soil as measured by sentinel insects and real-time PCR assays. *Ann. Appl. Biol.* **2011**, *158*, 55–68. [CrossRef]

61. Campos-Herrera, R.; Martín, J.A.; Escuer, M.; García-González, M.T.; Duncan, L.W.; Gutiérrez, C. Entomopathogenic nematode food webs in an ancient, mining pollution gradient in Spain. *Sci. Total Environ.* **2016**, *572*, 312–323. [CrossRef] [PubMed]

62. Porazinska, D.L.; Giblin-Davis, R.M.; Faller, L.; Farmerie, W.; Kanzaki, N.; Morris, K.; Powers, T.O.; Tucker, A.E.; Sung, W.; Thomas, W.K. Evaluating high-throughput sequencing as a method for metagenomic analysis of nematode diversity. *Mol. Ecol. Resour.* **2009**, *9*, 1439–1450. [CrossRef]

63. Jaffuel, G.; Mader, P.; Blanco-Perez, R.; Chiriboga, X.; Fliessbach, A.; Turlings, T.C.J.; Campos-Herrera, R. Prevalence and activity of entomopathogenic nematodes and their antagonists in soils that are subject to different agricultural practices. *Agric. Ecosyst. Environ.* **2016**, *230*, 329–340. [CrossRef]

64. Regaieg, H.; Ciancio, A.; Raouani, N.H.; Rosso, L. Detection and biocontrol potential of Verticillium leptobactrum parasitizing Meloidogyne spp. *World J. Microbiol. Biotechnol.* **2010**, *27*, 1615–1623. [CrossRef]
65. Ciancio, A.; Loffredo, A.; Paradies, F.; Turturo, C.; Sialer, M.F. Detection of Meloidogyne incognita and Pochonia chlamydosporia by fluorogenic molecular probes. *EPPO Bull.* **2005**, *35*, 157–164. [CrossRef]
66. Stirling, G.R. *Biological Control of Plant-Parasitic Nematodes*, 2nd ed.; CABI: Wallingford, UK, 2014. [CrossRef]

*Review*

# Taxonomy, Morphological and Molecular Identification of the Potato Cyst Nematodes, *Globodera pallida* and *G. rostochiensis*

John Wainer * and Quang Dinh

Agriculture Victoria Research, Department of Jobs, Precincts and Regions, Bundoora, VIC 3086, Australia; quang.dinh@agriculture.vic.gov.au
* Correspondence: john.wainer@agriculture.vic.gov.au

**Abstract:** The scope of this paper is limited to the taxonomy, detection, and reliable morphological and molecular identification of the potato cyst nematodes (PCN) *Globodera pallida* and *G. rostochiensis*. It describes the nomenclature, hosts, life cycle, pathotypes, and symptoms of the two species. It also provides detailed instructions for soil sampling and extraction of cysts from soil. The primary focus of the paper is the presentation of accurate and effective methods to identify the two principal PCN species.

**Keywords:** PCN; potato cyst nematode; *Globodera*; taxonomy; detection; morphology; molecular identification; PCR

**Citation:** Wainer, J.; Dinh, Q. Taxonomy, Morphological and Molecular Identification of the Potato Cyst Nematodes, *Globodera pallida* and *G. rostochiensis*. *Plants* **2021**, *10*, 184. https://doi.org/10.3390/plants10010184

Received: 25 December 2020
Accepted: 15 January 2021
Published: 19 January 2021

**Publisher's Note:** MDPI stays neutral with regard to jurisdictional claims in published maps and institutional affiliations.



## 1. Introduction

Potato cyst nematodes (PCN) are damaging soilborne quarantine pests of potato and other solanaceous crops worldwide [1,2]. The two most damaging species, *G. pallida* (Stone, 1973) Behrens, 1975, the pale or white cyst nematode, and *G. rostochiensis* (Wollenweber, 1923) Behrens, 1975, the golden cyst nematode, have proved to be highly adaptive at exploiting new environments, being passively transported, undetected across borders, in intimate association with tubers of their major host, the potato. *Globodera* species feeding on potato also include *G. ellingtonae*, restricted to Chile, Argentina, and two states in northwest USA [3–5] and *G. leptonepia* (Cobb and Taylor, 1953) Skarbilovich, 1959 found only once in a ship-borne consignment of potatoes [6,7].

Potato cyst nematodes are obligate sedentary endoparasites that can cause stunting of plants, reduce yields, and sometimes lead to complete crop failure. PCN causes losses of 9% of total potato production in Europe and can cause total losses in other parts of the world when no control strategies are employed [8]. When PCN populations are high in the field, potato yields can be less than the tonnage per unit area of the planted seed [9,10]. PCN presents formidable problems to farmers, advisors, and policy makers due to their small size and cryptic nature within large volumes of soil, their extreme specialization and intimate association with their host, and their adaptation for long-term survival in the soil in the absence of a suitable host. In fact, PCN is recognized throughout the temperate regions of the world as one of the most difficult crop pests to control [11].

As internationally recognized plant-quarantine organisms, efficient sampling and detection methods of PCN are critical to the effective management of these pests in both emergency response and on-going control situations [12–15]. Cysts are the dead remnants of female nematodes and contain hundreds of eggs; they can survive in soil without a host for 20 years or more [16]. *Globodera rostochiensis* and *G. pallida* are closely related species and difficult to be distinguished from each other solely based on morphology. The European and Mediterranean Plant Protection Organization has published a diagnostic protocol for the two species [17].

## 2. Nomenclature

Phylum: Nematoda Diesing, 1861, Order: Rhabditida Chitwood, 1933, Suborder: Tylenchina Chitwood, 1950, Family: Heteroderidae Filip'ev & Schuurmans Stekhoven, 1941, Genus: *Globodera* (Skarbilovich, 1959) Behrens, 1975.

*Globodera pallida* synonyms:
*Heterodera rostochiensis* Wollenweber, 1923 in partim
*Heterodera pallida* Stone, 1973
*Heterodera (Globodera) pallida* Stone, 1973
*Globodera pallida* (Stone, 1973) Mulvey and Stone, 1976.
*Globodera pallida* common names:
English: PCN, white potato cyst nematode, pale potato cyst nematode; French: nématode blanc de la pomme de terre; Spanish: nemátodo quiste blanco de la papa.

*Globodera rostochiensis* synonyms:
*Heterodera rostochiensis* Wollenweber, 1923
*Heterodera schachtii rostochiensis* Wollenweber, 1923
*Heterodera schachtii solani* Zimmermann, 1927
*Heterodera solani* Zimmermann, 1927
*Heterodera (Globodera) rostochiensis* (Wollenweber, 1923) Skarbilovich, 1959
*Heterodera pseudorostochiensis* Kirjanova, 1963
*Globodera pseudorostochiensis* (Kirjanova, 1963) Mulvey and Stone, 1976
*Globodera rostochiensis* (Wollenweber, 1923) Mulvey and Stone, 1976
*Globodera arenaria* Chizhov, Udalova and Nasonova, 2008.
*Globodera rostochiensis* common names:
English: PCN, yellow potato cyst nematode, golden potato cyst nematode, golden nematode, potato root eelworm; French: anguillule a kyste de la pomme de terre, anguillule des racines de la pomme de terre, nématode doré, nématode doré de la pomme de terre; German: kartoffelnematode; Spanish: nemátodo dorado.

## 3. Hosts

PCN hosts are restricted to the nightshade family Solanaceae. The most important host is *Solanum tuberosum* (potato) although other agronomic crops such as *Solanum lycopersicum* (tomato) and *Solanum melongena* (eggplant) are also attacked [18]. Up to 90 *Solanum* spp. And their hybrids can be PCN hosts including some weed species. These include *Datura* spp. (devil's trumpets), *Hyoscyamus niger* (henbane), *Nicotiana acuminata* (manyflower tobacco), *Physalis* spp. (husk tomatoes), *Physochlaina orientalis* (oriental physochlaina), *Salpiglossis* sp. (painted tongue), *Capsicum annuum* (chili pepper), and *Jaltomata procumbens* (creeping false holly) [19,20]. For a more complete list of PCN hosts, see [21–24].

Vermiform juveniles and adult PCN cysts can be found either in the soil or attached to roots or tubers, whereas adult males are found exclusively in the soil.

## 4. Life Cycle

The PCN cyst is the hardened dead body of a female and protects the eggs within. It is spheroid with a short neck. The female *G. rostochiensis* changes during maturation from white to yellow and then into brown cysts, whereas *G. pallida* changes from creamy white directly to brown. Cysts are highly resistant and long-lived and can be readily spread, mostly in association with soil, by human activities [25].

After infective juvenile nematodes hatch, they can disperse in the soil a distance of about 1 m and infect plants by entering a root near the growing tip. The nematode becomes sedentary, establishing a feeding site by modifying plant cells which then provide nutrients. Infested potato plants have a reduced root system and poor productivity [26]. Plant death can occur [27].

The lifecycle of PCN (Figure 1) can be described as follows:

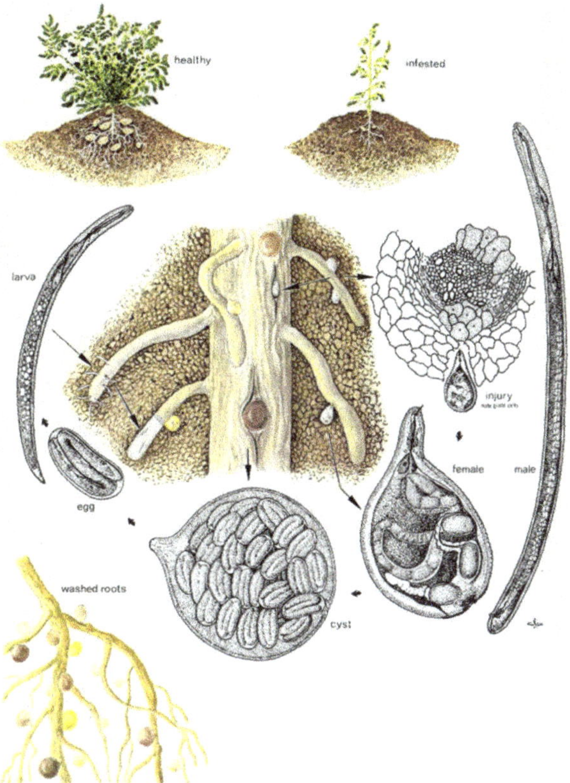


**Figure 1.** Illustration of the life cycle of *Globodera rostochiensis* (modified after Charles S Papp, Exclusion and Detection, Plant Pest Detection Manual 5:1, California Department of Food and Agriculture, Division of Plant Industry, USA).

A period of 38–48 days (depending on soil temperature) is required for PCN to complete its life cycle [28]. Nematodes reproduce sexually; males are attracted to females by a pheromone sex attractant. Nematodes may mate several times. After mating, each female produces approximately 200–500 eggs [29], dies, and the cuticle of the dead female forms a cyst. Eggs mostly remain dormant within the cyst until receiving a hatching stimulus (i.e., specific chemical released by host plant roots). PCN eggs can remain dormant and viable within the cyst for at least 30 years [16] and are resistant to nematicides [11].

When soil temperatures are warm enough (above 10 °C) [30], and hatching stimuli are received [31], second-stage juveniles hatch from the eggs, escape from the cyst, and migrate towards the host plant roots. Egg hatching is stimulated by host root diffusate, but not all eggs hatch (60–80%); by comparison only about 5% will hatch in water. Some eggs do not hatch until subsequent years [2].

Juveniles penetrate roots where they begin to feed. Host plant cells within the root cortex are stimulated to form specialized cells (syncytia) which transfer nutrients to the nematodes. After feeding commences, the juvenile grows and undergoes three more moults to become an adult. Females grow and become round, breaking through the roots and exposing the posterior portion of their body to the external environment.

Male juveniles remain active, feeding on the host plant until maturity, at which time they stop feeding, become vermiform, and seek females [32]. Adult males do not feed.

Sex is determined by food supply—more juveniles develop into males under adverse conditions and heavy infestations.

## 5. Pathotypes

Pathotypes (or virulence groups) of PCN are characterized by ability to multiply on certain clones and hybrids of *Solanum* spp. Both species of PCN have several pathotypes under several different schemes [1,33–36]. Under the European scheme [35], there are five pathotypes (Ro1–Ro5) for *G. rostochiensis* and three (Pa1–Pa3) for *G. pallida*. A wide range of commercial potato cultivars currently available carry the H1 gene that confers near complete resistance to the Ro1 and Ro4 pathotypes [37] and other genes (e.g., Gro1) confer resistance to all *G. rostochiensis* pathotypes. Although various genes confer a degree of resistance to *G. pallida*, complete resistance is not known, which means that some multiplication of the nematode is possible for most commercial cultivars [38]. The term "pathotype" is now considered too general, as many PCN populations cannot be assigned conclusively to pathotypes [17]. Any population showing signs of a new virulence should be tested as soon as possible.

## 6. Symptoms

Symptoms of PCN infestation are not specific and may not be apparent even when crop yield is significantly reduced. At high densities, patches of poor growth can occur in potato crops, sometimes with yellowing, wilting, or necrosis of the foliage. These symptoms may be caused by many other plant pathogens, including other nematodes, and should not be considered proof of PCN presence. If there are clear patches of stunting, plants should be lifted for a visual check for cysts on the roots. This is only possible for a short time at the appropriate stage of the crop; as young females mature into cysts they are easily detached when lifting plants.

When infested plants are lifted carefully, the swollen females or the cysts appear as small bead-like objects attached to the roots and can be easily seen with the naked eye. With severe infestations, cysts may be seen on the surface of tubers or stolons.

A cyst that changes during maturation from white to yellow and then into brown is *G. rostochiensis* while one which changes from creamy white directly to brown is *G. pallida*. Note that this feature can only be used at the appropriate stage of the life cycle: young cysts of both species are white or cream, and mature cysts of both species are brown.

## 7. Soil Sampling

Visual symptoms alone cannot be used to identify the presence of PCN in a potato crop. There are two methods available to sample fields for PCN: (1) taking soil samples and processing them in the laboratory; or (2) lifting plants and examining their roots for females or cysts in either the field or laboratory. The latter method has been used to detect low populations, which may have been undetectable by soil sampling [39]. However, plant sampling is extremely labor intensive, and plants are available only during a part of the year or cropping cycle, whereas viable cysts can remain in the soil for many years. Soil samples must be large enough to achieve the required accuracy and sensitivity and must also be derived from many points in the field to ensure that they are representative of the area. Been and Schomaker [40] emphasized the importance of sample point spacing to the probability of detection of PCN cysts in a field. To achieve a 90% average probability of detection, grid sampling at 5 m spacing with 52 g cores (total sample size 6.9 kg/0.33 ha) was recommended as being the best compromise for minimizing sample size and maximizing detection probability, while minimizing time needed to collect and process the samples. This recommendation is based on detecting the minimum abundance of cysts which will cause crop losses, rather than the presence or absence of PCN. The level of sampling depends on the aim: delimiting surveys for biosecurity reasons require a relatively high level of accuracy (i.e., a high probability of detection) whereas routine sampling, for example of seed potato crops, is generally done at a lower level of accuracy and probability of detection.

## 8. Cyst Extraction

*Globodera* spp. Juveniles and adult males can be extracted from soil by general nematode extraction methods such as Whitehead Trays or the more efficient differential flotation [41,42]. An additional cyst extraction on the soil is desirable because a combination of cyst and juvenile or male characteristics is better for identification.

To extract cysts from soil, the commonly used methods are flotation and elutriation. Flotation works on the principle that dried cysts will float. Standard methods include the Fenwick can and Schuiling centrifuge. Elutriation is based on cysts having lower density than soil particles and so can be used for wet soil.

The Fenwick can, as modified by Oostenbrink [19], is the most commonly used instrument for the extraction of cysts from soil samples using the principles of flotation [43,44]. Nematode cysts are relatively light in relation to the inorganic fraction of soil, have a waxy covering, and contain a pocket of air within, so it is possible to separate cysts in the lighter organic fraction of the soil for identification and assessment.

The can tapers toward the top, with a sloping collar around the outside of the rim which collects overflow and directs it towards an outlet. The can has a sloping internal base with a drain plug at its lowest point. Soil is placed at the bottom of the can. Water is then turned on and enters near the bottom of the can. As the can fills, lighter soil particles and cysts flow over the spout and onto sieves from which cysts are "backwashed" after at least 15 min and when the overflow water has become clear.

Soil samples should first be air dried at 37 °C for 48 h to ensure consistency of sample weight and to aid floatation of cysts, which improves efficiency of recovery. If relatively free of organic matter, put the sample of soil directly into the Fenwick can or into a funnel on top of the can. The recommended soil sample size for a smaller or standard-sized Fenwick can (height 30 cm, volume 2 L) is 300 g [45,46]. However, Bellvert [47] found that cyst extraction efficiency was stable in their Fenwick can using soil samples from 100 g up to the physical limits of the can (600 g). Collins et al. [48,49] achieved greater average cyst extraction efficiency using large-scale Fenwick cans (height 50 cm, soil sample size 2 kg) than with medium-sized Fenwick cans (87.5% and 76%, respectively) and concluded that a large Fenwick can is an effective tool for extraction of cysts from large soil samples. Fenwick [43] found very efficient cyst extraction from a can 60 cm high with a capacity of 19 L by using a soil sample size of 4.5 kg.

To achieve improved cyst recovery efficiency, very organic soils should be washed through an 850 μm sieve into the can to allow coarse organic material to be excluded. Fill the can with tap water from the inlet at the bottom, washing through the soil as the can fills. The organic matter with the cysts will rise and overflow onto the collar. Place two sieves with apertures of 850 and 250 μm under the collar outlet. The cysts are collected on the 250 μm sieve for further processing, as they are on average about 450 μm in diameter [46].

## 9. Taxonomic Descriptions

(After Golden and Ellington [50], Stone [29,51], Subbotin et al. [7])
*Globodera pallida*
Female. Body subspherical with projecting neck bearing head, pharynx corpus, isthmus, and anterior part of pharyngeal glands. White in color, some populations passing, after 4–6 weeks, through a cream stage, turning glossy brown when dead. Labial region with amalgamated lips and one or two prominent annuli, deep irregular annulations present on neck, changing to reticulate pattern of ridges over most of body surface. Head framework weakly developed, hexaradiate. Stylet knobs sloping backward. Very large median pharyngeal bulb, almost circular with large crescentic valve plates. Pharyngeal gland lobe broad, frequently displaced anteriad, three gland nuclei. Prominent excretory pore situated at base of neck. Internal structures in neck region often obscured by hyaline

secretions on cuticle surface. Vulva a transverse slit at posterior end, set in a slight circular depression or vulval basin. Cuticle surface between anus and vulval basin including about 12 parallel ridges with a few cross connections. Subsurface punctations irregularly arranged over much of body surface, may be confused with surface papillae on vulval crescents.

Cyst. White when first visible on root surface, changing to glossy brown with maturity, subspherical with protruding neck. Vulval region intact or fenestrated with single circumfenestrate opening occupying all or part of vulval basin. Abullate, but small darkened or thickened "vulval bodies" sometimes present in vulval region. Anus visible in most specimens, often at apex of a V-shape mark. Cuticular pattern as in female but more accentuated. Subcrystalline layer absent.

Male. Heat-relaxed specimens C- or S-shaped, posterior part twisted 90–180° about longitudinal axis. Cuticle with regular annulations and four incisures in lateral field, terminating on tail. Labial region offset, rounded with large oral disc, six irregular lips, six or seven annuli, and heavily sclerotized hexaradiate framework. Stylet well developed with posteriorly sloping basal knobs and cone forming ca 45% of total stylet. Ellipsoid pharyngeal median bulb with strong crescentic valve plates linked by a narrow isthmus encircled by a broad nerve ring, to a narrow, ventrally situated, pharyngeal gland lobe. Hemizonid two annuli long, situated two or three annuli posterior to excretory pore. One testis, commencing with single cap cell 40–65% of body length from head, terminating in a narrow vas deferens with glandular walls. Cloaca with small raised circular lip containing two stout arcuate spicules terminating distally in uni-pointed tips. Small dorsal gubernaculum without ornamentation, slightly wider in dorsoventral aspect. Tail short with bluntly rounded terminus of variable shape.

Juvenile (J2). Lateral field with four incisures but with three anteriorly and posteriorly, occasionally completely areolated. Cuticle thickened for first seven or eight body annuli. Labial region rounded, slightly offset with four to six annuli. Oral disc surrounded by two lateral lips bearing amphidial apertures, adjacent dorsal and ventral submedial lips often fused. Contour of lips and oral disc sub-rectangular [52]. Heavily sclerotized hexaradiate head framework, dorsal and ventral radii bifurcate at tips in 60% of specimens. Stylet well developed, basal knobs with distinct anterior projection as viewed laterally. Gland lobe extending posteriorly for ca 35% of body length. Excretory pore ca 20% of body length from anterior end. Distinct hemizonid two annuli long, located one annulus anterior to excretory pore; hemizonion five or six annuli posterior to excretory pore. Genital primordium at ca 60% of body length from anterior end. Tail tapering uniformly with a finely rounded point, hyaline region forming about half of tail region.

*Globodera rostochiensis*

Female. Pearly white, subspherical to ovate, with elongate, protruding neck. Color changing from white to yellow to light golden as female matures to cyst stage. Cuticle thick, with superficial, rugose, lace-like pattern, D-layer present, punctations resolved near or beneath surface. Labial region slightly offset, bearing two annuli. Labial framework weakly developed. Stylet fairly strong, straight to slightly curved, with well-developed rounded basal knobs, sloping posteriorly. Median bulb large, nearly spherical, with well-developed valve. Pharyngeal glands often obscured but appearing clustered. Excretory pore conspicuous, always at or near base of neck. Vulva terminal, slit of medium length. Vulval area circumfenestrate. No anal fenestration, but anus and vulva both lying in a "vulval basin", anal area not encircled by cuticular rings. Often beneath vulva, generally in a cluster, are vulval bodies of highly variable size and shape, large superficial tubercles clumped near vulva. Vulva ellipsoid in shape, anus shorter than vulva. All eggs retained in body, no egg mass.

Cyst. Yellow when first visible on root surface, eventually turning brown with age, ovate to spherical in shape with protruding neck, circumfenestrate, abullate, without distinct "vulval bodies" commonly seen in white females. Fenestra circular, anus conspicuous at apex of a V-shaped subsurface cuticular mark. Cyst wall pattern basically as in female but often more prominent, especially near mid-body, tending to form wavy lines going around body. Subcrystalline layer absent. Punctations generally present but variable in intensity and arrangement. Each cyst containing 200–1000 eggs.

Male. Body vermiform, slightly tapering at both anterior and posterior regions. Cuticle with prominent annulation. Labial region slightly offset, hemispherical, with six annuli. Labial framework heavily sclerotized. Stylet strong, with prominent knobs. Anterior and posterior cephalids present. Lateral fields with four equally spaced lines. Median bulb ellipsoidal. Excretory pore ca two annuli posterior to often distinct hemizonid. One testis. Spicules slightly arcuate, tips rounded, not notched. Tail short, variable in length and shape.

Juvenile (J2). Body tapering at both extremities but more at posterior end. Cuticular annulation well defined. Lateral fields with four lines extending for most of body length, outer two lines crenate but without areolation. Labial region slightly offset, bearing 4–6 annuli, considerably wider at base than anteriorly, presenting a rounded, though rather anteriorly flattened, appearance. Labial framework heavily sclerotized. Stylet well developed, with prominent rounded knobs as viewed laterally. Anterior and posterior cephalids present. Valve of median bulb prominent, ellipsoidal. Isthmus and pharyngeal glands typical for the genus. Excretory pore almost adjacent yet slightly posterior to hemizonid. Genital primordium slightly posterior to mid-body, with four cells commonly resolved. Tail tapering to small, rounded terminus. Phasmids generally difficult to see, when visible, located about halfway along tail.

## 10. Identification

*Globodera rostochiensis* and *G. pallida* are morphologically and morphometrically very similar [29,51,52]. Therefore, identification of as many stages as possible should be performed using a combination of morphological characters and molecular techniques.

Nematode cysts separated from soil organic matter must first be carefully inspected using moderate power (up to about 25×) of a dissecting microscope to exclude all non-globose cysts, including those of *Cactodera*, *Betulodera*, *Dolichodera*, *Heterodera*, and *Paradolichodera*.

Any remaining cysts should be considered as suspect PCN cysts. If the laboratory possesses positive control DNA of both species of PCN, single cyst sub-samples should be tested using the PCR protocol provided in Section 10.3.2.

When positive control DNA is not available, there are two potential courses of action, viz. molecular sequencing using the DNA sequencing protocol or morphological examination using the morphological protocol. Morphological identification of suspected *Globodera* cysts and juvenile nematodes to genus and species levels is difficult and requires an experienced nematologist. When a skilled nematologist is available, it is preferable to utilize both the DNA sequencing and morphological protocols to enhance the level of certainty of identification.

### 10.1. Morphological Identification to Genus

An early consideration is how to distinguish cysts of *Globodera* from those of other cyst-forming genera. There is the potential to confuse cysts of *Globodera* with those of the six other genera of the subfamily Heteroderinae, where all females turn into a hard-walled cyst. Cyst shape can be an important character to help distinguish *Globodera* from other genera: globose or spheroid in *Globodera* and generally elongate-ovoid in *Dolichodera* and *Paradolichodera*, and lemon-shaped or pear-shaped in *Betulodera*, *Cactodera*, and *Heterodera*. Occasionally, cysts of *Betulodera* and *Cactodera* tend towards the globose shape, and these specimens can be separated by the presence of a terminal cone, which is a posterior protrusion of the cyst encompassing the anus and vulva and is not present in *Globodera*.

*Punctodera* cysts lack a terminal cone and some species of the genus have globose cysts like *Globodera*, but all can be distinguished from other cyst-forming genera including *Globodera* by the formation of a fenestra in the anal region, of similar shape and size to the vulval fenestra. A fenestra is a terminal region of a cyst where the wall remains very thin and therefore can rupture to permit emergence of juveniles. The vulval slit of *Punctodera* is very short at <5 µm, whereas it is about 9 or 10 µm for *G. rostochiensis* [29,50] and about 11.5 µm for *G. pallida* [52]. In cysts of *Globodera*, the anus is at the apex of a conspicuous V-shaped subsurface cuticular mark not seen in *Punctodera*. Additionally, all members of the genus *Punctodera* are parasites of monocotyledonous plants.

Adult female root knot nematodes (*Meloidogyne* sp., family Meloidogynidae), like *Globodera* are swollen and sedentary plant root feeders, and can be distinguished from *Globodera* by their lack of cuticle thickening and pigmentation as a persistent container for the eggs, i.e., a cyst. The perineum of swollen adult female *Meloidogyne* retains its annulation in the form of fingerprint-like whorls, whereas this annulation is lost in *Globodera*. Unlike *Globodera*, female *Meloidogyne* create an egg-mass, which is a collection of extruded eggs embedded within a secreted gelatinous matrix. In addition, females of *Meloidogyne*, but not *Globodera*, are gall-inciting.

Second-stage juvenile specimens of *Globodera* are more robust than their *Meloidogyne* counterparts. The more conspicuous stylet is longer and thicker, and the tail terminus is hyaline (transparent), whereas it is non-hyaline in *Meloidogyne*. The phasmids (paired postanal lateral chemoreceptor sensory organs) of *Meloidogyne* are small and pore-like, whereas they are larger and lens-like in *Globodera*.

Male *Globodera* lack the distinctive lateral amphidial cheeks (outer part of the lateral lip of the head, adjacent the opening of the amphid sense organ) of *Meloidogyne*; they also have a long, slender esophageal isthmus in contrast to the very short, broad isthmus of *Meloidogyne*.

To identify suspected *Globodera* nematodes to genus level, refer to the key in Table 1. Additionally, to identify cysts within the family Heteroderidae, the keys of Hesling [53], Mulvey and Golden [54], Golden [55], Baldwin and Mundo-Ocampo [56], Brzeski [57], Wouts and Baldwin [58], Siddiqi [59], or Subbotin et al. [7] based on cyst form including characteristics of the vulva-anus region, should be consulted.

**Table 1.** Simplified dichotomous morphological key to genus *Globodera*.

| | | |
|---|---|---|
| 1 | Nematode with spear or stylet | 2 |
| | Nematode without spear or stylet | not *Globodera* |
| 2 | Three-part esophagus with a valvulated metacorpus (median bulb) followed by a slender isthmus and glandular basal bulb; stylet with basal knobs | 3 |
| | Two-part oesophagus, no valvulated apparatus; stylet usually without basal swelling | not *Globodera* |
| 3 | Dorsal oesophageal gland outlet in procorpus; metacorpus less than three-fourths body width | 4 |
| | Dorsal oesophageal gland outlet in metacorpus, anterior to valve; metacorpus large, often nearly as wide as body | not *Globodera* |
| 4 | Head without setae | 5 |
| | Head with setae | not *Globodera* |
| 5 | Metacorpus with sclerotized valve | 6 |
| | Metacorpus absent or without sclerotized valve | not *Globodera* |
| 6 | Mature female greatly enlarged | 7 |
| | Mature female vermiform | not *Globodera* |
| 7 | Mature female pyriform-saccate, spheroid, or lemon-shaped, usually without tail | 8 |
| | Mature female elongate-saccate or kidney-shaped, usually with tail | not *Globodera* |
| 8 | Female without irregular body annules around perineum; excretory pore posterior to median bulb; second-stage juvenile stylet usually >20 µm; well-developed labial framework | 9 |
| | Female with irregular body annules around perineum; excretory pore at level with stylet or close behind it; second-stage juvenile stylet <20 µm; weakly-developed labial framework | not *Globodera* |
| 9 | Vulva terminal or subterminal; cuticle with lacelike pattern | 10 |
| | Vulva subequatorial; cuticle annulated | not *Globodera* |
| 10 | Cyst stage present | 11 |
| | No cyst stage | not *Globodera* |
| 11 | Cyst generally lemon-shaped; vulva on terminal cone | not *Globodera* |
| | Cyst spherical or subspherical; vulva not on terminal cone | *Globodera* |

*10.2. Morphological Identification to Species*

Once all other genera are excluded and it is confirmed that cysts belong to the genus *Globodera*, the following procedure should be followed.

Use a combination of cyst and second-stage juvenile (J2) characteristics if possible. Both stages are normally present in most soil samples infested with PCN, but juveniles will not be extracted by the flotation methods that rely on dried cysts floating to the top of a column of water. Alternatively, to obtain larvae, a cyst can be broken open in a droplet of water on a microscope slide to release the contained eggs. During the process the delicate shells of some eggs will inevitably be broken, enabling the larvae to escape and unfold ready for identification. The most reliable characteristics for identification of second-stage juveniles (J2) within the genus *Globodera* are stylet length, stylet knob width, and stylet knob shape. In *G. pallida*, J2 stylet knobs are distinctly anteriorly directed to flattened anteriorly, and the mean J2 stylet length is >23 µm, whereas in *G. rostochiensis* J2 stylet knobs are rounded to flattened anteriorly, and the mean J2 stylet length is <23 µm (Figure 2).

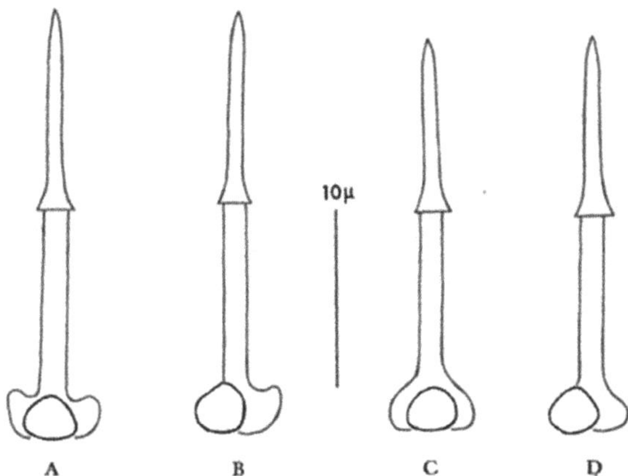


**Figure 2.** Stylets of second-stage juveniles of *G. pallida* (diagrams (**A**,**B**)) and *G. rostochiensis* (diagrams (**C**,**D**)) (after Stone [52]).

Cysts should be observed under a dissecting microscope directly on the filter paper used to catch the cysts during the extraction process, at low to moderate magnification (up to about 25×). For species identification, a 40× objective on a compound microscope is adequate to examine the perineal region after the cyst wall has been mounted on a slide. There are no clear differences in size, shape, or color of mature cysts of *G. rostochiensis* and *G. pallida*; the most important cyst differences can be obtained from examination of the perineal area, i.e., number of cuticular ridges between vulval basin and anus (Figure 3), and Granek's ratio (see Section 10.2.2), the distance from the anus to the nearest edge of the vulval basin divided by vulval basin diameter [60]. However, in some cases cuticular ridges are not visible or are very difficult to count, so Granek's ratio is considered a more reliable diagnostic tool, and when combined with the important second-stage juvenile measurements, a species diagnosis can be made. Confirmation with molecular techniques is also recommended.

A key to species of *Globodera* is presented in Table 2. Further keys to species can be found in Mulvey [61,62], Hesling [53], Wouts [63], Golden [55], Wouts and Baldwin [58], Subbotin et al. [7], and EPPO [17].

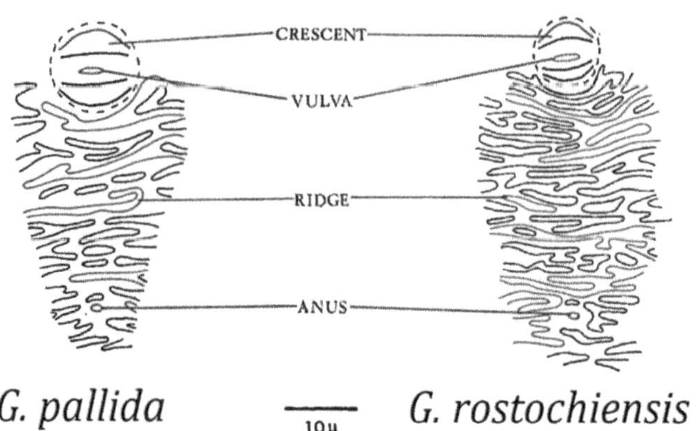


**Figure 3.** Vulval-anal ridge patterns for *G. pallida* and *G. rostochiensis* (after Stone [52]).

**Table 2.** Dichotomous morphological key to species of the genus *Globodera*. (after Subbotin et al. [7] and EPPO [17], with the addition of *G. agulhasensis* [64] and *G. sandveldensis* [65]).

| | | |
|---|---|---|
| 1 | Cuticle of cyst thin, transparent | *G. mali* |
| | Cuticle of cyst thick, dark in colour | 2 |
| 2 | Mean length of J2 stylet ≤26 µm | 3 |
| | Mean length of J2 stylet ≥27 µm | *G. zelandica* |
| 3 | Mean length of J2 stylet <19 µm | *G. leptonepia* |
| | Mean length of J2 stylet ≥19 µm | 4 |
| 4 | Hyaline tail region of J2 >31 µm | 5 |
| | Hyaline tail region of J2 ≤31 µm | 6 |
| 5 | Mean J2 body length <500 µm; mean J2 stylet length <24 µm; mean J2 DGO [1] <5 µm; mean J2 tail length >58 µm | *G. bravoae* |
| | Mean J2 body length >550 µm; mean J2 stylet length >26 µm; mean J2 DGO [1] >6 µm mean J2 tail length >62 µm | *G. sandveldensis* |
| 6 | Mean Granek's ratio usually >2, mostly parasites of Solanaceae | 7 |
| | Mean Granek's ratio ≤2, mostly parasites of Asteraceae | 12 |
| 7 | With a combination of the following characters: mean J2 DGO [1] ≥5.5 µm; mean Granek's ratio <3; J2 lip region with 4–6 annules; stylet knobs rounded to slightly anteriorly projected | 8 |
| | Not with the above combination of all characters; mean J2 DGO [1] <5.5 µm | 9 |
| 8 | Cyst wall lacking a network-like pattern, ridges close; mean number of cuticular ridges = 13 (10–18); male spicules with a pointed, thorn-like tip | *G. ellingtonae* |
| | Cyst wall exhibiting network-like or maze-like patterns; mean number of cuticular ridges = 7–8 (5–15); male spicules with a finely rounded tip | *G. tabacum* |
| 9 | Cysts with prominent bullae in the terminal region of most specimens; J2 lip region with 3 annules, mean hyaline tail region >28 µm | *G. capensis* |
| | Cyst abullate, at most with small vulval bodies in some specimens; J2 lip region with 4–6 annules, mean hyaline tail region <28 µm | 10 |
| 10 | J2 stylet knobs distinctly anteriorly directed to flattened anteriorly; mean J2 stylet length > 23 µm; Granek's ratio <3 | 11 |
| | J2 stylet knobs rounded to flattened anteriorly; mean J2 stylet length <23 µm; Granek's ratio ≥ 3 | *G. rostochiensis* |
| 11 | Mean Granek's ratio = 2.1–2.5 | *G. pallida* |
| | Mean Granek's ratio = 2.8 | *G. mexicana* |
| 12 | J2 lip region with 5–6 annules | 13 |
| | J2 lip region with 3–4 annules | 14 |
| 13 | Mean stylet ≥25 µm in J2, male gubernaculum [2] = 11.2–12.9 µm | *G. millefolii* |
| | Mean stylet <25 µm in J2, male gubernaculum = 6.0–9.9 µm | *G. artemisiae* |
| 14 | Mean stylet length 26.3 µm; mean DGO [1] 5.3 µm; mean hyaline tail region 29.5 µm; number of ridges between anus and vulval basin 6–20 | *G. capensis* |
| | Mean stylet length 23.5 µm; mean DGO [1] 4.4 µm; mean hyaline tail region 25 µm; number of ridges between anus and vulval basin 6–12 | *G. agulhasensis* |

[1] DGO = distance from anterior end to orifice of dorsal gland opening. [2] gubernaculum = grooved cuticular structure which guides the spicule or intromittent organ.

Three other *Globodera* species could cause confusion during identification of potato cyst nematodes: *G. achilleae* (Golden and Klindic, 1973) Behrens, 1975, *G. artemisiae* (Eroshenko and Kazachenko, 1972) Behrens, 1975, and *G. tabacum sensu lato*. None are parasitic on potato, although the *G. tabacum* species complex (*G. tabacum tabacum* (Lownsbery and Lownsbery, 1954) Skarbilovich, 1959; *G. tabacum solanacearum* (Miller and Gray, 1972) Behrens, 1975, and *G. tabacum virginiae* (Miller and Gray, 1972) Behrens, 1975) parasitizes *Nicotiana tabacum* (tobacco) and some other solanaceous plants (but not potato). To help resolve species determination, Table 3 shows morphometric and morphological comparisons between PCN and *G. achilleae*, *G. artemisiae* and *G. tabacum*. See also Baldwin and Mundo-Ocampo [56], Brzeski [57], Wouts and Baldwin [58], and Subbotin et al. [7,66] for more detailed information on other members of the Heteroderinae.

**Table 3.** Mean and range (in parentheses) values of some essential characters of *Globodera rostochiensis*, *G. pallida*, *G. tabacum (tabacum)*, *G. achilleae*, and *G. artemisiae*, as given in Baldwin and Mundo-Ocampo [56], Brzeski [57], Fleming and Powers [67], Manduric et al. [68], and Dobosz et al. [69].

| | J2 Measurements and Characteristics | | | | Cyst Measurements | |
|---|---|---|---|---|---|---|
| **Species** | **Body Length (μm)** | **Stylet Length (μm)** | **Stylet Knob Width (μm)** | **Stylet Knob Shape** | **Ridges between Anus and Vulval Basin** | **Granek's Ratio** |
| *G. rostochiensis* | 468 (425–505) | 21.8 (19–23) | (3.2–4.0) | anteriorly flattened to rounded, without forward projections | >14 (16–31) | >3 (1.3–9.5) |
| *G. pallida* | 484 (440–520) | 23.8 (22–24) | (4–5) | anterior surface flat to concave with forward projections | <14 (8–20) | <3 (1.2–3.5) |
| *G. tabacum (tabacum)* | 477 (410–527) | 23–24 | (4–5) | anterior surface rounded | (10–14) | <2.8 (1–4.2) |
| *G. achilleae* | 492 (472–515) | 25 (24–26) | (4–5) | anterior surface rounded to anchor shape | <10 (4–11) | 1.6 (1.3–1.9) |
| *G. artemisiae* | 413 (357–490) | 23 (18–29) | (3–5) | rounded, anteriorly flattened, sometimes slightly indented | (5–16) | 1.0 (0.8–1.7) |

10.2.1. Microscope Slide-Mounting of Cyst Wall

1. Place one cyst in a small droplet of water on a glass microscope slide.
2. Puncture the cyst wall with a new scalpel blade towards the neck area to release pressure so further cutting does not cause splits in the cyst wall (Figure 4).
3. Cut across the base of the cyst with a scalpel blade so that a small section containing the perineal region opposite the neck is detached (Figure 5). The smaller the section, the less likely that splits will occur when the wall is flattened in step 7. It is important to avoid creating splits as they can disfigure important diagnostic areas.
4. With a fine needle remove any eggs away from the excised section of cyst wall.
5. Pipette some eggs from the slide surface into an Eppendorf tube for PCR but leave some on the slide for hatching and measuring.
6. Make two more cuts in the cyst section with a scalpel blade as shown in Figure 6.
7. In a small droplet of glycerol on a fresh microscope slide, lay the excised section out flat with the outer surface uppermost (Figure 7). Place a cover slip over the section.
8. Seal the cover slip with clear nail polish, allow to dry, and examine with 40× objective on compound microscope and take measurements in microns.

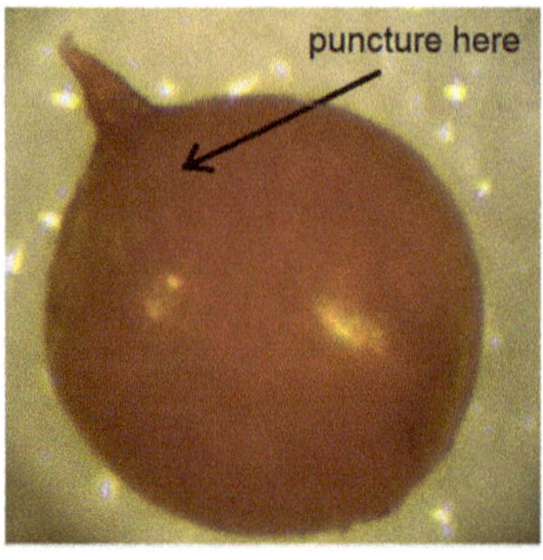



**Figure 4.** Puncture cyst wall to release pressure.

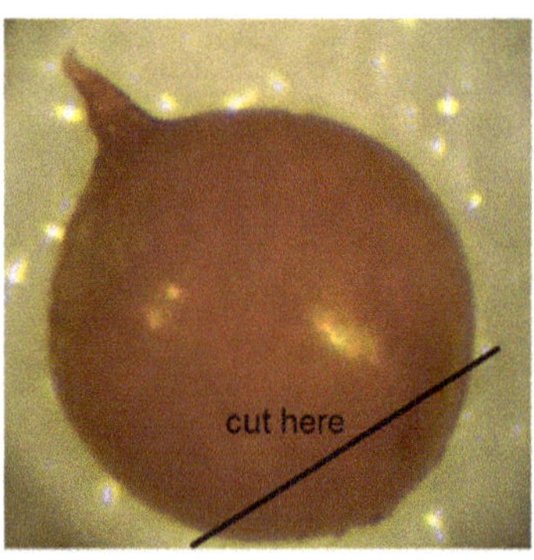



**Figure 5.** Cut across base of cyst to retain vulval and anal region.

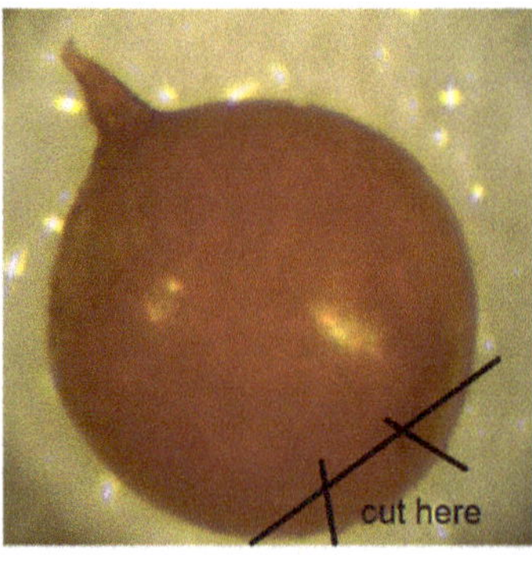



**Figure 6.** Make two more cuts to the excised section of cyst wall.

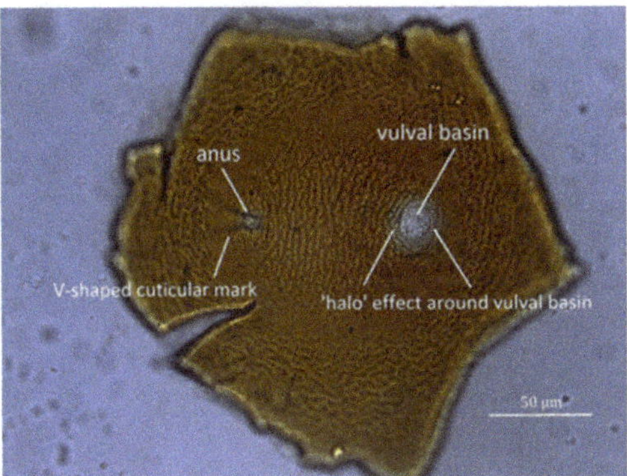


**Figure 7.** Light microscope image of cuticle surface of perineal region of potato cyst nematode (PCN) (*G. rostochiensis*) cyst laid flat on glass microscope slide.

### 10.2.2. Taking Measurements for Granek's Ratio

1.  Measure the distance from the anus at the apex of the V-shaped cuticular mark to the vulval basin (A) (Figure 8).
2.  Measure the diameter of the vulval basin (B). Note that there is a "halo" effect around the vulval basin (Figure 8a). The "halo" is not part of the vulval basin. Measurements are taken as shown in Figure 8b.
3.  Divide A by B.

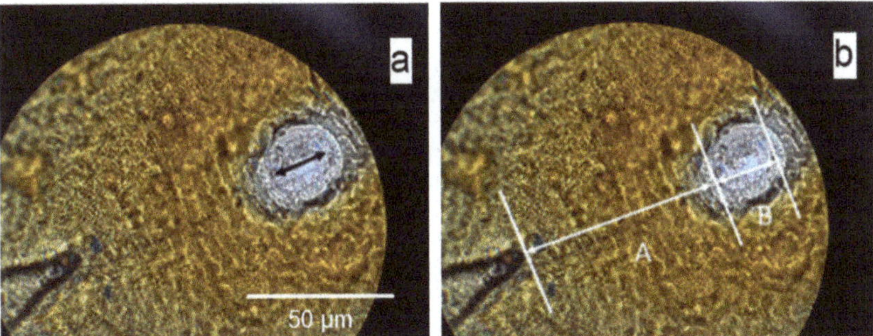


**Figure 8.** (**a**) Image of perineal region of *Globodera*, showing the "halo" effect around the vulval basin, the actual diameter of which is marked by an arrow. (**b**) Measurements taken to calculate Granek's ratio, where A is the distance from the anus (at the apex of the V-shaped cuticular mark) to the vulval basin, and B is the diameter of the vulval basin.

### 10.3. Molecular Identification

For potato growers to attain phytosanitary certification, and for a country's authorities to maintain official control of PCN, molecular techniques are often the preferred choice for regular routine soil testing. When new introductions are suspected, the identification of *G. pallida* and *G. rostochiensis* should combine molecular and morphological methods.

### 10.3.1. DNA Extraction from PCN Cysts

Cysts collected from soil can be washed/soaked in deionized water or briefly washed in 70% ethanol to avoid possible fungal/bacterial contamination.

The following DNA extraction method works with cysts that contain larvae or unhatched eggs; it does not work with empty cysts.

Material/equipment:

1. DNeasy Blood & Tissue Kit
2. Pipettes and tips
3. 1.5 mL centrifuge tubes
4. TissuLyser (Shaker) machine
5. Centrifuge tube rack/stand
6. Micro pestle
7. Balance
8. Centrifuge with $17,000\times g$ capacity
9. Shaker
10. Refrigerator
11. Gloves (nitrile)
12. References:

Operation manual of DNA extraction KIT Cat Nos 69504/69506, Protocol for Purification of Total DNA from Animal Tissues (Spin-Column Protocol)

Quader et al. [70]

Method:

1. Place a cyst into 1.5 mL Eppendorf tube and thoroughly crush it with a micro pestle.
2. Add 180 µL ATL (Tissue Lysis) buffer. Gently flick the tube with the pestle to shake off into the buffer as much as possible of the material stuck to the pestle. Close the tube lid and vortex for a few seconds.
3. Add 20 µL Proteinase K. Vortex and briefly spin to get all liquid off the tube lid.
4. Incubate the tube at 56 °C for at least 3 h, vortexing occasionally during this time.
5. Add 200 µL AL buffer. Vortex.
6. Add 200 µL 100% Ethanol. Vortex.
7. Pipette the mixture (including any precipitate) into the DNeasy Mini spin column placed in a 2 mL collection tube (provided). Centrifuge at $6000\times g$ for 1 min. Discard flow-through and collection tube.
8. Place the DNeasy Mini spin column in a new 2 mL collection tube (provided), add 500 µL Buffer AW1 (DNA wash buffer), and centrifuge for 1 min at $6000\times g$. Discard flow-through and collection tube.
9. Place the DNeasy Mini spin column in a new 2 mL collection tube (provided), then: (1) add 700 µL Buffer AW2, and centrifuge for 1 min at $6000\times g$. Discard flow-through; (2) add 200 µL Buffer AW2, and centrifuge for 2 min at $16,000\times g$. Discard flow-through and collection tube.
10. Place the DNeasy Mini spin column in a labelled 1.5 mL Eppendorf tube. Use 10 µL tips to pipette any leftover liquid (ca. 2–3 µL) inside the column wall corner immediately above the membrane (this is a poor design of Qiagen columns).
11. Pipette 100 µL Buffer AE (elution buffer) directly onto the DNeasy membrane. Incubate at room temperature for 1 min, and then centrifuge for 1 min at $6000\times g$ to elute.
12. Pipette 100 µL Buffer AE directly onto the DNeasy membrane. Incubate at room temperature for 1 min, and then centrifuge for 1 min at $6000\times g$ to elute.
13. Measure the extracted DNA concentration where possible.
14. The extracted DNA should be stored at $-20$ °C until required.

10.3.2. Multiplex PCR for the Identification of Species of PCN

Material/equipment:

1. Pipettes and tips
2. MyTaq™ Red Mix (Meridian Bioscience)
3. Primers for the nematodes at a concentration of 10 µM
4. DNA marker/ladder
5. SYBR™ Safe DNA Gel Stain
6. PCR grade water
7. Thermo-cycler PCR machine

8. Agarose gel and 0.5× TBE buffer
9. Ice
10. 0.2 mL Eppendorf/PCR tubes
11. Texta/marker pen
12. References:

    Bulman and Marshal [71]
    White et al. [72]
    PCR primer sequences:
    ITS5 5′-CGCGCGGATCCGGAAGTAAAAGTCGTAACAAGG-3′
    PIr3 5′-AGCGCAGACATGCCGCAA-3′
    PIp4, 5′-ACAACAGCAATCGTCGAG-3′
    ITS26 5′-TATATGGATCCATATGCTTAAGTTCAGCGGGT-3′

Primer ITS5 is used in combination with primer PITSr3 in a specific PCR to detect *G. rostochiensis* only. Primer ITS5 is used in combination with primer PITSp4 in a specific PCR to detect *G. pallida* only. Primer ITS5 is used in combination with PITSr3 and PITSp4 to detect both species from a mixed population.

Primers ITS5 and ITS26 should amplify both *G. pallida* and *G. rostochiensis*. These primers are used in a housekeeping nematode PCR to check the quality of DNA extracts. The PCR ensures that DNA is present or that there are no inhibitors in the DNA extracts that retard the activity of the DNA polymerase.

DNA barcoding based on the 18S rDNA gene and the internal transcribed spacer ITS1 region of rDNA (ITS) has been determined as suitable for species identification in *Globodera*. Bulman and Marshall [71] designed the PCR-based *G. pallida*-specific primer PITSp4 and *G. rostochiensis*-specific primer PITSr3 to be used in conjunction with the universal ITS5 primer. These can be used singly or in a multiplex PCR. Alternatively, the universal ITS5 and ITS26 primer pair can be used to amplify the barcoding region, and the resultant product sequenced and compared with verified reference sequences on the NCBS GenBank database.

Method:

1. Determine sample numbers and species of PCN to be tested.
2. Place ice in a suitable container, e.g., esky lid or disposable take away plastic container.
3. Label 0.2 mL PCR tubes according to the number of samples.
4. Make up a master mix of specific PCRs for PCN in a sterile Eppendorf tube by adding the ingredients described in Table 4. Vortex.
5. Make up a master mix of housekeeping PCRs in a sterile Eppendorf tube by adding the ingredients described in Table 5. Vortex.
6. Pipette the master mixes into the PCR reaction tubes.
7. Add PCN DNA templates into each PCR reaction tube. The volume of the DNA extracts can be varied to accommodate 50–100 ng DNA template per reaction.
8. Positive control 1: template DNA of the species *G. rostochiensis*.
9. Positive control 2: template DNA of the species *G. pallida*.
10. Negative control: Sterile distilled water.
11. Spin down all the liquid in the reaction PCR tubes before loading onto the PCR machine.

    PCR cycles:

1. Program cycles as showed in Table 6 in PCR machine for 50 μL reaction volumes in case the total reaction volumes are greater than 25 μL.
2. Transfer PCR tubes into PCR machine and start.

**Table 4.** Master mix of specific PCRs for both PCN species identification in one reaction (multiplex PCR).

| Items | Quantity (μL) |
|---|---|
| Water | 10 |
| MyTaq™ Red Mix | 12 |
| Primer ITS5 10 μM | 1.0 |
| * Primer PITSr3 10 μM | 1.0 |
| * Primer PITSp4 10 μM | 1.0 |

* PITSr3 for *G. rostochiensis* and PITSp4 for *G. pallida*.

**Table 5.** Master mix of housekeeping PCRs for both PCN species.

| Items | Quantity (μL) |
|---|---|
| Water | 10 |
| MyTaq™ Red Mix | 12 |
| Primer ITS5 10 μM | 1.0 |
| Primer ITS26 10 μM | 1.0 |

**Table 6.** PCR cycles for PCN species detection.

| Cycles | Temperature °C | Duration |
|---|---|---|
| ×1 cycle | 94 | 2 min |
| ×35 cycles | 94 | 30 s |
|  | 60 | 30 s |
|  | 72 | 30 s |
| ×1 cycle | 72 | 5 min |

Gel run and photograph:

1. Prepare 2% agarose gel (e.g., 2.0 g agarose in 100 mL 0.5× TBE) heated with a microwave oven until agarose is melted and let it cool down to about 70–80 °C
2. Add a 1/10,000 proportion of SYBR™ Safe (e.g., 1.0 μL SYBR safe in 100 mL melted agarose) and gently swirl agarose solution before pouring into a gel casting tray containing comb(s) and allow to set.
3. Pipette the 7.0–10 μL PCR products into wells.
4. Run PCR products on agarose gel at 100 V for 45 min to 1 h.
5. Visualize and photograph gel under UV light.

PCR product sizes:

1. For Globodera rostochiensis = 434 bp
2. For Globodera pallida = 256 bp
3. For housekeeping DNA = 1 kb

10.3.3. DNA Sequencing

For confirmation, the PCR products of the reactions using primers ITS5 and ITS26 for single cysts should be sequenced.

Sequencing reactions using BigDye™ Terminator v3.1 Cycle Sequencing Kit (Table 7) can be done in the laboratory using PCR products cleaned up by QIAquick PCR Purification Kit. Either cleaned PCR products or sequencing reaction products can be sent to Sanger sequencing services, e.g., Macrogen or Micromon, along with the primers to obtain forward and reverse sequences. Forward and reverse sequences of each sample should be de novo assembled and edited/corrected using a suitable computer program, e.g., Geneious. The consensus sequence should be subjected to a database search, e.g., GenBank or private sequence libraries, and phylogenetic analysis. Sequences should be compared with those in GenBank for accession numbers EF622513–EF622532 for *G. rostochiensis* and HQ260426–8,

FJ212165 for *G. pallida*. For a match to be positive, the sequence must have a similarity of greater than 99% with these GenBank sequences.

**Table 7.** Sequencing reaction mix.

| Items | Quantity (µL) |
|---|---|
| Water | 13 |
| Cleaned PCR product | 60 ng |
| 5× Buffer | 3.5 |
| BigDye | 1.0 |
| Primer ITS5 10 µM | 0.5 |
| Primer ITS26 10 µM | 0.5 |

### 10.3.4. Genotyping

It is possible to compare the genetic differentiation of PCN populations using polymorphic microsatellite DNA markers. DNA can be screened after extracting it from single larvae dissected from cysts. For methodology of this genotyping, see Boucher et al. [73], Alenda et al. [74], and Blacket et al. [75].

There have been many phylogenetic analyses of species within the genus *Globodera* (e.g., [5,73,76–86]). A recent study, based on a phylogenetic analysis of gene sequences of three molecular markers (455 ITS rRNA, 219 *COI*, and 164 *cytb*) of 11 valid and 2 undescribed species of *Globodera* [87], found that *Globodera* displayed two main clades in their phylogenetic trees: (i) *Globodera* from South and North America parasitizing plants from Solanaceae; and (ii) *Globodera* from Africa, Europe, Asia, and New Zealand parasitizing plants from Asteraceae and other families. They hypothesized that the split between solanaceous and non-solanaceous lineages occurred roughly 2.9 ± 0.5 Mya (million years ago), divergence dates of the solanaceous *Globodera* lineages started 2.7 ± 0.2 Mya and the nonsolanaceous *Globodera* lineages 1.6 ± 0.3 Mya, and dispersals of *Globodera* to Europe and New Zealand occurred 1.4 ± 0.3 and 0.9 ± 0.2 Mya, respectively.

**Author Contributions:** Conceptualization, J.W.; methodology, J.W. and Q.D.; validation, J.W. and Q.D.; formal analysis, J.W. and Q.D.; investigation, J.W. and Q.D.; resources, J.W and Q.D.; data curation, J.W. and Q.D.; writing—original draft preparation, J.W. and Q.D.; writing—review and editing, J.W.; visualization, J.W.; supervision, J.W.; project administration, J.W.; funding acquisition, J.W. All authors have read and agreed to the published version of the manuscript.

**Funding:** The preparation of this paper was funded by Agriculture Victoria Research, Department of Jobs, Precincts and Regions, State Government of Victoria.

**Institutional Review Board Statement:** Not applicable.

**Informed Consent Statement:** Not applicable.

**Acknowledgments:** This research received financial support from Plant Health Australia, the Subcommittee of Plant Health Diagnostics and the State Government of Victoria. Dolf De Boer and Jacky Edwards provided valuable suggestions on the manuscript.

**Conflicts of Interest:** The authors declare no conflict of interest.

## References

1. EPPO/CABI. *Globodera rostochiensis* and *Globodera pallida*. In *Quarantine Pests for Europe*, 2nd ed.; Smith, I.M., McNamara, D.G., Scott, P.R., Holderness, M., Eds.; CAB International: Wallingford, UK, 1997; pp. 601–606.
2. Moens, M.; Perry, R.N.; Jones, J.T. Cyst Nematodes—Life Cycle and Economic Importance. In *Cyst Nematodes*; Perry, R.N., Moens, M., Jones, J.T., Eds.; CAB International: Wallingford, UK, 2018; pp. 1–26. [CrossRef]
3. Lax, P.; Rondan Duenas, J.C.; Franco-Ponce, J.; Gardenai, C.N.; Doucet, M.E. Morphology and DNA sequence data reveal the presence of *Globodera ellingtonae* in the Andean region. *Contrib. Zool.* **2014**, *83*, 227–243. [CrossRef]
4. Phillips, W.S.; Kitner, M.; Zasada, I.A. Developmental dynamics of *Globodera ellingtonae* in field-grown potato. *Plant Dis.* **2017**, *101*, 1182–1187. [CrossRef] [PubMed]

5. Skantar, A.M.; Handoo, Z.A.; Zasada, I.A.; Ingham, R.E.; Carta, L.K.; Chitwood, D.J. Morphological and molecular characterization of *Globodera* populations from Oregon and Idaho. *Phytopathology* **2011**, *101*, 480–491. [CrossRef] [PubMed]

6. Cobb, G.S.; Taylor, A.L. *Heterodera leptonepia* n. sp., a cyst-forming nematode found in soil with stored potatoes. *P. Helm. Soc. Wash.* **1953**, *20*, 13–15. Available online: http://science.peru.edu/COPA/ProcHelmSocWash_V20_N1_1953I.pdf (accessed on 11 December 2020).

7. Subbotin, S.A.; Mundo-Ocampo, M.; Baldwin, G.B. *Systematics of Cyst Nematodes (Nematoda: Heteroderinae). Nematology Monographs and Perspectives, Volume 8A*; E.J. Brill: Leiden, The Netherlands, 2010; p. 351. [CrossRef]

8. Turner, S.J.; Subbotin, S.A. Cyst Nematodes. In *Plant Nematology*, 2nd ed.; Perry, R.N., Moens, M., Eds.; CAB International: Wallingford, UK, 2013; pp. 109–143. [CrossRef]

9. Mai, W.F. Worldwide distribution of potato-cyst nematodes and their importance in crop production. *J. Nematol.* **1977**, *9*, 30–34. Available online: https://www.ncbi.nlm.nih.gov/pmc/articles/PMC2620207/pdf/30.pdf (accessed on 24 December 2020).

10. Whitehead, A.G.; Turner, S.J. Management and Regulatory Control Strategies for Potato Cyst Nematodes (*Globodera rostochiensis* and *Globodera pallida*). In *Potato Cyst Nematodes, Biology, Distribution and Control*; Marks, R.J., Brodie, B.B., Eds.; CAB International: Wallingford, UK, 1998; pp. 135–152.

11. Spears, J.F. *The Golden Nematode Handbook: Survey, Laboratory, Control, and Quarantine Procedures. USDA Agriculture Handbook No. 353*; Agriculture Research Service, United States Department of Agriculture: Washington, DC, USA, 1968; p. 81. Available online: https://naldc.nal.usda.gov/download/CAT87208717/PDF (accessed on 24 December 2020).

12. DPIPWE. *Plant Biosecurity Manual Tasmania—2020 Edition*; Department of Primary Industries, Parks, Water and Environment, Tasmanian Government: Hobart, TAS, Australia, 2019. Available online: https://dpipwe.tas.gov.au/Documents/Plant%20 Biosecurity%20Manual%20Tasmania.pdf (accessed on 8 December 2020).

13. PBPI-BQ. *Queensland Biosecurity Manual—Version 12.0 Current as at September 2019*; Plant Biosecurity and Product Integrity sub-program of Biosecurity Queensland, Department of Agriculture and Fisheries, Queensland Government: Brisbane, QLD, Australia, 2019. Available online: https://www.daf.qld.gov.au/__data/assets/pdf_file/0004/379138/qld-biosecurity-manual.pdf (accessed on 8 December 2020).

14. PBPPIS. *Plant Quarantine Manual for New South Wales—Version 1.0*; Plant Product Integrity and Standards Unit, New South Wales Department of Primary Industries: Orange, NSW, Australia, 2016. Available online: https://vinehealth.com.au/media/NSW-PQM-Jan-2016.pdf (accessed on 8 December 2020).

15. PIRSA. *Plant Quarantine Standard South Australia—Version 13*; Primary Industries and Regions South Australia, Government of South Australia: Glenside, SA, Australia, 2017. Available online: https://vinehealth.com.au/media/PQS-Version-13-October-20 17.pdf (accessed on 24 December 2020).

16. Winslow, R.D.; Willis, R.J. Nematode Diseases of Potatoes. II. Potato Cyst Nematode, *Heterodera rostochiensis*. In *Economic Nematology*; Webster, J., Ed.; Academic Press: New York, NY, USA, 1972; pp. 18–34.

17. EPPO. Diagnostic protocol PM 7/40 (4) *Globodera rostochiensis* and *Globodera pallida*. *OEPP/EPPO Bull.* **2017**, *47*, 174–197. [CrossRef]

18. Mai, W.F.; Lownsbery, B.F. Studies on the host range of the golden nematode of potatoes, *Heterodera rostochiensis* Wollenweber. *Am. Potato J.* **1948**, *25*, 290–294. [CrossRef]

19. Oostenbrink, M. Het aardappelaaltje (Heterodera rostochiensis Wollenweber), een Gevaarlijke Parasite voor de Eenzijdige Aardappel-cultuur. Versl. Meded. Pliektenk. Dienst Wageningen. No. 115; 1950. Available online: https://library.wur.nl/ WebQuery/wurpubs/fulltext/176615 (accessed on 14 December 2020).

20. Sullivan, M.J.; Inserra, R.N.; Franco, J.; Moreno-Leheude, I.; Greco, N. Potato cyst nematodes: Plant host status and their regulatory impact. *Nematropica* **2007**, *37*, 193–201. Available online: https://journals.flvc.org/nematropica/article/view/64425 (accessed on 11 December 2020).

21. Handoo, Z.A.; Subbotin, S.A. Taxonomy, Identification and Principal Species. In *Cyst Nematodes*; Perry, R.N., Moens, M., Jones, J.T., Eds.; CAB International: Wallingford, UK, 2018; pp. 365–398. [CrossRef]

22. Ferris, H. *Nemaplex, the "Nematode-Plant Expert Information System": A Virtual Encyclopedia on Soil and Plant Nematodes*; Department of Entomology and Nematology, University of California: Davis, CA, USA, 2019; Available online: http://nemaplex.ucdavis.edu/ (accessed on 11 December 2020).

23. CABI. Invasive Species Compendium; Globodera rostochiensis (Yellow Potato Cyst Nematode), Datasheet. Available online: https://www.CABI.org/isc/datasheet/27034 (accessed on 11 December 2020).

24. CABI. Invasive Species Compendium; Globodera pallida (White Potato Cyst Nematode), Datasheet. Available online: https://www.CABI.org/isc/datasheet/27033 (accessed on 11 December 2020).

25. Banks, N.C.; Hodda, M.; Singh, S.K.; Matveeva, E.M. Dispersal of potato cyst nematodes measured using historical and spatial statistical analyses. *Phytopathology* **2012**, *102*, 620–626. [CrossRef]

26. Trudgill, D.L.; Evans, K.; Phillips, M.S. Potato Cyst Nematodes: Damage Mechanisms and Tolerance in the Potato. In *Potato Cyst Nematodes: Biology, Distribution and Control*; Marks, R.J., Brodie, B.B., Eds.; CAB International: Wallingford, UK, 1998; pp. 117–133.

27. Brodie, B.B.; Evans, K.; Franco, J. Nematode Parasites of Potatoes. In *Plant Parasitic Nematodes in Temperate Agriculture*; Evans, K., Trudgill, D.L., Webster, J.M., Eds.; CAB International: Wallingford, UK, 1993; pp. 87–132.

28. Chitwood, B.G.; Buhrer, E.M. The life history of the golden nematode of potatoes, *Heterodera rostochiensis* Wollenweber, under Long Island, New York, conditions. *Phytopathology* **1946**, *36*, 180–189. Available online: https://archive.org/details/in.ernet.dli. 2015.271513 (accessed on 16 December 2020).

29. Stone, A.R. Heterodera rostochiensis. In *CIH Descriptions of Plant-parasitic Nematodes*; Set 2, No. 16; CAB International: Wallingford, UK, 1973; p. 3.

30. Ferris, J.M. Effect of soil temperature on the life cycle of the golden nematode in host and non-host species. *Phytopathology* **1957**, *47*, 221–230.

31. Clarke, A.J.; Hennessy, J. Movement of *Globodera rostochiensis* (Wollenweber) juveniles stimulated by potato-root exudate. *Nematologica* **1984**, *30*, 206–212. [CrossRef]

32. Green, C.D.; Greet, D.N.; Jones, F.G.W. The influence of multiple mating on the reproduction and genetics on *Heterodera rostochiensis* and *H. schachtii. Nematologica* **1970**, *16*, 309–326. [CrossRef]

33. Kort, J. Identification of pathotypes of the potato cyst nematode. *OEPP/EPPO Bull.* **1974**, *4*, 511–518. [CrossRef]

34. Canto Saenz, M.; de Scurrah, M.M. Races of the potato cyst nematode in the Andean region and a new system of classification. *Nematologica* **1977**, *23*, 340–349. [CrossRef]

35. Kort, J.; Ross, H.; Rumpenhorst, H.J.; Stone, A.R. An international scheme for the identification of pathotypes of potato cyst nematodes *Globodera rostochiensis* and *G. pallida. Nematologica* **1977**, *23*, 333–339. [CrossRef]

36. Fleming, C.C.; Powers, T.O. Potato cyst nematodes: Species, pathotypes and virulence concepts. In *Potato Cyst Nematodes, Biology, Distribution and Control*; Marks, R.J., Brodie, B.B., Eds.; CAB International: Wallingford, UK, 1998; pp. 51–57.

37. Faggian, R.; Powell, A.; Slater, A.T. Screening for resistance to potato cyst nematode in Australian potato cultivars and alternative solanaceous hosts. *Austral. Plant Pathol.* **2012**, *41*, 453–461. [CrossRef]

38. Dalamu, V.B.; Umamaheshwari, R.; Sharma, R.; Kaushik, S.K.; Joseph, T.A.; Singh, B.P.; Gebhardt, C. Potato cyst nematode (PCN) resistance: Genes, genotypes and markers—An update. *SABRAO J. Breed. Genet.* **2012**, *44*, 202–228. Available online: https://www.researchgate.net/publication/285649776 (accessed on 12 December 2020).

39. Wood, F.H.; Foot, M.A.; Dale, P.S.; Barker, C.J. Relative efficiency of plant sampling and soil sampling in detecting the presence of low potato cyst nematode infestations. *N. Z. J. Exp. Agric.* **1983**, *11*, 271–273. [CrossRef]

40. Been, T.H.; Schomaker, C.H. A new sampling method for the detection of low population densities of potato cyst nematodes (*Globodera pallida* and *G. rostochiensis*). *Crop Prot.* **1996**, *15*, 375–382. [CrossRef]

41. Whitehead, A.G.; Hemming, J.R. A comparison of some quantitative methods of extracting small vermiform nematodes from soil. *Ann. Appl. Biol.* **1965**, *55*, 25–38. [CrossRef]

42. De Grisse, A.T. Redescription ou modification de quelques techniques utilissée dans l'étude des nematodes phytoparasitaires. *Mededelingen Rijksfaculteit Landbouwwetenschappen Gent* **1969**, *34*, 351–369.

43. Fenwick, D.W. Methods for the recovery and counting of cysts of *Heterodera schachtii* from soil. *J. Helminthol.* **1940**, *18*, 155–172. [CrossRef]

44. EPPO Diagnostic protocol PM 7/119 (1) Nematode extraction. *OEPP/EPPO Bull.* **2013**, *43*, 471–495. [CrossRef]

45. Shepherd, A.M. Extraction and Estimation of Cyst Nematodes. In *Laboratory Methods for Work with Plant and Soil Nematodes. Technical Bulletin Number 2*; Southey, J.F., Ed.; Her Majesty's Stationary Office: London, UK, 1986; pp. 31–49.

46. Turner, S.J. Sample Preparation, Soil Extraction and Laboratory Facilities for the Detection of Potato Cyst Nematodes. In *Potato Cyst Nematodes: Biology, Distribution and Control*; Marks, R.J., Brodie, B.B., Eds.; CAB International: Wallingford, UK, 1998; pp. 75–90.

47. Bellvert, J.; Crombie, K.; Horgan, F.G. Effect of sample size on cyst recovery by flotation methods: Recommendations for sample processing during EU monitoring of potato cyst nematodes (*Globodera* spp.). *OEPP/EPPO Bull.* **2008**, *38*, 205–210. [CrossRef]

48. Collins, S.A.; Marshall, J.M.; Zhang, X.A.; Vanstone, V.A. Area freedom from *Globodera rostochiensis* in Western Australia. *Asp. Appl. Biol.* **2010**, *103*, 55–62.

49. Collins, S.; Vanstone, V.; Zhang, X. *PCN "Area Freedom" for WA: Evaluation of the Current Status of Potato Cyst Nematode (Globodera rostochiensis) in Western Australia*; Final Report for Project PT04004 and MT04000; Horticulture Australia Limited: Sydney, Australia, 2010; 108p.

50. Golden, A.M.; Ellington, D.M.S. Redescription of *Heterodera rostochiensis* (Nematoda: Heteroderidae) with key and notes on related species. *P. Helm. Soc. Wash.* **1972**, *39*, 64–78. Available online: http://science.peru.edu/COPA/ProcHelmSocWash_V39_N1_1972I.pdf (accessed on 19 December 2020).

51. Stone, A.R. Heterodera pallida. In *CIH Descriptions of Plant-Parasitic Nematodes*; Set 2, No. 17; CAB International: Wallingford, UK, 1973; p. 2.

52. Stone, A.R. *Heterodera pallida* n. sp. (Nematoda: Heteroderidae), a second species of potato cyst nematode. *Nematologica* **1973**, *18*, 591–606. [CrossRef]

53. Hesling, J.J. Cyst Nematodes: Morphology and Identification of *Heterodera, Globodera* and *Punctodera*. In *Plant Nematology*; Southey, J.F., Ed.; Her Majesty's Stationery Office: London, UK, 1982; pp. 125–155.

54. Mulvey, R.H.; Golden, A.M. An illustrated key to the cyst-forming genera and species of Heteroderidae in the Western Hemisphere with species morphometrics and distribution. *J. Nematol.* **1983**, *15*, 1–59.

55. Golden, A.M. Morphology and Identification of Cyst Nematodes. In *Cyst Nematodes*; Lamberti, F., Taylor, C.E., Eds.; Plenum Press: New York, NY, USA, 1986; pp. 23–45. [CrossRef]

56. Baldwin, J.G.; Mundo-Ocampo, M. Heteroderinae, Cyst- and Non-Cyst-Forming Nematodes. In *Manual of Agricultural Nematology*; Nickle, W.R., Ed.; Marcel Dekker: New York, NY, USA, 1991; pp. 275–362.

57. Brzeski, M.W. *Nematodes of Tylenchina in Poland and Temperate Europe*; Muzeum i Instytut Zoologii Polska Akademia Nauk: Warsaw, Poland, 1998; 397p.

58. Wouts, W.M.; Baldwin, J.G. Taxonomy and Identification. In *The Cysts Nematodes*; Sharma, S.B., Ed.; Kluwer: Dordrecht, The Netherlands, 1998; pp. 83–122.

59. Siddiqi, M.R. Key to Genera of Heteroderinae. In *Tylenchida: Parasites of Plants and Insects*; Siddiqi, M.R., Ed.; CABI Publishing: Wallingford, UK, 2000; pp. 395–396. [CrossRef]

60. Granek, I. Additional morphological differences between the cysts of *Heterodera rostochiensis* and *Heterodera tabacum*. *Plant Dis. Rep.* **1955**, *39*, 716–718.

61. Mulvey, R.H. Identification of *Heterodera* cysts by terminal and cone top structures. *Can. J. Zool.* **1972**, *50*, 1277–1292. [CrossRef]

62. Mulvey, R.H. Morphology of the terminal areas of white females and cysts of the genus *Heterodera* (s.g. *Globodera*). *J. Nematol.* **1973**, *5*, 303–311. [PubMed]

63. Wouts, W.M. *Globodera zelandica* n. sp. (Nematoda: Heteroderidae) from New Zealand, with a key to the species of the genus *Globodera*. *N. Z. J. Zool.* **1984**, *11*, 129–135. [CrossRef]

64. Knoetze, R.; Swart, A.; Wentzel, R.; Tiedt, L.R. Description of *Globodera agulhasensis* n. sp. (Nematoda: Heteroderidae) from South Africa. *Nematology* **2017**, *19*, 305–322. [CrossRef]

65. Knoetze, R.; Swart, A.; Wentzel, R.; Tiedt, L.R. Description of *Globodera sandveldensis* n. sp. (Nematoda: Heteroderidae) from South Africa. *Nematology* **2017**, *19*, 805–816. [CrossRef]

66. Subbotin, S.A.; Mundo-Ocampo, M.; Baldwin, G.B. *Systematics of Cyst Nematodes (Nematoda: Heteroderinae). Nematology Monographs and Perspectives, Volume 8B*; E.J. Brill: Leiden, The Netherlands, 2010; p. 512. [CrossRef]

67. Fleming, C.C.; Powers, T.O. Potato Cyst Nematode Diagnostics: Morphology, Differential Hosts and Biochemical Techniques. In *Potato Cyst Nematodes, Biology, Distribution and Control*; Marks, R.J., Brodie, B.B., Eds.; CAB International: Wallingford, UK, 1998; pp. 91–114.

68. Manduric, S.; Olsson, E.; Englund, J.E.; Andersson, S. Separation of *Globodera rostochiensis* and *G. pallida* (Tylenchida: Heteroderidae) using morphology and morphometrics. *Nematology* **2004**, *6*, 171–181. [CrossRef]

69. Dobosz, R.; Obrepalska-Steplowska, A.; Kornobis, S. *Globodera artemisiae* (Eroshenko et Kazachenko, 1972) (Nematoda: Heteroderidae) from Poland. *J. Plant Prot. Res.* **2006**, *46*, 403–407. Available online: http://www.plantprotection.pl/Globodera-artemisiae-Eroshenko-et-Kazachenko-1972-Nematoda-Heteroderidae-from-Poland,90324,0,2.html (accessed on 23 November 2020).

70. Quader, M.; Nambiar, L.; Cunnington, J. Conventional and real-time PCR based species identification and diversity of potato cyst nematodes (*Globodera* spp.) from Victoria, Australia. *Nematology* **2008**, *10*, 471–478. [CrossRef]

71. Bulman, S.R.; Marshall, J.W. Differentiation of Australasian potato cyst nematode (PCN) populations using the polymerase chain reaction (PCR). *N. Z. J. Crop Hort. Sci.* **1997**, *25*, 123–129. [CrossRef]

72. White, T.J.; Bruns, T.; Lee, S.; Taylor, J. Amplification and Direct Sequencing of Fungal Ribosomal RNA Genes for Phylogenetics. In *PCR Protocols: A Guide to Methods and Applications*; Innes, M.A., Gelfard, D.H., Sninsky, J.J., White, T.J., Eds.; Academic Press: San Diego, CA, USA, 1990; pp. 315–322. [CrossRef]

73. Boucher, A.C.; Mimee, B.; Montarry, J.; Bardou-Valette, S.; Belair, G.; Moffett, P.; Grenier, E. Genetic diversity of the golden cyst nematode *Globodera rostochiensis* and determination of the origin of populations in Quebec, Canada. *Can. Mol. Phylogenet. Evol.* **2013**, *69*, 75–82. [CrossRef]

74. Alenda, C.; Montarry, J.; Grenier, E. Human influence on the dispersal and genetic structure of French *Globodera tabacum* populations. *Infect. Genet. Evol.* **2014**, *27*, 309–317. [CrossRef] [PubMed]

75. Blacket, M.J.; Agarwal, A.; Wainer, J.; Triska, M.D.; Renton, M.; Edwards, J. Molecular assessment of the introduction and spread of potato cyst nematode, *Globodera rostochiensis*, in Victoria, Australia. *Phytopathology* **2019**, *109*, 659–669. [CrossRef] [PubMed]

76. Blok, V.C.; Malloch, G.; Harrower, B.; Phillips, M.S.; Vrain, T.C. Intraspecific variation in ribosomal DNA in populations of the potato cyst nematode *Globodera pallida*. *J. Nematol.* **1998**, *30*, 262–274. Available online: https://www.ncbi.nlm.nih.gov/pmc/articles/PMC2620292/pdf/262.pdf (accessed on 18 November 2020). [PubMed]

77. Subbotin, S.A.; Halford, P.D.; Warry, A.; Perry, R.N. Variations in ribosomal DNA sequences and phylogeny of *Globodera* parasitising Solanaceae. *Nematology* **2000**, *2*, 591–604. [CrossRef]

78. Subbotin, S.A.; Cid Del Prado Vera, I.; Mundo-Ocampo, M.; Baldwin, J.G. Identification, phylogeny and phylogeography of circumfenestrate cyst nematodes (Nematoda: Heteroderidae) as inferred from analysis of ITS-rDNA. *Nematology* **2011**, *13*, 805–824. [CrossRef]

79. Madani, M.; Subbotin, S.A.; Ward, L.J.; Li, X.; De Boer, S.H. Molecular characterization of Canadian populations of potato cyst nematodes, *Globodera rostochiensis* and *G. pallida* using ribosomal nuclear RNA and cytochrome b genes. *Can. J. Plant Pathol.* **2010**, *32*, 252–263. [CrossRef]

80. Hoolahan, A.H.; Blok, V.C.; Gibson, T.; Dowton, M. A comparison of three molecular markers for the identification of populations of *Globodera pallida*. *J. Nematol.* **2012**, *44*, 7–17.

81. Picard, D.; Sempere, T.; Plantard, O. A northward colonisation of the Andes by the potato cyst nematode during geological times suggests multiple host-shifts from wild to cultivated potatoes. *Mol. Phylogenet. Evol.* **2007**, *42*, 308–316. [CrossRef]

82. Picard, D.; Sempere, T.; Plantard, O. Direction and timing of uplift propagation in the Peruvian Andes deduced from molecular phylogenetics of highland biotaxa. *Earth Planet. Sci. Lett.* **2008**, *271*, 326–336. [CrossRef]

83. Plantard, O.; Picard, D.; Valette, S.; Scurrah, M.; Grenier, E.; Mugniéry, D. Origin and genetic diversity of Western European populations of the potato cyst nematode *Globodera pallida* inferred from mitochondrial sequences and microsatellite loci. *Mol. Ecol.* **2008**, *17*, 2208–2218. [CrossRef]

84. Pylypenko, L.A.; Phillips, M.S.; Blok, V.C. Characterisation of two Ukrainian populations of *Globodera pallida* in terms of their virulence and mtDNA, and the biological assessment of a new resistant cultivar Vales Everest. *Nematology* **2008**, *10*, 585–590. [CrossRef]

85. Hockland, S.; Niere, B.; Grenier, E.; Blok, V.; Phillips, M.; den Nij, L.; Anthoine, G.; Pickup, J.; Viaene, N. An evaluation of the implications of virulence in non-European populations of *Globodera pallida* and *G. rostochiensis* for potato cultivation in Europe. *Nematology* **2012**, *14*, 1–13. [CrossRef]

86. Grenier, E.; Bossis, M.; Fouville, D.; Renault, L.; Mugniéry, D. Molecular approaches to the taxonomic position of Peruvian potato cyst nematodes and gene pool similarities in indigenous and imported populations of *Globodera*. *Heredity* **2001**, *86*, 277–290. [CrossRef] [PubMed]

87. Subbotin, S.A.; Franco, J.; Knoetze, R.; Roubtsova, T.V.; Bostock, R.M.; Cid Del Prado Vera, I. DNA barcoding, phylogeny and phylogeography of the cyst nematode species from the genus *Globodera* (Tylenchida: Heteroderidae). *Nematology* **2020**, *22*, 269–297. [CrossRef]

*Review*

# Taxonomy and Identification of Principal Foliar Nematode Species (*Aphelenchoides* and *Litylenchus*)

Zafar Handoo *, Mihail Kantor and Lynn Carta

Mycology and Nematology Genetic Diversity and Biology Laboratory, USDA, ARS, Northeast Area, Beltsville, MD 20705, USA; mihail.kantor@usda.gov (M.K.); lynn.carta@usda.gov (L.C.)
* Correspondence: zafar.handoo@ars.usda.gov

Received: 25 September 2020; Accepted: 2 November 2020; Published: 4 November 2020

**Abstract:** Nematodes are Earth's most numerous multicellular animals and include species that feed on bacteria, fungi, plants, insects, and animals. Foliar nematodes are mostly pathogens of ornamental crops in greenhouses, nurseries, forest trees, and field crops. Nematode identification has traditionally relied on morphological and anatomical characters using light microscopy and, in some cases, scanning electron microscopy (SEM). This review focuses on morphometrical and brief molecular details and key characteristics of some of the most widely distributed and economically important foliar nematodes that can aid in their identification. *Aphelenchoides* genus includes some of the most widely distributed nematodes that can cause crop damages and losses to agricultural, horticultural, and forestry crops. Morphological details of the most common species of *Aphelenchoides* (*A. besseyi, A. bicaudatus, A. fragariae, A. ritzemabosi*) are given with brief molecular details, including distribution, identification, conclusion, and future directions, as well as an updated list of the nominal species with its synonyms. *Litylenchus* is a relatively new genus described in 2011 and includes two species and one subspecies. Species included in the *Litylenchus* are important emerging foliar pathogens parasitizing trees and bushes, especially beech trees in the United States of America. Brief morphological details of all *Litylenchus* species are provided.

**Keywords:** foliar nematodes; taxonomy; *Aphelenchoides*; *Litylenchus*

---

## 1. Introduction

Foliar nematodes are mostly pathogens of ornamental crops in greenhouses, nurseries, and forest trees, as well as field crops [1]. Foliar nematodes include several nematode genera among which *Aphelenchoides, Anguina, Ditylenchus,* and *Litylenchus*. Foliar nematodes have been documented as associated with more than 1100 different species of plants, belonging to 126 botanical families, to include dicots, monocots, gymnosperms and angiosperms, ferns and mosses [2]. *Aphelenchoides,* as well as nematodes of genus *Litylenchus,* are phytoparasites known to infect leaves, stems, and buds [3]. The damage caused by the foliar nematodes can cause marketability problems in ornamentals because they interfere with the appearance of the plant or they can reduce yield in food crops [2].

## 2. General Techniques

For morphological observation, adult specimens of foliar nematodes can be extracted from fresh leaves. The best method for extracting nematodes from fresh leaves is by using the Baermann Funnel method. Another simple extraction method of nematodes from rice seeds was described by Hoshino and Togashi [4]. They cut the rice seeds longitudinally in two, then transferred the pieces into single plastic pipette tips, which were placed upright in glass vials with water. The pipet tips are transferred to new vials 2, 4, 8, and 24 h later the rate of nematode extraction can be observed. Nematodes are transferred to Syracuse watch glasses and counted. The final step includes dissection of seeds and counting the remaining nematodes after additional 24 h.

For light microscopic observation, fresh specimens are fixed using different methods, such as the ones described by Golden [5] and Hooper [6]. Another method of fixing nematodes was described by Ryss et al. [7] in which nematodes are placed into cold 4% formalin and 1% glutaraldehyde in 0.01 M phosphate buffer at pH 7.3, and then stored at 48 °C for light and scanning electron microscopy (SEM).

For SEM observations, nematode specimens can be fixed in phosphate-buffered aldehyde and transferred to special chambers [8], rinsed for 15 min in distilled water, transferred for 2 h in 1% aqueous osmium tetroxide, rinsed again in distilled water and dehydrated in increasing concentrations of ethanol (10% to 100%) in 10% increments for 30 min each, followed by three changes of 100% ethanol. Alcohol is removed using a critical point dryer and the dried specimens stored under vacuum over silica gel. Dried specimens can be mounted on double-sided adhesive tape placed on SEM stubs, sputter-coated with 30 nm of gold, and photographed [7]. To observe the nematode stylets, one individual (alive) specimen is placed in a 1 µL drop of 45% lactic acid on a 12-mm-round, glass cover slip. A small sliver of a broken cover slip, approximately 1 × 3 mm, is placed over the specimen and pressure is applied to it with a needle until the nematode ruptures and the stylet and guiding apparatus extrude. After 24 h, small triangles of filter paper are applied to the edge of the broken sliver to remove the lactic acid, which is exchanged with 2% formalin, followed by three changes of 50% ethanol. The sliver is then floated by adding 50% ethanol and removed with forceps. Stylets adhering to the glass cover slips are air-dried and prepared for SEM as described previously.

Other methods to prepare nematodes for low temperature SEM observations were described by Kantor et al. [9] and Carta et al. [10]. Nematodes can be placed in 1.5 Eppendorf tubes filled with a fixative composed of 2% Paraformaldehyde, 2.5% Glutaraldehyde, 0.05 M Na Cacodylate, and 0.005 M $CaCl_2$ for at least 12 h. After 12 h, specimens are rinsed in distilled water and individual nematodes placed onto ultra-smooth, round (12 mm diameter), carbon adhesive tabs (Electron Microscopy Sciences, Inc., Hatfield, PA, USA) secured to 15 mm × 30 mm copper plates. The nematode specimens are frozen conductively, in a Styrofoam box, by placing the plates on the surface of a pre-cooled (−196 °C) brass bar whose lower half is submerged in liquid nitrogen. After 20–30 s, the brass plate containing the frozen sample is transferred to the Quorum PP2000 cryo transfer system (Quorum Technologies, East Sussex, UK), attached in this case to an S-4700 field emission scanning electron microscope (Hitachi High Technologies America, Inc., Dallas, TX, USA). The specimens are freeze- etched inside the cryotransfer system to remove any surface contamination (condensed water vapor) by raising the temperature of the stage to −90 °C for 10–15 min. Following etching, the temperature inside the chamber is lowered below −130 °C, and the specimens coated with a 10 nm layer of platinum using a magnetron sputter head equipped with a platinum target. The specimens are transferred to a pre-cooled (−130 °C) cryostage in the SEM for observation. An accelerating voltage of 5 kV is used to view the specimens.

DNA extraction from live specimens can be performed using the freeze-thaw lysis with a single live nematode in a 0.2 mL PCR tube containing 25 µL of extraction buffer (10 mM Tris pH 8.2, 2.5 mM $MgCl_2$, 50 mM KCl, 0.45% TWEEN 20 and 0.05% gelatin). Next, the PCR tube is submerged in liquid nitrogen for 10 to 15 s and then placed at 95 °C for 2 min in a thermal cycler. The tube is submerged one more time in liquid nitrogen for 10 to 15 sec and then slow-thawed at room temperature. After thawing, the sample is lysed with 1 µL of proteinase K (800 U/mL, Sigma-Aldrich, St. Louis, MO, USA) at 60 °C for 60 min, followed by 95 °C for 15 min to deactivate the proteinase K. It is recommended to use at least three single nematodes for the individual DNA extraction. The lysates can be stored at −20 °C until needed [11,12]. After extraction, the DNA fragments can be amplified using SSU rDNA (18S), D2D3 (28S) expansion region of the LSU rDNA and cytochrome oxidase subunit 1 of the mitochondrial DNA (mtCOI) markers [13]. The primers used for the 18S fragment amplification are 1813F (CTGCGTGAGAGGTGAAAT) and 2646R (GCTACCTTGTTACGACTTTT) and were first published by Holterman et al. [14]. Primers used for the amplification of the 28S region are D2A (ACAAGTACCGTGAGGGAAAGTTG) and D3B (TCCTCGGAAGGAACCAGCTACTA) [15].

The mtCOI fragment can be amplified using COI-F1(CCTACTATGATTGGTGGTTTTGGTAA TTG) and COI-R2 (GTAGCAGCAGTAAA ATAAGCACG) primers [16].

## 3. Genus *Aphelenchoides* Fischer, 1894

After Goodey [17] the genus *Aphelenchoides* Fischer, 1894 is characterized as follows:

- Six fused, non-annulated, similar lips, slightly offset from body;
- Male tail without bursa, with one pair of approximately adanal and two pairs of postanal, ventro-submedian, caudal papillae;
- Spicules paired and shaped like rose thorns;
- Tails of both sexes never elongate filiform but short, tapering, conical, and frequently ending in one or more mucrones.

A polytomous key was developed and tested on 14 populations by Hockland [18] and the primary key characters were identified as:

- The length of the post-vulval sac;
- The shape of the tail terminus and tail;
- Body length;
- Ratios 'a' and 'c'.

A more detailed characterization of the genus was first given by Allen [19]:

- Cuticle marked by fine transverse striae;
- Lateral field marked as longitudinal incisures;
- Lip region set off from body;
- Six lips supported by six radial internal sclerotization;
- Lips not annulated;
- Stylet with or without basal knobs;
- Medial esophageal bulb well developed;
- Intestine joining esophagus immediately behind bulb;
- Nerve ring encircling anterior ends of intestine and the esophageal glands;
- Esophageal glands free in the body cavity;
- Single anteriorly directed ovary, oocytes on tandem or multiple;
- Male tail without bursa or gubernaculum;
- Three pairs of ventro-submedian papillae usually present on male tail;
- Spicules paired, ventrally arcuate.

Female and male tail never elongate filiform. A recent characterization of *Aphelenchoides* was given by Wheeler and Crow [20]:

- Stylet with small basal knobs;
- Males are common;
- Vulva located near 2/3 the body length from the anterior;
- Prodelphic (anteriorly outstretched) ovary and a post-uterine sac;
- Males have prominent, thorn-shaped spicules (paired, cuticularized copulatory structures).

According to Hunt [21] members of the *Aphelenchoides* genus can be diagnosed by the following morphological characteristics:

- Body length between 0.4 to 1.2 mm (commonly from 0.4 to 0.8 mm);
- Females become straight to ventrally arcuate when heat relaxed while males assume a "walking-stick shape";

- Cuticle finely annulated, two to four (rarely six) incisures in the lateral field;
- Stylet slender with basal knobs (sometimes indistinct), length between 10–12 μm;
- Long and slender procorpus; well-developed spherical to rounded-rectangular shaped metacorpus, with central valve plates; esophageal gland lobe long, with dorsal overlap of the intestine;
- Vulva usually between 60 and 75% of the body length;
- Ovary monoprodelphic, typically outstretched, but may reflex;
- Post-vulval sac present most of the times;
- Oocytes in one or more rows;
- Post-uterine sac present (sometimes absent) and most of the times contains spermatozoa;
- Tail shape is conoid to variable; males have a tail more strongly curved ventrally and papillae variable;
- Tail terminus with one or more mucros or without mucros;
- Spicules well-developed, thorn-shaped, paired and separate without bursa.

## 4. Genus *Aphelenchoides* Fischer, 1894

Emended Diagnosis [22]

These nematodes are small and slender, averaging around one millimeter in length and a width less than 20 microns. One characteristic of thee Aphelenchidae nematodes family is that they have a larger median bulb as compared to other plant parasitic nematodes in the order Tylenchida. The dorsal esophageal gland orifice connects to the esophageal lumen at the base of the stylet in most plant-parasitic nematodes, but in Aphelenchida, this duct empties into the esophageal lumen within the median bulb. In *Aphelenchoides*, males are more common, and they reproduce primarily by amphimixis. In most species, the vulva of the female is located near 2/3 the body length from the anterior. Females have a single, prodelphic (anteriorly outstretched) ovary and a post-uterine sac, while males have prominent, thorn-shaped spicules (paired, cuticularized copulatory structures). There is a considerable variation in the shape of the tail terminus within populations of species of the genus *Aphelenchoides*. The tail terminus can be used to divide *Aphelenchoides* species into four groups [22]. The four groups are:

a. Tail without any outgrowth or mucro;
b. Tail with one or sometimes two mucronate structures on tail end;
c. Star shaped tail with four mucronate structures;
d. Tail end with outgrowth other than spine or star.

## 5. Systematic Position

The number of valid nominal species in the Aphelenchoidea is still debatable. However, modern molecular technology may help solve this problem soon. According to Hunt [23] there are 453 'valid' species in Aphelenchoidea, of which 33 belong to the Aphelenchidae and 420 to the Aphelenchoididae. From Aphelenchoididae family, *Aphelenchoides* genera has the most species, namely 153 [23]. A more recent 2015 study conducted by Sánchez-Monge et al. [2] assigned approximately 200 species to the genus. However, after conducting a through literature review, the authors have identified 182 valid nominal species assigned to the *Aphelenchoides* genus.

## 6. Diagnostic Characters

Some diagnostic characteristics of *Aphelenchoides* are presented below [3]:

- Slender body, length variable;
- Lips often slightly offset;
- Stylet with basal knobs;
- Oocytes in one or more rows;

- Post-uterine sac usually well-developed, with variable length;
- Spicules paired, rose thorn-shaped, not fused, rostrum usually prominent;
- Male tail without caudal alae or gubernaculum; with three pairs of ventro-submedian papillae;
- Tails of both sexes never elongate-filiform, but usually more or less tapering, conical, and frequently ending in one or more mucrons.

## 7. Genus Synonyms

Emended list of *Aphelenchoides* species and synonyms:

**Type species:**

1. *A. kuehnii* Fischer, 1894 = *A.* (*Aphelenchoides*) *kuehnii* Fischer, 1894 (Filipjev, 1934)

**Other species:**

1. *A. absari* Husain and Khan, 1967
2. *A. abyssinicus* (Filipjev, 1931) Filipjev, 1934 = *Aphelenchus abyssinicus* Filipjev, 1931
3. *A. aerialis* Chanu, Mohilal, Victoria and Shah, 2015
4. *A. africanus* Dassonville and Heyns, 1984
5. *A. agarici* Seth and Sharma, 1986
6. *A. aligahriensis* Siddiqi, Hussain and Khan, 1967
7. *A. andrassyi* Husain and Khan, 1967
8. *A. angusticaudatus* Eroshenko, 1968
9. *A. appendurus* Singh, 1967
10. *A. arachidis* = *Robustodorus arachidis* Bos, 1977
11. *A. arcticus* Sanwal, 1965
12. *A. asterocaudatus* Das, 1960
13. *A. asteromucronatus* Eroshenko, 1967
14. *A. baguei* Maslen, 1979
15. *A. besseyi* Christie, 1942 = *Aphelenchoides oryzae* Yokoo, 1948 *Asteroaphelenchoides besseyi* (Christie 1942) Drozdovski, 1967
16. *A. bicaudatus* (Imamura, 1931) Filipjev and Schuurmans Stekhoven, 1941 = *Aphelenchus bicaudatus* (Imamura, 1931)
17. *A. bimucronatus* Nesterov, 1985
18. *A. blastophthorus* Franklin, 1952
19. *A. brassicae* Edward and Misra, 1969
20. *A. brevicaudatus* Das, 1960
21. *A. brevionchus* Das, 1960
22. *A. breviuteralis* Eroshenko, 1967
23. *A. brushimucronatus* Bajaj and Walia, 1999
24. *A. capsuloplanus* = *Paraphelenchoides capsuloplanus* Haque, 1967
25. *A. centralis* Thorne and Malek, 1968
26. *A. chalonus* Chawla and Khan, 1979
27. *A. chamelocephalus* (Steiner, 1926) Filipjev, 1934
28. *A. chauhani* Tandon and Singh, 1974
29. *A. chinensis* Husain and Khan, 1967
30. *A. cibolensis* Riffle, 2011
31. *A. citri* Andrássy, 1957
32. *A. clarolineatus* Baranovskaya, 1958
33. *A. clarus* Thorne and Malek, 1968

34. *A. composticola* Franklin, 1957
35. *A. confusus* Thorne and Malek, 1968
36. *A. conimucronatus* Bessarabova, 1966
37. *A. conophthori* Massey, 1974
38. *A. curiolis* Gritsenko, 1971
39. *A. cyrtus* Paesler, 1957
40. *A. dactylocercus* Hooper, 1958
41. *A. dalianensis* Cheng, Hou and Lin, 2009
42. *A. daubichaensis* Eroshenko, 1968
43. *A. delhiensis* Cwala, Bhamburkar, Khan and Prasad, 1968
44. *A. dhanachandhi* Chanu, Mohilal and Shaw, 2012
45. *A. dubitus* Ebsary, 1991
46. *A. echinocaudatus* Haque, 1968
47. *A. eldaricus* Esmaeili, Heydari, Golhasan and Kanzaki, 2017
48. *A. editocaputis* Shavrov, 1967
49. *A. eltayebi* Zeidan and Geraert, 1991
50. *A. emiliae* Romaniko, 1966
51. *A. ensete* Swart, Bogale and Tiedt, 2000
52. *A. eradicitus* Eroshenko, 1968
53. *A. fluviatilis* Andrassy, 1960
54. *A. fragariae* (Ritzema Bos, 1891) Christie, 1932 = *Aphelenchoides olesistus* (Ritzema Bos, 1893) Steiner, 1932 *Aphelenchoides olesistus* var. *longicollis* (Schwartz, 1911) Goodey, 1933 *Aphelenchoides pseudolesistus* (Goodey, 1928) Goodey, 1933 *Aphelenchus fragariae* Ritzema Bos, 1891 *Aphelenchus olesistus* Ritzema Bos, 1893 *Aphelenchus olesistus* var. *longicollis* Schwartz, 1911 *Aphelenchus pseudolesistus* Goodey, 1928
55. *A. franklini* Singh, 1969
56. *A. fuchsi* Esmaeili, Heydari, Ziaie and Gu, 2016
57. *A. fujianensis* Zhuo, Cui, Ye, Luo, Wang, Hu, and Liao, 2010
58. *A. giblindavisi* Aliramaji, Pourjam, Alvarez-Ortega, Afshar and Pedram, 2017
59. *A. goeldii* (Steiner, 1914) Filipjev, 1934 = *Aphelenchus goeldii* Steiner, 1914 *Aphelenchoides* (A.) *goeldii* (Steiner, 1914) Filipjev, 1934
60. *A. goldeni* Suryawanshi, 1971
61. *A. goodeyi* Siddiqi and Franklin, 1967
62. *A. gorganensis* Miraeiz, Heydari and Bert, 2017
63. *A. graminis* Baranovskaya and Haque, 1968
64. *A. gynotylurus* Timm and Franklin, 1969
65. *A. haguei* Maslen, 1978
66. *A. hamatus* Thorne and Malek, 1968
67. *A. heidelbergi* Carta, Li, Skantar, and Newcombe, 2016 = *Laimaphelenchus heidelbergi* Zhao, Davies, Riley, and Nobbs, 2007
68. *A. helicosoma* Maslen, 1978
69. *A. helicus* Heyns, 1964
70. *A. helophilus* (de Man, 1880) Goodey, 1933 = *Aphelenchus helophilus* le Man, 1880 *Aparietinus var. helophilus* de Man, 1880 *Aphelenchoides (A.) helophilus* (de Man, 1880) Goodey, 1933 *Aphelenchus elegans* Micoletzky, 1913
71. *A. heterophallus* Steiner, 1934
72. *A. huntensis* Esmaeili, Fang, Li and Heydari, 2016

73. *A. hunti* Steiner, 1935
74. *A. hylurgi* Massey, 1974
75. *A. indicus* Chawla, Bhamburkar, Khan and Prasad, 1968
76. *A. involutus* Minegawa, 1992
77. *A. iranicus* Golhasan, Heydari, Alvarez-Ortega and Palomares-Rius, 2016
78. *A. jacobi* Husain and Khan, 1967
79. *A. jodhpurensis* Tikyani, Khera and Bhatnagar, 1970
80. *A. jonesi* Singh, 1977
81. *A. kheirii* Golhasan, Heydari, Esmaeili and Kanzaki, 2018
82. *A. kungradensis* Karimova, 1957
83. *A. lanceolatus* Tandon and Singh, 1974
84. *A. lagenoferrus* Baranovskaya, 1963
85. *A. lanceolatus* Tandon and Singh, 1974
86. *A. lichenicola* Siddiqi and Hawksworth, 1982
87. *A. lilium* Yokoo, 1964
88. *A. limberi* Steiner, 1936 = *Paraphelenchoides limberi* (Steiner, 1936) Hague, 1967
89. *A. longiurus* Das, 1960
90. *A. longiuteralis* Eroshenko, 1967
91. *A. loofi* Kumar, 1982
92. *A. lucknowensis* Tandon and Singh, 1973
93. *A. macromucrons* Slankis, 1967
94. *A. macronucleatus* Baranovskaya, 1963
95. *A. macrospica* Golhasan, Heydari, Esmaeili and Miraeiz, 2017
96. *A. marinus* Timm and Franklin, 1969
97. *A. martinii* Ruhm, 1955
98. *A. medicagus* Wang, Bert, Gu, Couvrer and Li, 2019
99. *A. meghalayensis* Bina and Mohilal, 2017
100. *A. menthae* Lisetzkaya, 1971
101. *A. microsylus* Kaisa, 2000
102. *A. minor* Seth and Sharma, 1986
103. *A. myceliophagus* Seth and Sharma, 1986
104. *A. nechaleos* Hooper and Ibrahim, 1994
105. *A. neocomposticola* Seth and Sharma, 1986
106. *A. neoechinocaudatus* Chanu, Mohilal and Shah, 2012
107. *A. nonveilleri* Andrassy, 1959
108. *A. obtusicaudatus* Eroshenko, 1967
109. *A. obtusus* Thorne and Malek, 1968
110. *A. orientalis* Eroshenko, 1968
111. *A. pannocaudus* Massey, 1966
112. *A. paradalianensis* Cui, Zhuo, Wang and Liao, 2011
113. *A. paramonovi* Eroshenko and Kruglik, 2004
114. *A. paranechaleos* Hooper and Ibrahim, 1994
115. *A. parasaprophilus* Sanwal, 1965
116. *A. parasexalineatus* Kalinich, 1984
117. *A. montanus* Singh, 1967
118. *A. panaxi* Skarbilovich and Potekhina, 1959
119. *A. parabicaudatus*, Shavrov, 1967

120. *A. parascalacaudatus* Chawla, Bhamburkar, Khan and Prasad, 1968
121. *A. parasubtenuis* Shavrov, 1967
122. *A. paraxui* Esmaeili, Heydari, Fang and Li, 2017
123. *A. parietinus* (Bastian, 1865) Steiner, 1932
124. *A. petersi* Tandon and Singh, 1970
125. *A. pinusi* Bajaj and Walia, 1999
126. *A. pityokteini* Massey, 1974
127. *A. platycephalus* Eroshenko, 1968
128. *A. polygraphi* Massey, 1974
129. *A. primadentus* Esmaeili, Heydari, Golhasan and Kanzaki, 2018
130. *A. pseudogoodeyi* Oliveira, Subbotin, Alvarez-Ortega, Desaeger, Brito, Xavier, Freitas, Vau and Inserra, 2019
131. *A. pusillus* (Thorne, 1929) Filipjev, 1934
132. *A. rarus* Eroshenko, 1968
133. *A. rhytium* Massey, 1971
134. *Aphelenchoides ritzemabosi* (Schwartz, 1911) Steiner and Buhrer = *Aphelenchoides ribes* (Taylor, 1917) Goodey, 1933; *Aphelenchus phyllophagus* Stewart, 1921; *Aphelenchus ribes* (Taylor, 1917) Goodey, 1923; *Aphelenchus ritzemabosi* (Schwartz, 1911); *Pathoaphelenchus ritzemabosi* (Schwartz, 1911) Steiner, 1932; *Pseudaphelenchoides ritzemabosi* (Schwartz, 1911) Drozdovski, 1967; *Tylenchus ribes* Taylor, 1917
135. *A. rosei* Dmitrenko, 1966
136. *A. rotundicaudatus* Fang, Wang, Gu and Li, 2014
137. *A. rutgersi* Hooper and Myers, 1971
138. *A. sacchari* Hooper, 1958
139. *A. sanwali* Chaturvedi and Khera, 1979
140. *A. saprophilus* Franklin, 1957
141. *A. salixae* Esmaeili, Heydari, Tahmoures and Ye, 2017
142. *A. scalacaudatus* Sudakova, 1958
143. *A. seiachicus* Nesterov, 1973
144. *A. sexlineatus* Eroshenko, 1967
145. *A. shamimi* Khera, 1970
146. *A. siddiqii* Fortuner, 1970
147. *A. silvester* Andrassy, 1968
148. *A. sinensis* (Wu and Hoeppli, 1929) Andrassy, 1960
149. *A. singhi* Das, 1960
150. *A. sinodendroni* Ruhn, 1957
151. *A. smolae* Cai, Gu, Wang, Fang and Li, 2020
152. *A. solani* Steiner, 1935
153. *A. spasskii* Eroshenko, 1968
154. *A. sphaerocephalus* Goodey, 1953
155. *A. spicomucronatus* Truskova, 1973
156. *A. spinosus* Paesler, 1957
157. *A. spinohamatus* Bajaj and Walia, 1999
158. *A. spinosus* Paesler, 1957
159. *A. stammeti* Korner, 1954
160. *A. steineri* Ruhm, 1956
161. *A. stellatus* Fang, Gu, Wang and Li, 2014

162.  *A. submersus* Truskova, 1973

163.  *A. subparietinus Sanwal,* 1961

164.  *A. subtenuis = Robustodorus subtenuis* (Cobb, 1926) Steiner and Buhrer, 1932

165.  *A. suipingensis* Feng and Li, 1986

166.  *A. swarupi* Seth and Sharma, 1986

167.  *A. tabarestanensis* Golhasan, Fang, Li, Maadi and Heydari, 2019

168.  *A. tagetae* Steiner, 1941

169.  *A. taraii* Edward and Misra, 1969

170.  *A. tsalolikhini* Ryss, 1993

171.  *A. trivialis* Franklin and Siddiqi, 1963

172.  *A. tumulicaudatus* Truskova, 1973

173.  *A. turnipi* Israr, Shahina and Nasira, 2017

174.  *A. tuzeti* B'Chir, 1978

175.  *A. unisexus* Jain and Singh, 1984

176.  *A. varicaudatus* Ibrahim and Hooper, 1994

177.  *A. vaughani* Maslen, 1978

178.  *A. vigor* Thorne and Malek, 1968

179.  *A. wallacei* Singh, 1977

180.  *A. xui* Wang, Wang, Gu, Wang and Li, 2013

181.  *A. zeravschanicus* Tulaganov, 1948

## 8. Principal Species

The following four species have been selected for further discussion because of their commonality, economic importance, and/or worldwide distribution:

- *Aphelenchoides besseyi* Christie, 1942;
- *Aphelenchoides bicaudatus* (Imamura, 1931) Filipjev and Schuurmans Stekhoven;
- *Aphelenchoides fragariae* (Ritzema Bos, 1891) Christie, 1932;
- *Aphelenchoides ritzemabosi* (Schwartz, 1911) Steiner and Buhrer, 1941.

Each species is illustrated below (Figures 1–11). Data were obtained from various sources, including Allen [19]; Christie [24] De Jesus et al. [25], 2016; Xu et al. [26]; Siddiqi [27–30]; Shahina [22]; Siddiqui and Taylor [31]; Jen et al. [32]; Khan et al. [33]; Chizhov et al. [34]; Zhao et al. [35]; Khan et al. [36]; Hunt [21], Kanzaki et al. [37] 2019 and Carta et al. [11], and original descriptions and/or re-descriptions.

Because *Aphelenchoides besseyi* Christie, 1942, *Aphelenchoides fragariae* (Ritzema Bos, 1891) Christie, 1932, *Aphelenchoides ritzemabosi* (Schwartz, 1911) Steiner and Buhrer, *Aphelenchoides bicaudatus* (Imamura, 1931) Filipjev and Schuurmans Stekhoven, 1941 are of major economic importance and widely distributed all over the world, they will be discussed in detail.

## 9. Rice White-Tip Nematode (*Aphelenchoides besseyi* Christie, 1942)

*Aphelenchoides besseyi* (Figure 1) is an economically important pathogen of rice and has been reported from many countries. However, it is not commonly found in ornamentals [38,39], with the exception of some reports on tuberose [36], begonia [40], gerbera [41], hydrangea [27], tuberose [42], and even on bird nest fern [43]. *A. besseyi* distribution is mostly in warmer climates, whereas *A. ritzemabosi* and *A. fragariae* are more commonly associated with temperate climates, while found in both tropical and temperate localities [1].

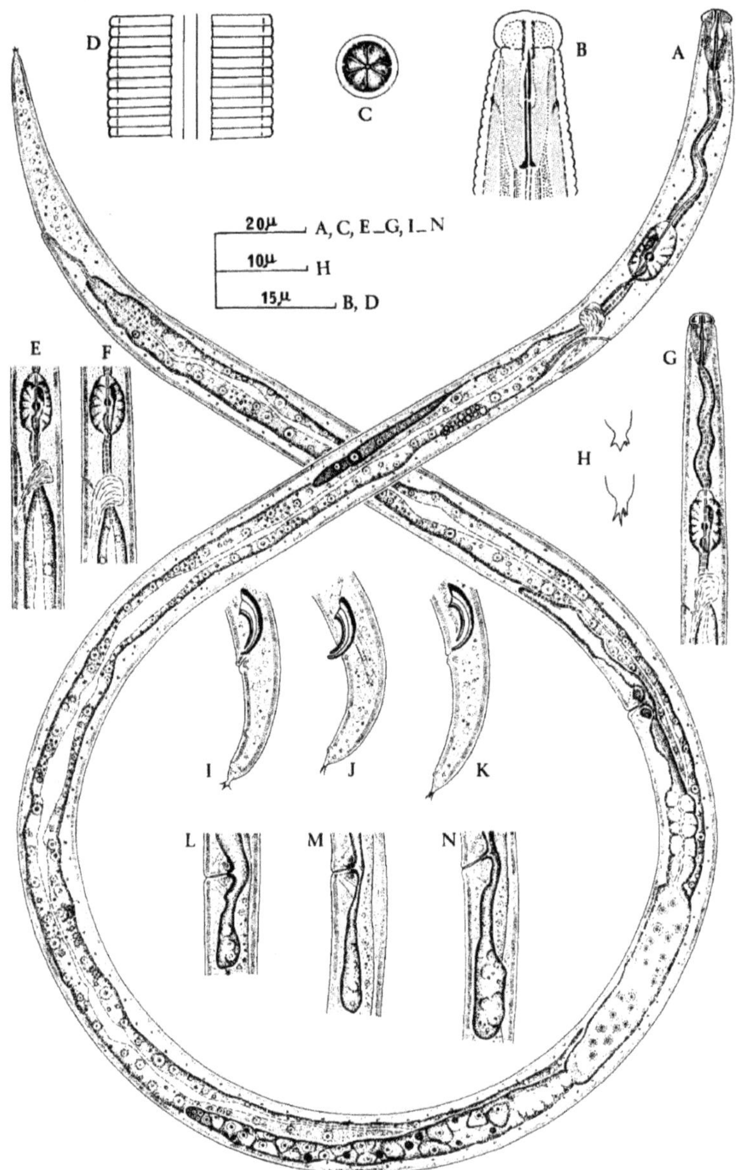

**Figure 1.** *Aphelenchoides besseyi* Christie (**A**) female; (**B**) female head end; (**C**) female en face view; (**D**) Lateral field; (**E,F**) variation in female esophageal bulb and position of excretory pore with respect to nerve ring; (**G**) male anterior end; (**H**) female tail termini showing variation in shape mucro; (**I–K**) male tail ends; (**L–N**) variation in post-vulval uterine sac (**B** and **D** original, the rest after Fortuner, 1970) after Franklin and Siddiqi [27]. Courtesy of *Commonwealth Institute of Helminthology*.

Measurements

After Christie [24].

Females ($n$ = 10): length = 0.66–0.75 mm; a = 32–42 (width = 17–22); b = 10.2–11.4 (esophagus = 64–68 μm); c = 17–21 (tail = 36–42 μm); V = 68–70%.

Males ($n$ = 10): length = 0.54–0.62 mm; a = 36–39 (width = 14–17 μm); b = 8.6–8.8 (esophagus = 63–66 μm); c = 15–17 (tail = 34–37 μm); T = 44–61%

After Allen [19].

Females: length = 0.62–0.88 mm; a = 38–58; b = 9–12; c = 15–20; V = 66–72

Males: length = 0.44–0.72 mm; a = 36–47; b = 9–11; c = 14–19; T = 50–65%.

After De Jesus et al. [20]

Females: length = 0.65–0.75 mm; a = 42.8–49; c = 15.6–17.5; c' = 4.0–4.5.

Males: length = 0.65–0.75 mm; a = 42.8–49; c = 15.6–17.5; c' = 4.0–4.5; spicule = 14.1–18.3 μm.

After Xu et al. [44]

Body length ($n$ = 11) = 0. 656 ± 18.5 (0.546–0.729) mm; body width = 14.4 ± 0.32 (12.4–15.9) μm; pharynx = 124 ± 2.53 (111.0–137.8) μm; stylet = 12.5 ± 0.21 (10.6–13.3) μm; median bulb end to anterior end 69.7 ± 1.07 (65.7–75.3) μm; tail length 36.9 ± 0.38 (35.3–38.9) μm; anus/cloacal width 9.27 ± 0.47(7.5–12.1) μm.

Description

Female: female specimens share a slender body, slightly arcuate ventrally when relaxed, anteriorly tapering from the level of esophageal glands to the head, which is one half of the body width. Four lateral lines (occasionally six noted) are present in the lateral field (Figure 2). In en face view, the pore-like amphids are on outer margins of lateral lips; four papillae, one on each submedian lip (Figure 2). Lip region is non-striated and set off from body by a constriction as wide as or slightly wider than adjacent body; labial framework weakly developed; cheilorhabdions well sclerotized. Basal knobs of spear distinct, 2 μm across. Procorpus cylindrical; median esophageal bulb one and a half times to twice as long as wide, with refractive valvular apparatus slightly posterior to center. Esophageal glands extending over intestine 5 to 8 body widths. Excretory pore at 58 to 83 μm from anterior end, level with or slightly anterior to nerve ring. Hemizonid distinct in specimens from rice seeds (but not from cultured specimens), 11 to 15 μm behind excretory pore; hemizonion 20 to 30 μm behind hemizonid, usually difficult to see. Tail straight, slender, regularly tapering to a narrowly rounded end, 3–5 to 5 anal body diameters long; mucro with 3 to 4 processes. Ovary not extending to esophageal glands; oocytes in 2 to 4 rows; spermatheca very conspicuous, elongate oval, full of rounded sperms showing a central nucleolus usually surrounded by a circle of black dots of unknown nature. Post-vulval uterine sac short, slender and extending up to one fourth of the distance from vulva to anus (2.5 to 3 body diameters) often found empty and collapsed but more conspicuous and rounded in nematodes from cultured specimens. Vulval lips slightly protruding after Fortuner [45].

Male: tail end usually curved by 90° (a greater curvature has also been found) in specimens killed in 3% formaldehyde; mucro of diverse shape, with 2 to 4 processes. Spicule length between 17 to 21 μm along dorsal limb. Different morphometric characters, such as the shape of the head, the position of the excretory pore in relation to the nerve ring and the shape and length of the post-vulval uterine sac were found to be variable between populations [45].

Distribution

According to Devran et al. [46], *A. besseyi* was on the quarantine lists of nine countries in 1982 and up to 70 countries in 2002. Centre for Agriculture and Biosciences International (CABI), Invasive Species Compendium [47] lists *A. besseyi* being present in 75 countries around the world. The quarantine pests lists *A. besseyi* as the second most prevalent nematode after *Globodera rostochiensis* [39].

*Aphelenchoides bicaudatus* (Imamura, 1931) Filipjev and Schuurmans Stekhoven, 1941 *Aphelenchoides bicaudatus* (Imamura, 1931) Filipjev and Schuurmans Stekhoven, 1941 was originally described from a paddy field in Japan and previously considered a primarily mycophagous species. Since then, it has been reported to parasite more than 200 plant species [31,48].

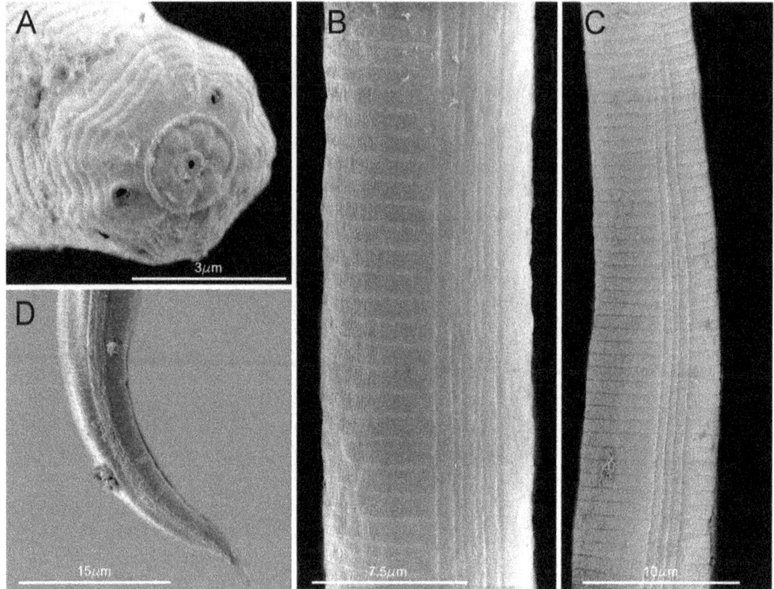

**Figure 2.** SEM photomicrographs of *Aphelenchoides besseyi* female (**A**) head end; (**B,C**) lateral fields; (**D**) tail end, after Khan et al. [36]. Courtesy of *Journal of Nematology*.

Measurements

After Imamura [49].

Female ($n$ = 18): L = 0.38–0.47 (0.43) mm; a = 31.3–31.7 (31.5); b = 6.8–8.4 (7.4); c = 9.4–12.6 (10.6); V%= 61.7–90.2 (0.4).

After Siddiqui and Taylor [31].

Female ($n$ = 50): L = 0.41–0.55 (0.46) mm; a = 25–31 (28.0); b = 7.3–9.6 (8.2); c = 9.8–13.7 (11.4); V% = 65–70 (67.5); stylet = 10–12 (11.2) μm.

Male: L = 0.385 mm; a = 22.6; b = 7.5; c = 11.4; stylet = 10 μm.

After Jen et al. [32].

Female ($n$ = 50): L = 499.12 ± 67.95 (0.376–0.637) mm; maximum body width = 15.24 ± 2.69 (11–22) μm; a = 33.03 ± 2.42 (27.00–38.64); b = 9.0 ± 0.7 (7.5–10.0); b′ = 5.13 ± 0.76 (3.61–7.94); c = 11.94 ± 0.93 (10.16–14.80); c′ = 5.41 ± 0.56 (4.13–7.14); V% = 68.53 ± 1.20 (64.90–71.83); stylet = 10.38 ± 0.63 (9–12) μm; length of post-uterine sac expressed as % of length from vulva to anus = 18.98 ± 4.54 (9.23–33.80) μm.

After Israr et al. [50].

Female ($n$ = 2): L = 0.36 mm; a = 30.1, 32.7; b = 8.8,7.2; b′ = 5.6, 5.8; c = 11.3, 12; c′ = 2.9, 3.7; V% = 66.8–67.2; $G_1$% = 25, 26.2; body diameter 12, 12,5; stylet = 10, 11 μm, median bulb length 10, 10 μm; median bulb width 7, 8 μm; median bulb length/ width 1.4, 1.3; distance anterior end to distal end of median bulb 51, 52 μm; anterior end to excretory pore 50, 51 μm; anterior end to nerve ring 55, 56 μm; anterior end to vulva 242, 248 μm; ovary length 95, 84 μm; distance from vulva to anus 85, 84 μm; post uterine sac length 24, 22 μm; post uterine sac length/vulva anus distance% 22.4, 24; esophageal length 90, 92 μm; esophageal intestinal junction 62, 64 μm; tail length 31, 30 μm; anal body width 31, 30 μm, anal body width 11, 8.

Male ($n$ = 1): L = 0.40 mm; a = 30.7; b = 4.3; b′ = 6.2; c = 10; c′ = 3.9; T% = 52; body diameter 13; stylet = 10 μm, median bulb length 12 μm; median bulb width 9 μm; median bulb length/width

1.3; distance anterior end to distal end of median bulb 54 µm; anterior end to excretory pore 62 µm; anterior end to nerve ring 60 µm.

Description

Female: have a slender body, attenuated slightly anteriorly, and more prominently toward posterior end (Figure 3). When relaxed by gentle heat the position of the body is straight and only the tail region is slightly curved. Cuticle is finely striated, with annuli measuring between 0.47–0.58 µm wide and 0.39–0.51 µm thick. Lateral field has two lateral lines. Head distinctly set off from body. Lip region rounded, offset with no annules. Stylet weak, with small basal swellings. Metacorpus rounded, occupying approximately 73% of body width. Nerve ring is located about 1/2 body width behind metacorpus. Excretory pore opposite anterior margin of nerve ring. Vulva a transverse slit and slightly protruding, about 66% of body length from anterior end. Post-vulvar uterine sac extending for one-fifth of distance from vulva to end of tail. Rectum prominent, straight, near ventral body wall, and in length approximately three-fourths of anal body width. Tail gradually tapering to terminus, which is unevenly bifurcated with one prong longer than the other.

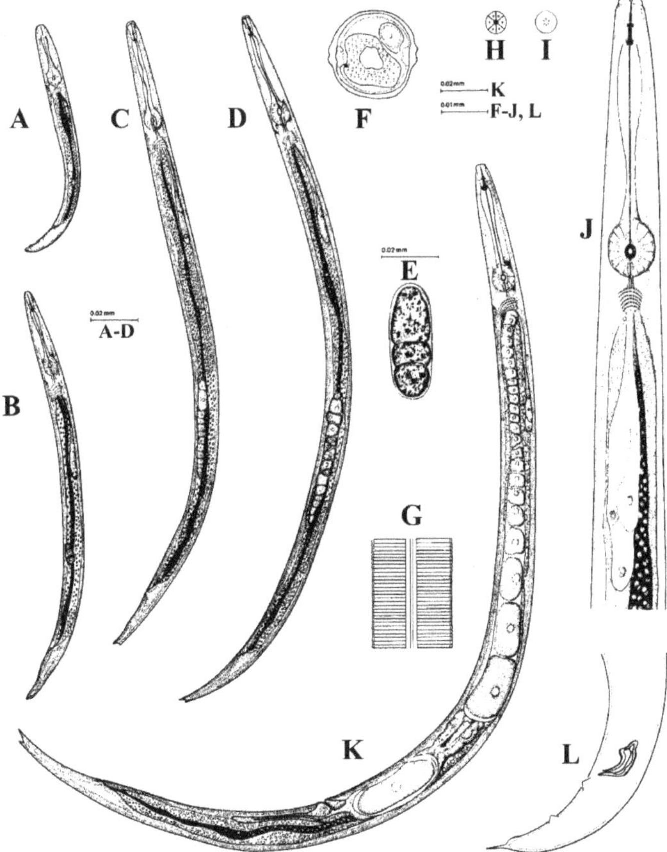

**Figure 3.** *Aphelenchoides bicaudatus* (Imamura) Filip. and Sch. Stek (**A–D**) Larvae, first of fourth stages; (**E**) egg; (**F**) cross section of female at mid-body; (**G**) lateral field; (**H**) face view; (**I**) framework around oral opening; (**J**) esophageal region in dorsal view; (**K**) whole female; (**L**) male tail after Siddiqi [28]. Courtesy of *Commonwealth Institute of Helminthology*.

Females of *A. bicaudatus* (Figure 4) can be differentiated from other members of the genus by having an unevenly bifurcated tail tip with prongs of different lengths [51].

Male: extremely rare.

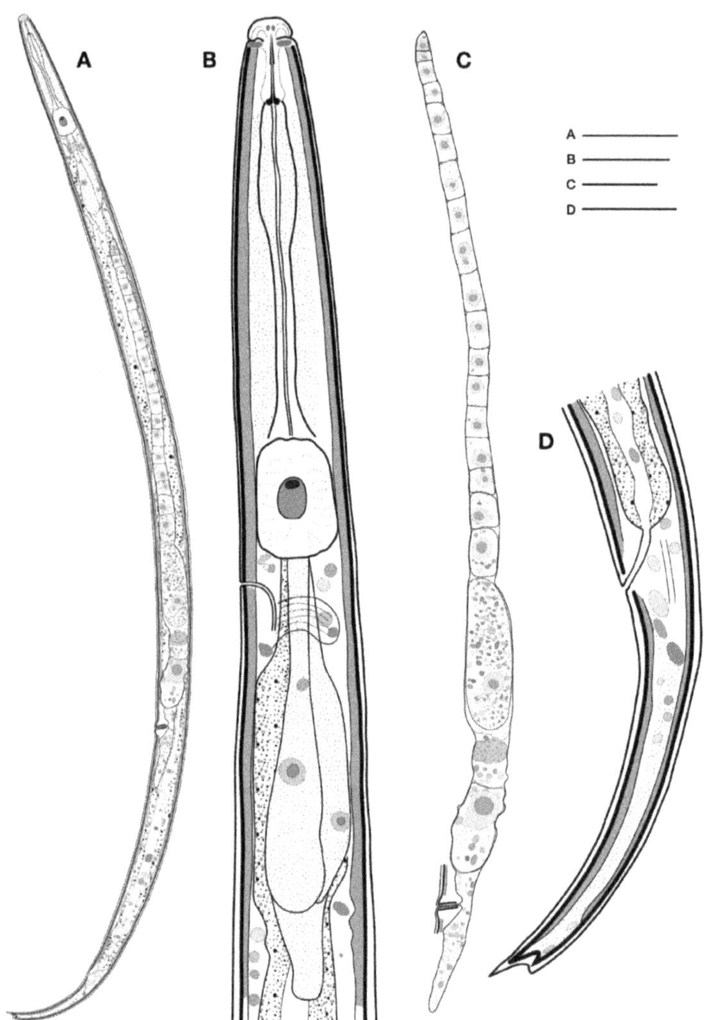

**Figure 4.** *Aphelenchoides bicaudatus* (Imamura, 1931) Filipjev and Schuurmans Stekhoven, 1941. (**A**) Entire female; (**B**) neck region; (**C**) female reproductive system; (**D**) female posterior region. Scale bars: (**A**) = 50 μm, (**B,D**) = 10 μm, (**C**) = 20 μm after Kim et al. [52]. Courtesy of *Animal Systematics Evolution and Diversity Journal*.

Distribution

*A. bicaudatus* was recorded in most of the tropical and subtropical regions of the world as well as some warmer temperate areas [21]. More specifically, it was reported in the following countries: Australia, Brunei, France, Japan, USA, Russia, Venezuela [28], South Korea [52], Taiwan [32].

## 10. Strawberry Crimp Nematode (*Aphelenchoides fragariae* (Ritzema Bos, 1891) Christie, 1932)

*Aphelenchoides fragariae* was originally described by Ritzema Bos (1891) in specimens recovered from strawberry plants sent to him from England (Figure 5). When compared to all the *Aphelenchoides* species mentioned previously, it has the widest distribution as well as hosts range (more than 600 species), to include ferns, herbaceous perennials and bedding plants [2,33,53]. *A. fragariae* is an ecto- and endo-parasite of the above ground parts of a plant, but it can also be mycetophagous [2,21,33]. The nematodes enter the plant leaves through stomata or wounds [1,47]. In the leaves, nematodes feed on mesophyll cells which causes characteristic vein delimited lesions [1,47]. *A. fragariae* survives overwinter in soil, dormant buds, dry leaves, but not in roots [18,47]. Research showed that *A. fragariae* nematodes can tolerate temperature as high as 40 °C and as low as −80 °C once in leaf tissues [18].

Measurements

After Allen [19].
Females: length = 0.45–0.80 mm; a = 45–60; b = 8–15; c = 12–20; V%= 64–71. Males: length = 0.48–0.65 mm; a = 46–63; b = 9–11; c = 16–19; T% = 44–61.
After Franklin [54].
Females: length = 0.552–0.886 (0.796) mm; a = 36–63 (53); body width = 12–17 (15) μm.
Males: length = 0.573–0.864 mm; a = 40–63; body width= 12–17 (14) μm.
After Khan et al. [33].
Females (*n* = 7): length = 0.620–0.895 mm; a = 46.2–64.5; b = 9.0–13.2; c = 13.4–20.3, V% = 66.5–72.2; stylet = 10.0–11.5 μm.
Males (*n* = 7): length = 0.480–0.623 mm; a = 45.7–61.7; b = 9.3–10.8; c = 15.7–18.5, T% = 45.6–60; stylet = 10.0–11.2 μm; spicules = 16.9–19.0 μm.
After Chizhov et al. [34].
Females (*n* = 25): length = 0.525–0.685 (0.579 ± 0.043) mm; a = 37.1–59.8 (48.7 ± 4.8); b = 7.6–9.1 (8.1 ± 0.3); c = 15.2–20.6 (17.0 ± 1.2), c′ = 3.6–5.7 (4.7 ± 0.3); V = 65.0–74.0% (69.0 ± 2.0); stylet = 8.0–11.0 (9.0) μm; head region width = 4.0–5.0 μm; head region high = 3.0 μm; distance from anterior end to: medial bulb base = 52.0–64.0 (58.0) μm, nerve ring = 63.0–78.0 (72.0) μm, excretory pore = 68.0–85.0 (76.0) μm and esophageal gland base = 100.0–150.0 (128.0) μm; post uterine sac length = 58.0–98.0 (77.0) μm; tail length = 28.0–40.0 (34.0) μm; body width at vulva level = 10.0–16.0 (12.0) μm and anus level = 6.0–8.0 (7.0) μm.
Males (*n* = 24): length = 0.435–0.562 (0.493 ± 0.037) mm; a = 41.2–54.8 (46.8 ± 3.1); b = 6.5–8.1 (7.2 ± 0.4); c = 15.9–24.1 (18.5 ± 1.8); stylet = 8.0–10.0 (9.0) μm; head region width = 4.0–5.0 μm; head region height = 3.0 μm; spicule length = 10.0–13.0 (12.0) μm; distance from anterior end to: medial bulb base = 52.0–62.0 (57.0) μm and esophageal gland base = 100.0–135.0 (118.0) μm; nerve ring = 68.0–77.0 (71.0) μm; excretory pore = 70.0–82.0 (76.0) μm; testis length = 204.0–289.0 (250.0) μm; maximal body width = 10.0–13.0 (11.0) μm; tail length = 21.0–33.0 (27.0) μm.

Description

Body very slender (a = 45–63 μm), straight or arcuate when relaxed. Cuticle marked by fine transverse striae about 0.9 μm apart; lateral field with two incisures, 1/7th of body-width. Cephalic region, smooth, anteriorly flattened with straight to curved side margins, almost continuous with neck contour. Lips without annulation. Stylet slender, approximately 10 μm long, with small but distinct basal knob. Median esophageal bulb well developed, oval. Nerve ring about one body width behind median bulb. Excretory pore level at or close behind nerve ring. Esophageal glands stretched five body widths behind the medium bulb, joining esophagus immediately behind the medium bulb. Tail elongate-conoid, terminus bearing a terminal peg which is simple, spike-like.
Female: vulva a transverse slit, at approximately 64–71% of body. Spermatheca elongate-oval. Posterior uterine sac more than half the vulva-anus distance, often containing sperm. Ovary single, with oocytes in a single row. Tail terminus with a single mucronate points point enlarged at the base.

Male: abundant. Male tail curved to about 45–90 degrees. Three pairs of ventro-submedian copulatory papillae (1st slightly post-anal, 2nd midway, and 3rd near the end). Testis single, outstretched; sperm large-sized, rounded, in a row. Spicules large and prominent, ventrally curved, rose-thorn-shaped, with moderately developed dorsal and ventral processes (apex and rostrum) at proximal end; dorsal limb 14–17 μm long.

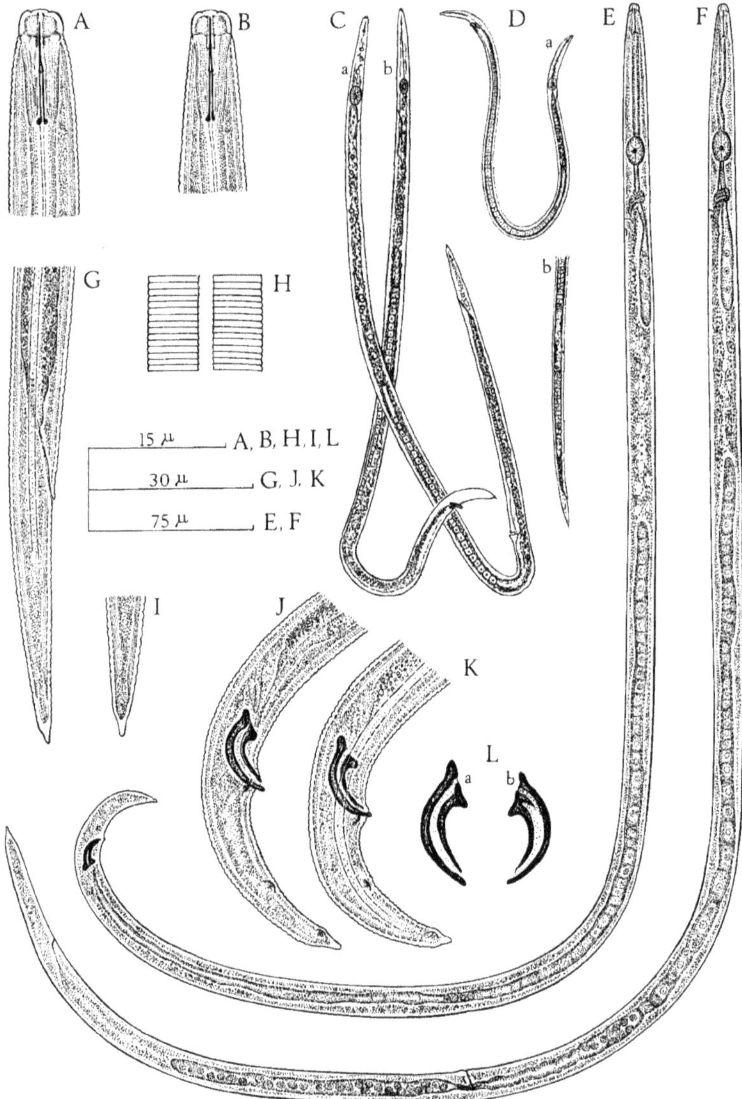

**Figure 5.** *Aphelenchoides fragariae*. (**A**) Female head end. (**B**) Male head end; (**C**) a, female; b male of *A. olesistus* Ritzema Bos, 1893 (= *A. fragariae*); (**D**) a, male; b, posterior portion of female, of *Aphelenchus fragariae* Ritzema Bos, 1891; (**E**) male; (**F**) female; (**G**) female tail; (**H**) lateral field; (**I**) female tail tip (**J,K**) male tails. (**L**) Spicules a, drawn from paratypes of Allen (1952); b, from specimens ex *Cornus canadensis* from Surrey, England after Siddiqi [29]. Courtesy of *Commonwealth Institute of Helminthology*.

Distribution

*A. fragariae* has a widespread distribution in Europe, Russia, Japan and North America [21]. According to the CABI Invasive Species Compendium [47], *A. fragariae* is currently reported to be present in 37 countries.

## 11. Chrysanthemum Nematode (*Aphelenchoides ritzemabosi* (Schwartz, 1911) Steiner and Buhrer)

*Aphelenchoides ritzemabosi* (Schwartz, 1911) Steiner and Buhrer, also known as the Chrysanthemum foliar nematode, is a common plant-parasite infecting more than 300 plant species, second only to *A. fragariae* [2] in the *Aphelenchoides* genus based on the number plants they parasitize.

Measurements

After Allen [13].
Females: length = 0.77–1.2 mm; a = 40–54; b = 10–13; c = 18–24; V% = 66–75.
Males: length = 0.70–0.93 mm; a = 31–50; b = 10–14; c = 16–30; T% = 35–64.
After Chizhov et al. [29].
Females ($n$ = 15): length = 0.768–1.027 (0.916 ± 0.067) mm; a = 43.4–60.5 (51.2 ± 3.7); b = 8.1–9.5 (9.1 ± 0.3); c = 16.8–21.2 (19.3 ± 1.1); c′ = 4.0–5.1 (4.6 ± 0.2); V% = 68–71 (69 ± 0.2); stylet = 9.0–11.0 (10.0) μm; head region width = 6.0–7.0 μm; head region height = 3.0 μm; distance from anterior end to: medial bulb base = 71.0–77.0 (74.0) μm; nerve ring= 95.0–108.0 (100.0) μm; excretory pore= 108.0–130.0 (121.0) μm and esophageal gland base = 145–185 (170) μm; postuterine sac length= 105.0–160.0 (134) μm; tail length= 41.0–54.0 (48.0) μm; body width at vulva level = 16.0–23.0 (18.0) μm and anus level= 8.0–12.0 (10.0) μm.
Males ($n$ = 15): length = 0.625–0.852 (0.721 ± 0.053) mm; a = 36.9–53.3 (46.3 ± 3.3); b = 6.5–9.4 (7.9 ± 0.6); c = 17.3–22.4 (19.9 ± 1.1); stylet = 9.0–11.0 (10.0) μm; head region width = 6.0–7.0 μm; head region height= 3.0 μm; spicule = 15–18 (16) μm; distance from anterior end to: medial bulb base = 67.0–72.0 (69.0) μm, nerve ring = 85.0–108.0 (93.0) μm; excretory pore = 92.0–118.0 (105.0) μm and esophageal gland base = 156.0–180.0 (169.0) μm; testis length = 353.0–512.0 (442.0) μm; tail length = 34.0–39.0 (36.0) μm.

Description (Figure 6)

Female: nematodes with slender body, with fine transverse striae on the cuticle. Four lines present in the lateral field. Lip region set off, wider than neck at base of lips with no annulations. Hexaradiate framework weakly sclerotized. Stylet approximately 12 μm long, with small but well-developed basal knobs. Median esophageal bulb well developed, oval in shape. Nerve ring 1.5 body widths behind median bulb. Excretory pore located behind nerve ring, approximately 0.5–2 body widths posterior to nerve ring. Esophageal glands extending 4 body widths over the intestine, joining esophagus immediately behind median bulb. Oocytes in multiple rows, several in a cross-section at middle of ovary. Posterior uterine branch extending for more than half the vulva-anus distance, usually containing sperms. Tail elongated-conoid. Terminus peg-like armed with two-four small mucronate points pointing posteriorly.

Male: males are common, having a tail curvature at about 180 degrees when relaxed. Testis single. Three pairs of ventro-submedian papillae. First pair adanal, second midway on tail, third near end. Spicules smoothly ventrally curved, the ventral piece without a ventral process at the distal end; dorsal limb 20–22 μm long. Terminus peg-like armed with two-four small mucronate points.

Distribution

*Aphelenchoides ritzemabosi* is a major pest of chrysanthemum in Europe, Russia, North America, South Africa, New Zealand, Australia, and Brazil [25]. According to the CABI Invasive Species Compendium [55], *A. ritzemabosi* is currently reported to be present in 35 countries around the world.

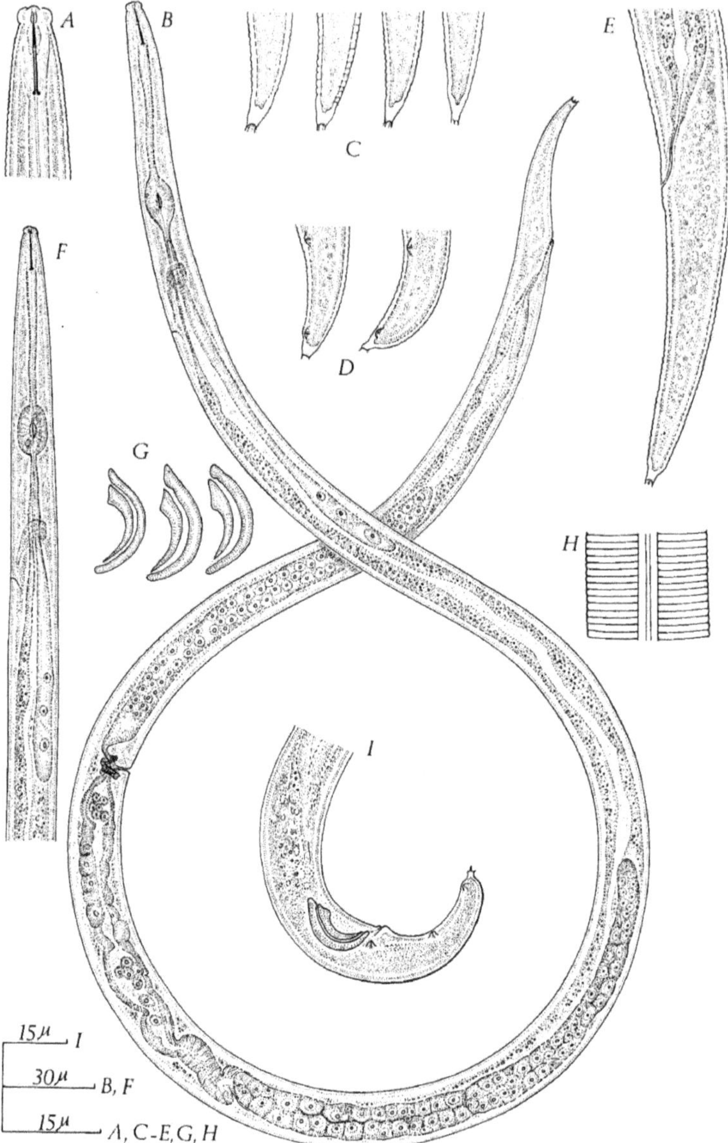

**Figure 6.** *Aphelenchoides ritzemabosi.* (**A**) Female head; (**B**) female; (**C**) female tail ends; (**D**) male tail ends; (**E**) female tail; (**G**) spicules; (**H**) lateral field; (**I**) male tail region. (**A**, **E**, and **F** syntypes; **B**, **C**, and **H** Specimens from chrysanthemum, Stockholm; I Specimen from *Buddleia* leaf, Sussex, England) after Siddiqi [25]. Courtesy of *Commonwealth Institute of Helminthology*.

Identification

Accurate identification of foliar nematodes (*Aphelenchoides* spp.) is crucial for effective disease control. Major efforts should be geared towards rapid and accurate classification of the pathogens so that appropriate control measures could be taken. In addition, timely and accurate diagnosis is also needed to make sound decisions regarding quarantine of imported and exported plant material and commodities. Nevertheless, the identification of foliar nematodes to species level remains a challenging endeavor. The diagnosis and/or relationship between conserved morphology, variable morphometrics, host effects, intraspecific variation, existence of cryptic species, and the ever-increasing number of described species, still vary significantly. To add to the confusion, there is verification of mixed populations and/or detection of rare species which require(s) identification techniques, including morphology of adult females; male, and labial region shape, and stylet morphology; V% age, body length, and shape of tail and tail terminus, and, in some cases, biochemical or molecular methodologies. Because of an increasing number of described species, the value of many of these characters often show large intraspecific variation. Isozyme electrophoresis has discriminated a number of these otherwise cryptic species. Currently used PCR-based molecular methodologies offer hope for a future relying on bigger genebanks that could be used by scientists for a more accurate specie identification. Integrated morphology and molecular approaches are essential to future improved identification of Anguinata nematodes. Detailed diagnostic characters differentiating various species of foliar nematodes have been given by authors such as Allen [19], Hunt [21], Shahina [22].

## 12. Genus *Litylenchus* Zhao, Davies, Alexander and Riley, 2011

Genus *Litylenchus* Zhao, Davies, Alexander and Riley, 2011 is a new genus with much smaller number of species when compared to *Aphelenchoides* genus. *Litylenchus crenatae* Kanzaki, 2019, *Litylenchus crenatae mccannii* Carta 2020, are emerging foliar pathogens of major economic importance. Nematodes from this genus parasitize trees (*Fagus grandifolia*) and bushes (*Coprosma repens*). *Litylenchus crenatae mccannii* described by Carta et al. [11] seems to be a very aggressive subspecies with devastating effects on beech trees (*Fagus grandifolia*). Even though *Litylenchus crenatae mccannii* was initially found infesting beech trees in Ohio [11], it was also reported in several other states and provinces, to include Pennsylvania, New York, Ontario, Canada [56], Connecticut [57], New Jersey, Rhode Island, and West Virginia (unpublished data).

After Zhao [35] the genus *Litylenchus* Zhao, Davies, Alexander and Riley, 2011 is characterized as follows:

- Adults and juveniles of *Litylenchus* gen. from within leaves not forming galls;
- Lacking obese females with a spiral form;
- Slender to semi-obese, cylindrical nematodes, barely curved around ventral axis;
- Lack of sexual dimorphism in head, pharyngeal, and tail characters;
- Cuticle with fine annulations, head offset;
- Stylet short (9–12 μm), robust, with rounded knobs;
- Pharynx with non-muscular fusiform median bulb, valve may be present;
- Pharyngeal glands contained in a large terminal bulb abutting intestine and three large nuclei present;
- Secretory/excretory pore opening 1–1.5 body diameter posterior to nerve ring;
- Female with mono-prodelphic gonad with quadricolumella and post-uterine sac;
- Male with arcuate spicules and simple gubernaculum;
- Bursa arising 1–2 cloacal body diameter anterior to cloacal aperture, extending nearly to tail tiptail medium, conoid, tip shape variable, usually bluntly rounded in male, more variable in female.

## 13. Systematic Position

Based on phylogenetic analyses, *Litylenchus* genus [35] is close to *Subanguina*. However, the two genera have many morphological differences as highlighted below:

- *Litylenchus* genus. does not induce typical galls like *Anguina* and *Nothanguina*;

- Lack of obese females with a spiral form in *Anguina* and *Nothanguina* and lack of semi-obese females in *Ditylenchus*;
- Stylet of *Litylenchus* genus is more robust and the stylet knobs are rounded compared to *Ditylenchus*;
- Excretory pore situated posterior to nerve ring;
- Tails of *Litylenchus* genus are conoid rather than elongate conoid to filiform in *Ditylenchus*, and elongate conoid in *Nothotylenchus* gen.;
- Males have a shorter bursa compared to those of *Nothotylenchus* gen.

List of *Litylenchus* species and synonyms:

Type species:

1. *Litylenchus coprosma*

Other species

1. *Litylenchus crenatae*
2. *Litylenchus crenatae mccannii*

## 14. *Litylenchus coprosma* Zhao, Davies, Alexander and Riley, 2011

Measurements

After Zhao et al. [35].
Slender female ($n$ = 13): L = 743 ± 50 (649–816) μm; a = 55.2 ± 4.0 (51.5–63.3); b = 4.4 ± 0.6 (3.9–5.8); c = 18.7 ± 1.3 (16.3–21.3); V %= 81.5 ± 2.4 (76.5–85.3); stylet = 10.8 ± 0.9 (8.9–11.7) μm.
Obese female ($n$ = 15): L = 856 ± 72 (710–940) μm; a = 32.8 ± 3.7 (24.9–37.7); b = 5.1 ± 0.6 (4.2–6.8); c = 19.4 ± 2.5 (15.4–25.0); V%= 82.2 ± 1.6 (78.8–84.7); stylet = 10.9 ± 0.3 (10.2–11.4) μm.
Male ($n$ = 11): L = 899 ± 66 (768–994) μm; a = 52.0 ± 4.4 (44.5–60.2); b = 5.4 ± 0.4 (4.8–6.2); c = 21.1 ± 1.9 (18.2–24.1); stylet = 10.5 ± 0.5 (9.7–11.3) μm; spicule= 16.2 ± 0.7 (14.9–17.0) μm.

Description

*Litylenchus coprosma* has adult females with two distinct forms, one described as semi-obese (a = 20–40) and the other slender (a = 45–65) (Figures 7 and 8).
Semi-obese female: when killed by heat body is almost straight, semi-obese. Maximum body width is at mid-body. Body cuticle finely striated, almost smooth. Four lines can be observed in lateral field extending almost to tail terminus. Head offset, cephalic framework, and stylet as described for male. Excretory pore located ca 3–3.5 body diameter from anterior, opening near anterior end of terminal bulb, duct with obvious cuticular lining. Hemizonid, pharynx, pharyngeal glands, and pharyngo-intestinal junction as described for male. Nerve ring is located approximately 100 μm from anterior extremity. Deirids and phasmids not seen. Gonads are monodelphic, prodelphic, outstretched, crustaformeria forming a quadricolumella. Oocytes arranged in single row. Oviduct with several cells forming a valve just anterior to elongate, sac-like spermatheca. Vulva located 7–11 anal body diameter anterior to anus (80–85% of body length). Vulval slit occupying almost half body diameter when viewed laterally, vagina almost perpendicular to body wall. Post-uterine sac extending 20–70% of distance from vulva to anus, approximately 2.7 anal body long, sometimes with sperms, lacking cellular relicts of posterior ovary. Rectum difficult to see, anus pore-like, opening in a cuticular depression. Tail approximately 4–5 anal body diameter long, conoid, straight, with a variable tail terminus, may be bluntly rounded, more or less bifurcate, or appear bilobed. Mucro not observed.
Slender female: very similar to the semi-obese females, but slender. Head capsule is a little bit bigger, 59–77% of body diameter at level of stylet knobs compared to the semi-obese females, where the head capsule is between 48–62%. Quadricolumella cells are smaller than in semi-obese female.
Male: when killed, the nematodes assume a smoothly ventrally arcuate shape, body cylindrical, narrowing to a bluntly rounded conoid tail. Body cuticle smooth with three incisures in the lateral field

visible in the region of procorpus increasing to four incisures at mid-body and extending almost to tail tip. Head is set off from the body, smooth, and not annulated. Lightly sclerotized cephalic framework with six sectors.

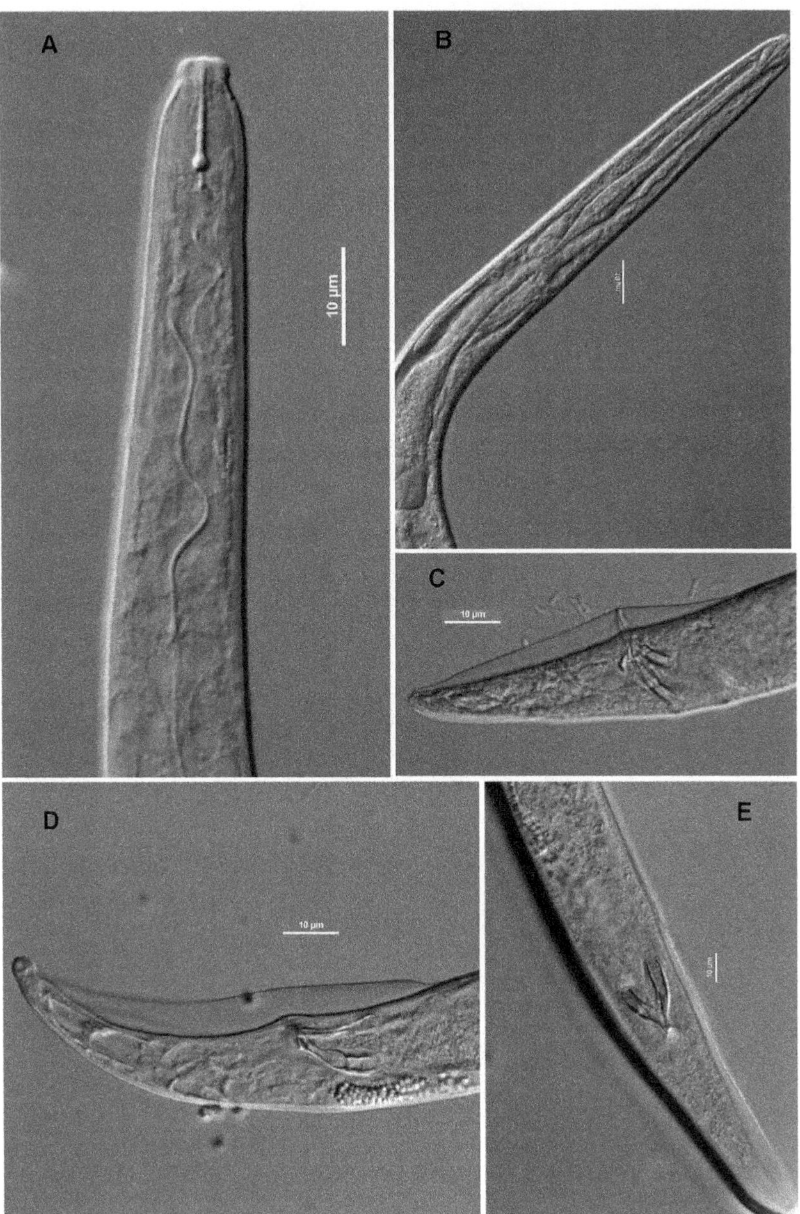

**Figure 7.** *Litylenchus coprosma.* All males in lateral view, except E which is ventral. (**A**) Anterior region; (**B**) pharynx showing median bulb; (**C**) tail with bursa; (**D**) tail showing spicules and variation in shape of tail tip; (**E**) spicules. (Scale bars = 10 μm) after Zhao et al. [35]. Courtesy of *Nematology.*

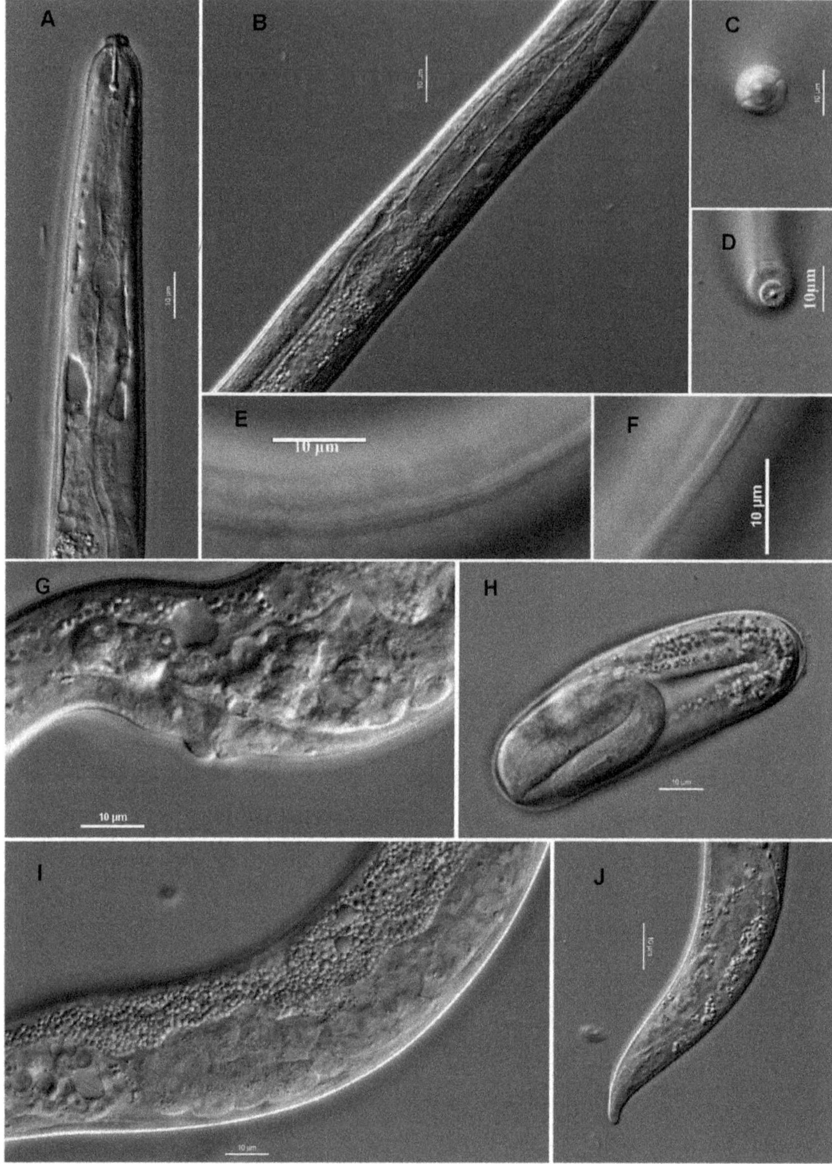

**Figure 8.** *Litylenchus coprosma*. All in lateral view, except C, D en face. (**A**) Head of mature, semi-obese female; (**B**) terminal pharyngeal bulb; (**C**) sub-terminal head showing amphidial apertures; (**D**) apical view of head; (**E**) lateral fields at mid-body showing four incisures; (**F**) lateral fields at pharyngeal region showing three incisures; (**G**) vulva and post-uterine sac; (**H**) second-stage juvenile within egg; (**I**) quadricolumella; (**J**) female tail. (Scale bars = 10 μm) after Zhao et al. [35]. Courtesy of *Nematology*.

En-face view shows amphidial apertures appearing as small lateral slits. Stylet robust, with well-developed rounded knobs, conus comprising ca 40% of stylet length, diameter narrowing sharply to be distinctly less than that of shaft. The opening of dorsal esophageal gland is located just posterior to stylet knobs. Nerve ring is located 70–110 μm from anterior extremity, surrounding isthmus, *ca* one body diameter long. Excretory pore is located ca 5–6 body diameter from anterior end, opening posterior to nerve ring. Hemizonid located immediately anterior to excretory pore. Procorpus cylindrical, fusiform, non-muscular median bulb which is approximately one body diameter long and narrowing sharply to isthmus which is slender, cylindroid, marked off from terminal bulb, pharyngeal glands enclosed in a pyriform terminal bulb containing three large nuclei. Esophago-intestinal junction is immediately posterior to terminal bulb and covered by it in some specimens, valve present, without hyaline cells. Deirids and phasmids were not observed. Testis outstretched, reflexed in some specimens, reaching to nerve ring in some specimens, with spermatocytes arranged in a single row. Spicule paired, similar, arcuate, 2–3 μm wide at anterior end, gradually narrowing towards tip. Capitulum absent. Gubernaculum simple and arcuate. Tail conoid with a variable in shape tail terminus, usually bluntly rounded, but may have terminal process; no mucron observed. Bursa membranous, crenate in some, arising ca 1–2 cloacal body diameter anterior to cloacal aperture, extending nearly (90–95% of tail length) to tail tip.

Distribution

*Litylenchus coprosma* was reported in New Zealand from *Coprosma repens* [35] and from *Coprosma robusta* [26].

### 15. *Litylenchus crenatae* Kanzaki, Ichihara, Aikawa, Ekino, and Masuya, 2019

Measurements

After Kanzaki et al. [37].
Mature female (*n* = 10): L = 816 ± 32 (758–870) μm; a = 35.9 ± 3.4 (30.2–41.1); b = 6.6 ± 0.4 (6.1–7.6); c = 24.5 ± 1.9 (21.8–28.1); V%= 81.5 ± 1.0 (79.4–83.2); stylet = 10.6 ± 0.5 (9.9–11.3) μm.
Immature female (*n* = 10): L = 868 ± 33 (837–915) μm; a = 67.5 ± 5.8 (60.7–74.4); b = 4.3 ± 0.3 (3.9–4.8); c = 15.7 ± 0.7 (14.4–16.7); V% = 77.4 ± 0.5 (76.6–78.3); stylet = 8.0 ± 0.4 (7.4–8.5) μm.
Mature male (*n* = 9): L = 805 ± 21 (766–840) μm; a = 41.0 ± 2.4 (37.4–44.4); b = 6.4 ± 0.4 (5.9–7.3); c = 24.8 ± 2.5 (21.4–30.3); stylet = 10.5 ± 0.4 (9.9–11.3) μm; spicule = 18.3 ± 1.0 (16.7–20.2) μm; gubernaculum = 8 ± 0.4 (7.1–8.5) μm.
Immature male (*n* = 8): L = 707 ± 41 (642–773) μm; a = 57.2 ± 4.7 (48.9–61.9); b = 5.3 ± 0.6 (4.5–6.3); c = 21.1 ± 2.0 (18.5–25.1); stylet = 10.2 ± 0.4 (9.9–11.0) μm; spicule = 15.6 ± 1.2 (14.2–17.7) μm; gubernaculum = 6.5 ± 0.4 (6.0–7.1) μm.

Description

Female (Figure 9a): when killed, the nematodes assume a smoothly ventrally arcuate shape, body cylindrical, vermiform to semi-obese. Anterior part and cuticular morphology similar to mature male. Female gonad single, anteriorly outstretched reaching to level of pharyngeal glands. Oocytes are arranged in single row in entire ovary. Oviduct is short and spermatheca is elongated oval filled with large sperm, posteriorly connected to crustaformeria, which consists of four rows of four large and rounded cells, i.e., forming a quadricolumella, posteriorly connected to uterus by a cluster of small cells. Uterus, a thick-walled tube, sometimes containing an egg. Vagina at right angles to body axis or slightly inclined anteriorly. Vulva, a horizontal slit. Post uterine sac present, well-developed, with a thin wall and a short appendage comprising several rounded cells at distal end. Rectum is about less than one anal body diameter in length, with muscular constriction at intestine-rectal junction. Tail is short and broad, abruptly narrowing at the end with a conoid and bluntly pointed terminus, sometimes appearing like a conical blunt mucron.

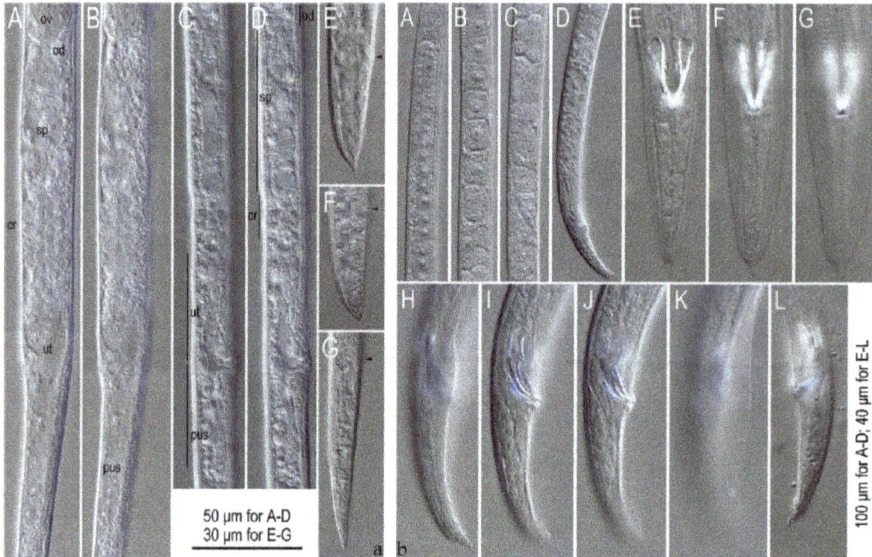

**Figure 9.** Males and Females of *Litylenchus crenatae*. (a) Female reproductive system and tail of *Litylenchus crenatae*; (**A,B**) posterior part of gonad of mature female in different focal planes; (**C,D**) posterior part of gonad of immature female in different focal planes; (**E,F**) tail of mature female; (**G**) tail of immature female. Ovary (ov), oviduct (od), spermatheca (sp), crustaformeria (cr), uterus (ut), and post-uterine sac (pus) are shown in (A–D), and anal opening is indicated by arrowheads in (**E–G**). (b) Male reproductive system of *Litylenchus crenatae* (**A–K**) are mature individuals, (**L**) is an immature individual. (**A**) Anterior end of testis; (**B**) middle part of mature testis; (**C**) posterior part of testis; (**D**) posterior end of testis and *vas deferens*; (**E,G**) ventral view of tail in different focal planes; (**H,K**) right lateral view of tail in different focal planes; (**L**) left lateral view of tail of immature individual. After Kanzaki et al. [32]. Courtesy of Nematology.

Male (Figure 9b): when killed, the nematodes assume a smoothly ventrally arcuate shape, body cylindrical, not clearly obese or semi-obese. Body cuticle annulated with six incisures in the lateral field at the anterior part of body, 6–8 incisures around mid-body, and posteriorly connected to bursa. Deirids present in middle of lateral field slightly posterior to hemizonid and excretory pore. Lip region slightly offset from body, with a truncated shape, separated by a very shallow constriction.

Stylet with narrow lumen and a shaft with prominent rounded basal knobs (3.6 µm in diameter). Dorsal esophageal gland is located posterior to stylet knobs. Procorpus is cylindrical. Median esophageal bulb is weakly developed, with small metacarpal valve at mid-bulb length. Isthmus is cylindrical, but narrower than the procorpus, enveloped by the nerve ring in its mid-length. Broad and glandular gland lobe with three large nuclei were observed (Figure 10). Hemizonid found at level the beginning of expansion of pharynx. Excretory pore located slightly posterior to hemizonid, with clear secretory-excretory duct. Nuclei of the esophageal overlap observed between hemizonid and pharyngo-intestinal junction, two being just anterior to the third, and latter located slightly anterior to junction. Gonad single, anteriorly outstretched reaching to level of pharyngeal glands. Testis outstretched with spermatocytes arranged in single row from anterior to middle part of testis and in multiple rows in posterior section. *Vas deferens* is visible, consisting of rounded cells, sometimes containing well-developed sperm. Spicules paired, smoothly arcuate ventrally, forming a smoothly curved horn-like blade with bluntly pointed distal end in lateral view (V-shaped). Gubernaculum simple, crescent or bow-shaped in lateral view. Bursa peloderan, well developed arising three cloacal body diameter anterior to cloacal opening and terminating near tail tip. Tail is conoid, bluntly pointed in lateral view.

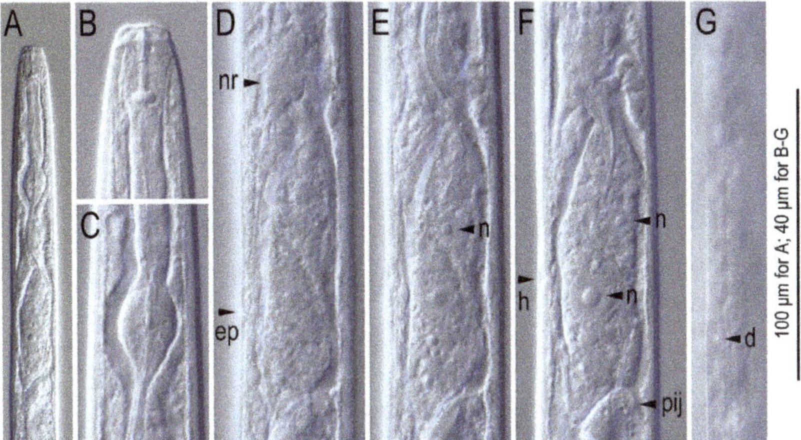

**Figure 10.** Anterior region of mature adults of *Litylenchus crenatae*; (**A**) anterior end to pharyngo-intestinal junction; (**B**) lip region; (**C**) metacorpus (median bulb); (**D–G**) pharyngeal gland region in different focal planes. Nerve ring (nr), excretory pore (ep), pharyngeal gland nuclei (n), hemizonid (h), pharyngo-intestinal junction (pij) and deirid (d) are indicated in (**D–G**) after Kanzaki et al. [37]. Courtesy of *Nematology*.

Distribution

*Litylenchus crenatae* was reported so far from Japan from *Fagus crenata* [37].

The phylogenetic relationships among anguinid nematodes inferred from three ribosomal RNA loci were provided by Kanzaki et al. [37]. The marker sequences derived from *Litylenchus crenatae* specimens, LC383723 (SSU), LC383725 (D2-D3 LSU), and LC383724 (ITS) were deposited to GenBank.

**16. *Litylenchus crenatae* Kanzaki et al., 2019 *mccannii* ssp. Carta, Handoo, Li, Kantor, Bauchan, McCann, Gabriel, Yu, Reed, Koch, Martin, Burke 2020**

Measurements

After Carta et al. [11].

Immature female ($n$ = 10): L = 823 ± 61 (750–947) μm; a = 72.9 ± 3. (61.0–86.0); b = 5.4 ± 0.7 (4.5–6.6); c =17.4 ± 3.3 (13.0–25.0); V%= 76.9 ± 1.2 (75.0–79.0); stylet = 9.7 ± 0.9 (8.5–11.2) μm.

Mature male ($n$ = 4): L = 548 ±16.7 (534.5–566.7) μm; a = 36.1 ± 5.4 (33.4–44.1); b = 4.8 ± 0.2 (4.6–4.9); c = 15.5 ± 0.2 (15.3–15.9); stylet = 11.1 ± 0.5 (10.5–11.4) μm; spicule= 16.3 ± 1.4 (14.9–17.6) μm; gubernaculum = 5.3 ± 0.8 (4.3–6.1) μm.

Description

Females have long and slender bodies, a lip region slightly offset with 5 annules. Stylet measures 9.7 ± 0.9 μm in young females with 5% of the pharynx length, and 7–10% of the pharynx length in males. Median bulb is weak without an obvious valve. The vulval region is kinked and irregular and the anterior gonad is relatively long, nearly five times the length of the post uterine sac. The post uterine sac is about three times the vulval body width and one quarter of the vulval anal distance. The rectum is approximately one quarter of the tail length and the anus is pore-like and obscure in most specimens. Tail is conical, slender and asymmetrically pointed, with a gradually tapering and the tail tip often with mucronate extension (Figure 11). There is a shape variation in tails of immature and mature females.

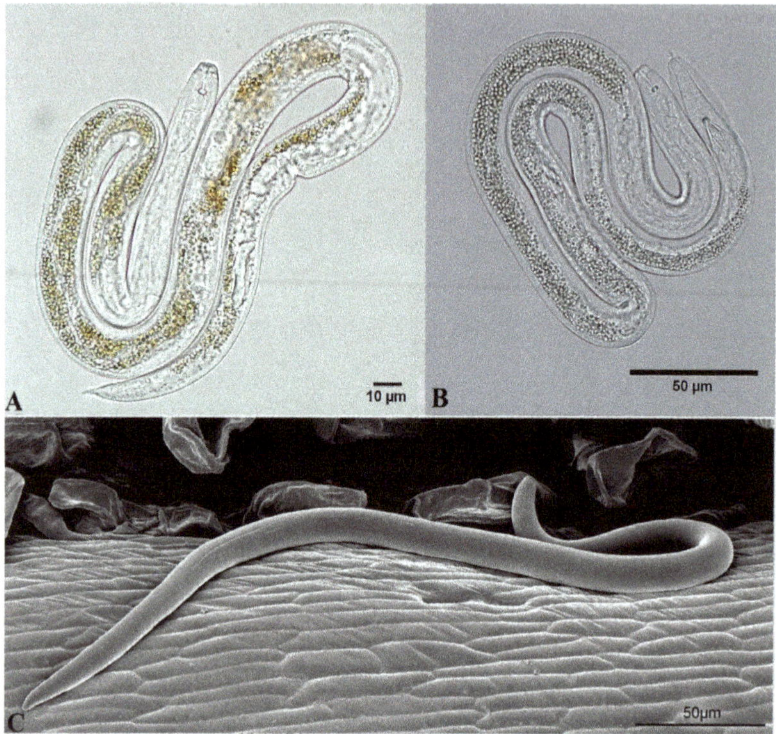

**Figure 11.** Males and Females of *Litylenchus crenatae mccannii*. (**A**) Mature Female; (**B**) male; (**C**) LT-SEM of young Female. Courtesy of Gary Bauchan and Shiguang Li of Electron and Confocal Microscopy and Mycology and Nematology Genetic Diversity and Biology Laboratory (MNGDBL), USDA, ARS, Beltsville, MD, respectively.

Female: *Litylenchus crenatae mccannii* ssp. n. young female population from North America can be differentiated from the *Litylenchus crenatae* described from Japan by:

- Having longer stylet 9.7 ± 0.9 µm (8.6–11.2) vs. 8.0 ± 0.4 (7.4–8.5) and longer stylet conus 4.6 µm (3.6–5.2) vs. 3.1 ± 0.2 (2.8–3.5);
- The post- uterine sac in mature females was shorter (36.9 ± 9.4 vs. 68 ± 7.4);
- Tail was shorter in the fixed immature female populations (48.3 ± 6.2 vs. 55 ± 3.8) but it was longer in the mature populations (43.7 ± 11.3 vs 33 ± 2.3) which was also reflected in different c (16.8 ± 1.4 vs 24.5 ± 1.9) and c′ (5.3 ± 1.2 vs. 2.9 ± 0.3) ratios;
- The body width in mature females was narrower in all populations (16.2 ± 2.4 vs. 22.9 ± 2.6).

Male: males of *Litylenchus crenatae mccannii* ssp. n. are very similar to *Litylenchus crenatae* males described from Japan. Carta et al. [11] noted some differences between the North America and the Japan population such as:

- Longer stylet (11.2 (10.6–12) vs. 10.2 (9.9–11)) µm and stylet conus (4.8 (4.4–5.3) vs. 3.6 (3.5–4.3)) µm;
- A wider body (16.7 (13.5–20.3)) µm than the fixed type population from Japan.

Molecularly, *Litylenchus crenatae mccannii* from Ohio, Pennsylvania, and the neighboring province of Ontario, Canada, showed some differences in morphometric averages among females when compared to the Japanese population described by Kanzaki et al. [32]. Ribosomal DNA marker sequences were nearly identical to the population from Japan [11]. The 18S rDNA and internal transcribed spacer (ITS) rDNA sequences for *Litylenchus crenatae* from Japan are 99.9% and 99.7% similar, respectively, to *Litylenchus crenatae mccannii* from North America. A sequence for the COI marker was also generated, although it was not available in the Japanese population [11]. The marker sequences derived from *Litylenchus crenatae mccannii* specimens, 104H78 and 104H82 were deposited to GenBank with accession numbers for rDNA (MK292137, MK292138) and COI (MN524968, and MN524969).

Phylogenetic trees for 18S rDNA of *Aphelenchoides* and *Litylenchus* are shown in Figures 12 and 13.

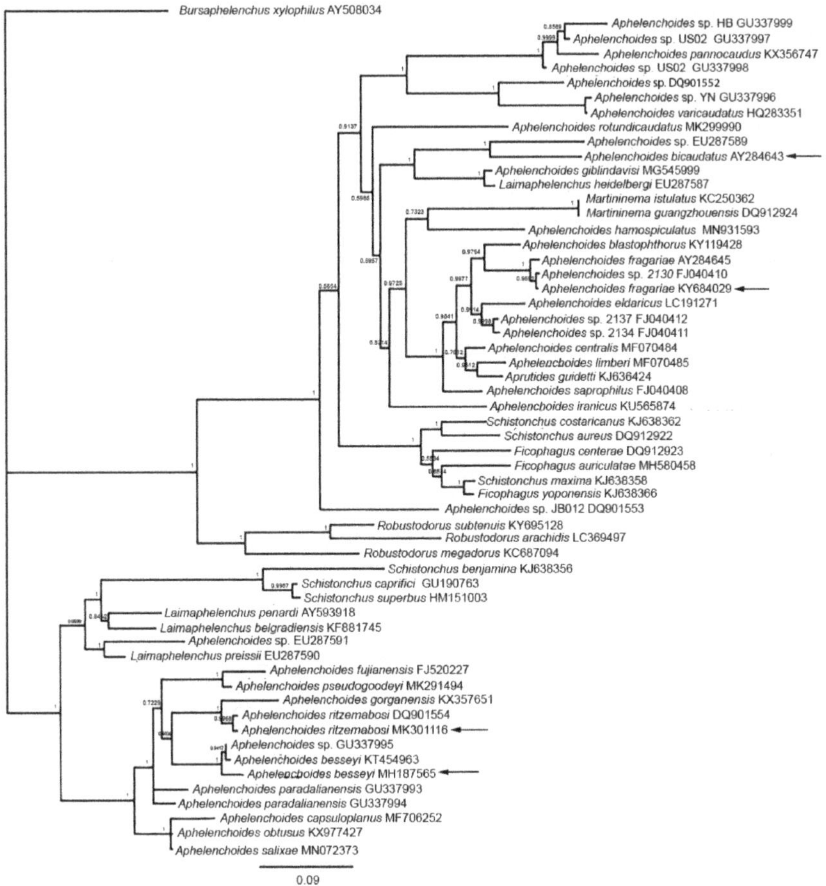


**Figure 12.** Phylogenetic Bayesian tree of 18S rDNA sequences for *Aphelenchoides* and related genera from multiple sequence alignment made with Clustal Omega (EMBL-EBI, https://www.ebi.ac.uk/Tools/msa/clustalo/); tree processed from 1,100,000 iterations in MrBayes version 3.2.6 [58] within Geneious Prime Version 2020.2.4 (Biomatters, Ltd., Auckland, NZ). Pathogenic species are indicated by arrows.

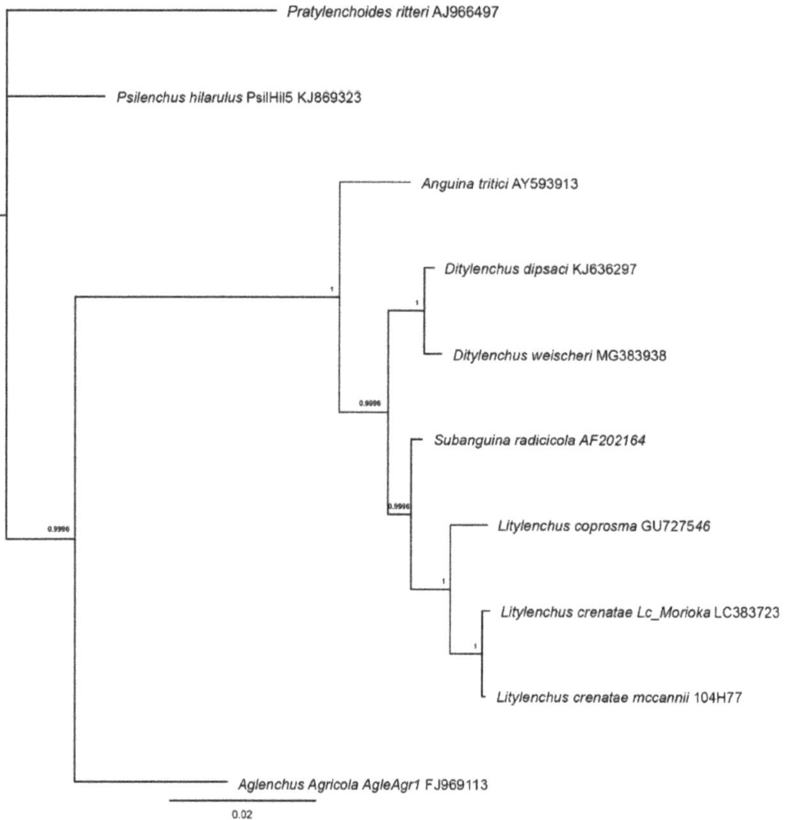



**Figure 13.** Phylogenetic Bayesian tree from 1,100,000 iterations created in MrBayes version 3.2.6 [59] from multiple sequence alignment made with Clustal Omega (EMBL-EBI, https://www.ebi.ac.uk/Tools/ msa/clustalo/) within Geneious Prime Version 2020.2.4 (Biomatters, Ltd., Auckland, NZ).

## 17. Conclusions and Future Prospects

Until recently, morphology used to be the only way to differentiate nematodes. With recent developments of molecular approaches in taxonomy gaining more widespread use, molecular identification has the potential to become an indispensable tool in the near future. As the GenBank continues to expand, molecular identification can become a reliable resource for nematode identification. Classical morphology continues to play a very important role in nematode identification, being reliable, cheap and quick. Molecular approaches can complement classical morphology and are crucial for species with similar morphological characters. A blend of both morphological (including SEM), morphometric, and molecular data is essential for future new foliar nematode species. The prospects in foliar nematode taxonomy and diagnostics are dependent on molecular-based methodologies that will discriminate not only species but also at the level of host races and pathotypes. This finer discrimination provides opportunities for more focused management strategies. These techniques can provide rapid diagnostics and help resolve the present problems associated with morphologically conservative organisms. When widely employed, these characterization techniques will allow differentiation between nominal species, also enhancing our understanding of the phylogeny of the genus and its relationship with other plant-parasitic nematodes.

**Author Contributions:** Conceptualization, Z.H. and M.K.; methodology, Z.H., M.K., L.C.; software, Z.H., M.K., L.C.; validation, Z.H., M.K. and L.C.; formal analysis, Z.H., M.K.; investigation, M.K., Z.H.; resources, Z.H. M.K.; data curation, Z.H., M.K., L.C.; writing—original draft preparation, M.K.; writing—review and editing, Z.H., M.K., L.C.; visualization, Z.H., M.K., L.C.; supervision, Z.H.; project administration, Z.H.; funding acquisition, Z.H. All authors have read and agreed to the published version of the manuscript.

**Funding:** This research received no external funding.

**Acknowledgments:** Mihail Kantor was supported in part by an appointment to the Research Participation Program at the Mycology and Nematology Genetic Diversity and Biology Laboratory USDA, ARS, Northeast Area, Beltsville, MD, administered by the Oak Ridge Institute for Science and Education (ORISE) through an interagency agreement between the U.S. Department of Energy and USDA-ARS. This research was funded by ORISE ARS Research Participation Program Outgoing Interagency Agreement number (60-8042-0-057). Mention of trade names or commercial products in this publication is solely for the purpose of providing specific information and does not imply recommendation or endorsement by the U.S. Department of Agriculture. USDA is an equal opportunity provider and employer. We would also like to thank Brill Publishers for allowing us to use some of the images published in this review.

**Conflicts of Interest:** The authors declare no conflict of interest.

## References

1.    Kohl, L.M. Foliar nematodes: A summary of biology and control with a compilation of host range. *Plant Health Prog.* **2011**, *12*, 23. [CrossRef]

2.    Sánchez Monge, G.A.; Flores, L.; Salazar, L.; Hockland, S.; Bert, W. An updated list of the plants associated with plant-parasitic *Aphelenchoides* (Nematoda: Aphelenchoididae) and its implications for plant-parasitism within this genus. *Zootaxa* **2015**, *4013*, 207–224. [CrossRef] [PubMed]

3.    Nickle, W.R. A taxonomic review of the genera of the Aphelenchoidea (Fuchs, 1937) thorne, 1949 (Nematoda: Tylenchida). *J. Nematol.* **1970**, *2*, 375. [PubMed]

4.    Hoshino, S.; Togashi, K. A simple method for determining *Aphelenchoides besseyi* infestation level of *Oryza sativa* seeds. *J. Nematol.* **1999**, *31*, 641. [PubMed]

5.    Golden, A.M. Preparation and mounting nematodes for microscopic observations. In *B. Plant Nematology Laboratory Manual*; Zuckerman, M., Mai, W.F., Krusberg, L.R., Amherst, M.A., Eds.; University of Massachusetts Agricultural Experiment Station: Amherst, MA, USA, 1990; pp. 197–205.

6.    Hooper, D.J. Handling, fixing, staining, and mounting nematodes. In *Laboratory Methods for Work with Plant and Soil Nematodes*, 5th ed.; Southey, J.F., Ed.; Her Majesty's Stationery Office: London, UK, 1970; pp. 39–54.

7.    Ryss, A.Y.; McClure, M.A.; Nischwitz, C.; Dhiman, C.; Subbotin, S.A. Redescription of *Robustodorus megadorus* with molecular characterization and analysis of its phylogenetic position within the family Aphelenchoididae. *J. Nematol.* **2013**, *45*, 237.

8.    McClure, M.A.; Stowell, L.J. A simple method of processing nematodes for electron microscopy. *J. Nematol.* **1978**, *10*, 376. [PubMed]

9.    Kantor, M.; Handoo, Z.A.; Skantar, A.M.; Hult, M.N.; Ingham, R.E.; Wade, N.M.; Ye, W.; Bauchan, G.R.; Mowery, J.D. Morphological and molecular characterisation of Punctodera mulveyi n. sp. (Nematoda: Punctoderidae) from a golf course green in Oregon, USA, with a key to species of Punctodera. *Nematology* **2020**, in press.

10.    Carta, L.K.; Bauchan, G.; Hsu, C.-Y.; Yuceer, C. Description of *Parasitorhabditis frontali* n. sp. (Nemata: Rhabditida) from *Dendroctonus frontalis* Zimmermann (Coleoptera: Scolytidae). *J. Nematol.* **2010**, *42*, 46–54.

11.    Carta, L.K.; Handoo, Z.A.; Li, S.; Kantor, M.R.; Bauchan, G.; McCann, D.; Gabriel, C.K.; Yu, Q.; Reed, S.E.; Koch, J.; et al. Beech leaf disease symptoms caused by newly recognized nematode subspecies *Litylenchus crenatae mccannii* (Anguinata) described from *Fagus grandifolia* in North America. *For. Path.* **2020**, *50*, e12580. [CrossRef]

12.    Carta, L.K.; Li, S. PCR amplification of a long rDNA segment with one primer pair in agriculturally important nematodes. *J. Nematol.* **2019**, *51*, 1–8. [CrossRef] [PubMed]

13.    Rybarczyk-Mydłowska, K.; Mooyman, P.; van Megen, H.; van den Elsen, S.; Vervoort, M.; Veenhuizen, P.; van Doorn, J.; Dees, R.; Karssen, G.; Bakker, J.; et al. Small subunit ribosomal DNA-based phylogenetic analysis of foliar nematodes (Aphelenchoides spp.) and their quantitative detection in complex DNA backgrounds. *Phytopathology* **2012**, *102*, 1153–1160. [CrossRef]

14. Sánchez-Monge, A.; Janssen, T.; Fang, Y.; Couvreur, M.; Karssen, G.; Bert, W. mtCOI successfully diagnoses the four main plant-parasitic Aphelenchoides species (Nematoda: Aphelenchoididae) and supports a multiple origin of plant-parasitism in this paraphyletic genus. *Eur. J. Plant Pathol.* **2017**, *148*, 853–866. [CrossRef]

15. Holterman, M.; Van Der Wurff, A.; Van Den Elsen, S.; Van Megen, H.; Bongers, T.; Holovachov, O.; Bakker, J.; Helder, J. Phylum-wide analysis of SSU rDNA reveals deep phylogenetic relationships among nematodes and accelerated evolution toward crown clades. *Mol. Biol. Evol.* **2006**, *23*, 1792–1800. [CrossRef]

16. Nunn, G.B. Nematode Molecular Evolution. An Investigation of Evolutionary Patterns among Nematodes Based upon DNA Sequences. Ph.D. Thesis, University of Nottingham, Nottingham, UK, 1992.

17. Goodey, J.B. The classification of the Aphelenchoidea Fuchs, 1937. *Nematologica* **1960**, *5*, 111–126. [CrossRef]

18. Jagdale, G.B.; Grewal, P.S. Infection behavior and overwintering survival of foliar nematodes, *Aphelenchoides fragariae*, on *Hosta*. *J. Nematol.* **2006**, *38*, 130.

19. Allen, M.W. Taxonomic status of the bud and leaf nematodes related to *Aphelenchoides fragariae* (Ritzema Bos, 1891). *Proc. Helminthol. Soc. Wash.* **1952**, *19*, 109–120.

20. Wheeler, L.; Crow, W.T. Foliar Nematode, Aphelenchoides (spp.). Available online: http://entnemdept.ufl.edu/creatures/NEMATODE/foliar_nematode.html (accessed on 28 July 2020).

21. Hunt, D.J. *Aphelenchida, Longidoridae and Trichodoridae: Their Systematics and Bionomics*; CABI International: Wallingford, UK, 1993; p. 352.

22. Shahina, F. A diagnostic compendium of the genus Aphelenchoides Fischer, 1894 (Nematoda: Aphelenchida) with some new records of the group from Pakistan. *Pak. J. Nematol.* **1996**, *14*, 1–32.

23. Hunt, D.J. A checklist of the Aphelenchoidea Nematoda:Tylenchina). *J. Nematode Morphol. Syst.* **2008**, *10*, 99–135.

24. Christie, J.R. A description of *Aphelenchoides besseyi* n.sp., the summer- dwarf nematode of strawberries, with comments on the identity of *Aphelenchoides subtenuis* (Cobb, 1929) and *Aphelenchoides hodsoni* Goodey, 1935. *Proc. Helminth Soc. Wash.* **1942**, *9*, 82–84.

25. De Jesus, D.S.; Oliveira, C.M.G.; Roberts, D.; Blok, V.; Neilson, R.; Prior, T.; de Lima Oliveira, R.D.A. Morphological and molecular characterisation of *Aphelenchoides besseyi* and *A. fujianensis* (Nematoda: Aphelenchoididae) from rice and forage grass seeds in Brazil. *Nematology* **2016**, *18*, 337–356. [CrossRef]

26. Xu, Y.M.; Li, D.; Alexander, B.J.; Zhao, Z.Q. First report of *Litylenchus coprosma* on *Coprosma robusta*. *Australas. Plant Dis. Notes* **2017**, *12*, 17. [CrossRef]

27. Franklin, M.T.; Siddiqi, M.R. Aphelenchoides besseyi. In *CIH Descriptions of Plant-Parasitic Nematodes*; Commonwealth Institute of Helminthology: St. Albans, UK, 1972; 3p.

28. Siddiqui, I.A. Aphelenchoides bicaudatus. In *CIH Descriptions of Plant-Parasitic Nematodes*; Commonwealth Institute of Helminthology: St. Albans, UK, 1976; 3p.

29. Siddiqui, M.R. Aphelenchoides fragariae. In *CIH Description of Plant-Parasitic Nematodes*; Commonwealth Institute of Helminthology: St. Albans, UK, 1975; 4p.

30. Siddiqui, M.R. Aphelenchoides ritzemabosi. In *CIH Description of Plant-Parasitic Nematodes*; Commonwealth Institute of Helminthology: St. Albans, UK, 1974; 4p.

31. Siddiqui, I.A.; Taylor, D.P. A Redescription of *Aphelenchoides bicaudatus* (Imamura, 1931) Filipjev & Schuurmans Stekhoven, 1941 (Nematoda: Aphelenchoididae), with a description of the previously undescribed male. *Nematologica* **1967**, *13*, 581–585. [CrossRef]

32. Jen, F.Y.; Tsay, T.T.; Chen, P. *Aphelenchoides bicaudatus* from ornamental nurseries in Taiwan and its relationship with some agricultural crops. *Plant Dis.* **2012**, *96*, 1763–1766. [CrossRef]

33. Khan, Z.; Son, S.H.; Moon, H.S.; Kim, S.G.; Shin, H.D.; Jeon, Y.H. Description of a foliar nematode, *Aphelenchoides fragariae* (Nematoda: Aphelenchida) with additional characteristics from Korea. *J. Asia-Pac. Entomol.* **2007**, *10*, 313–315. [CrossRef]

34. Chizhov, V.N.; Subbotin, S.A.; Chumakova, O.A.; Baldwin, J.G. Morphological and molecular characterization of foliar nematodes of the genus *Aphelenchoides*: *A. fragariae* and *A. ritzemabosi* (Nematoda: Aphelenchoididae) from the main botanical garden of the Russian Academy of Sciences, Moscow. *Russ. J. Nematol.* **2006**, *14*, 179–184.

35. Zhao, Z.Q.; Davies, K.; Alexander, B.; Riley, I.T. *Litylenchus coprosma* gen. n., sp. n. (Tylenchida: Anguinata), from Leaves of *Coprosma repens* (Rubiaceae) in New Zealand. *Nematology* **2011**, *13*, 29–44.

36. Khan, M.R.; Handoo, Z.A.; Rao, U.; Rao, S.B.; Prasad, J.S. Observations on the foliar nematode, *Aphelenchoides besseyi*, infecting tuberose and rice in India. *J. Nematol.* **2012**, *44*, 391.

37. Kanzaki, N.; Ichihara, Y.; Aikawa, T.; Ekino, T.; Masuya, H. *Litylenchus crenatae* n. sp. (Tylenchomorpha: Anguinidae), a laf gall nematode parasitizing *Fagus crenata* Blume. *Nematology* **2019**, *21*, 5–22. [CrossRef]

38. Daughtrey, M.L.; Wick, R.L.; Peterson, J.L. *Compendium of Flowering Potted Plant Diseases*; American Phytopathological Society: St. Paul, MN, USA, 1995.

39. Kepenekci, I. Rice white tip nematode (*Aphelenchoides besseyi*) in rice growing areas of Turkey. *Nematropica* **2013**, *43*, 181–185.

40. Oliveira, C.M.G.; Kubo, R.K. Foliar nematodes (*Aphelenchoides* spp.) on *Begonia* in Brazil. *Rev. Bras. Hortic. Ornam.* **2006**, *12*, 134–137.

41. Perez, A.; Fernandez, E. New hosts of *Aphelenchoides besseyi* (Christie, 1942) in Cuba. *Fitosanidad* **2004**, *8*, 45–46.

42. Hockland, S. *A Pragmatic Approach to Identifying Aphelenchoides Species for Plant Health Quarantine and Pest Management Programmes*; University of Reading: Reading, UK, 2001.

43. Wu, G.L.; Kuo, T.H.; Tsay, T.T.; Tsai, I.J.; Chen, P.J. Glycoside hydrolase (GH) 45 and 5 Candidate Cellulases in *Aphelenchoides besseyi* Isolated from Bird's-Nest Fern. *PLoS ONE* **2016**, *11*, e0158663. [CrossRef] [PubMed]

44. Xu, X.; Qing, X.; Xie, J.L.; Yang, F.; Peng, Y.L.; Ji, H.L. Population structure and species delimitation of rice white tip nematode, *Aphelenchoides besseyi* (Nematoda: Aphelenchoididae), in China. *Plant Pathol.* **2020**, *69*, 159–167. [CrossRef]

45. Fortuner, R. On the morphology of *Aphelenchoides besseyi* christie, 1942 and *A. siddiqii* n. sp. (Nematoda, Aphelenchoidea). *J. Helminthol.* **1970**, *44*, 141–152. [CrossRef]

46. Devran, Z.; Tülek, A.; Mıstanoğlu, İ.; Çiftçiğil, T.H.; Özalp, T. A rapid molecular detection method for *Aphelenchoides besseyi* from rice tissues. *Australas. Plant Path.* **2017**, *46*, 43–48. [CrossRef]

47. Fu, Z.; Agudelo, P.; Wells, C.E. Detoxification-related gene expression accompanies anhydrobiosis in the foliar nematode (Aphelenchoides fragariae). *J. Nematol.* **2020**, *52*, 1. [CrossRef]

48. Tsay, T.T. Quarantine of plant-parasitic nematodes. *Plant Pathol. Bull.* **1995**, *4*, 43–59.

49. Imamura, S. Nematodes in the paddy field, with notes on their population before and after irrigation. *J. Coll. Agric. Imp. Univ. Tokyo* **1931**, *11*, 193–240.

50. Israr, M.; Shahina, F.; Nasira, K. Description of *Aphelenchoides turnipi* n. sp. and Redescription of *A. siddiqii* with Notes on *A. bicaudatus* (Nematoda: Aphelenchoididae) from Pakistan. *Pak. J. Nematol.* **2017**, *35*, 3–12. [CrossRef]

51. Sanwal, K.C. A key to the species of the nematode genus *Aphelenchoides* fischer, 1894. *Can. J. Zool.* **1961**, *35*, 143–148. [CrossRef]

52. Kim, J.; Kim, T.; Park, J.K. First report of *Aphelenchoides bicaudatus* (Nematoda: Aphelenchoididae) from South Korea. *Anim. Syst. Evol. Divers.* **2016**, *32*, 253. [CrossRef]

53. Dunn, R.A. *Foliar Nematodes as Pests of Ornamental Plants*; Institute of Food and Agricultural Science, University of Florida: Gainesville, FL, USA, 2005; 3p.

54. Franklin, M.T. Two species of *Aphelenchoides* associated with strawberry bud disease in Britain. *Ann. Appl. Biol.* **1950**, *37*, 1–10. [CrossRef]

55. CABI. Invasive Species Compendium. Available online: https://www.cabi.org/isc/datasheet/6378 (accessed on 28 July 2020).

56. Reed, S.E.; Greifenhagen, S.; Yu, Q.; Hoke, A.; Burke, D.J.; Carta, L.K.; Handoo, Z.A.; Kantor, M.R.; Koch, J.L. Foliar nematode, *Litylenchus crenatae* ssp. *mccannii*, population dynamics in leaves and buds of beech leaf disease-affected trees in Canada and the US. *For. Path.* **2020**, *50*, e12599. [CrossRef]

57. Marra, R.E.; LaMondia, J. First Report of beech leaf disease, caused by the foliar nematode, *Litylenchus crenatae mccannii*, on American Beech (*Fagus grandifolia*) in Connecticut. *Plant Dis.* **2020**. [CrossRef]

58. Kanzaki, N.; Futai, K. A PCR primer set for determination of phylogenetic relationships of *Bursaphelenchus* species within the *xylophilus* group. *Nematology* **2002**, *4*, 35–41. [CrossRef]

59. Huelsenbeck, J.P.; Ronquist, F. MRBAYES: Bayesian inference of phylogenetic trees. *Bioinformatics* **2001**, *17*, 754–755. [CrossRef]

**Publisher's Note:** MDPI stays neutral with regard to jurisdictional claims in published maps and institutional affiliations.

*Review*

# Nematode Identification Techniques and Recent Advances

**Mesfin Bogale, Anil Baniya and Peter DiGennaro ***

Department of Entomology and Nematology, University of Florida, Gainesville, FL 32611, USA;
mazene@ufl.edu (M.B.); anilbaniya1@ufl.edu (A.B.)
* Correspondence: pdigennaro@ufl.edu; Tel.: +1-352-273-3959

Received: 28 July 2020; Accepted: 22 September 2020; Published: 24 September 2020

**Abstract:** Nematodes are among the most diverse but least studied organisms. The classic morphology-based identification has proved insufficient to the study of nematode identification and diversity, mainly for lack of sufficient morphological variations among closely related taxa. Different molecular methods have been used to supplement morphology-based methods and/or circumvent these problems with various degrees of success. These methods range from fingerprint to sequence analyses of DNA- and/or protein-based information. Image analyses techniques have also contributed towards this success. In this review, we highlight what each of these methods entail and provide examples where more recent advances of these techniques have been employed in nematode identification. Wherever possible, emphasis has been given to nematodes of agricultural significance. We show that these alternative methods have aided nematode identification and raised our understanding of nematode diversity and phylogeny. We discuss the pros and cons of these methods and conclude that no one method by itself provides all the answers; the choice of method depends on the question at hand, the nature of the samples, and the availability of resources.

**Keywords:** nematode identification; morphology-based methods; DNA-based methods; protein-based methods; image analysis

---

## 1. Introduction

Comprising over a million species [1], nematodes are likely the most diverse and numerous metazoans in soil and aquatic sediments. Despite this, nematodes are among the least studied organisms with less than 0.01% of their species diversity described to date [2]. Among some 26,000 described species, about 4100 are plant parasitic, which cause drastic economic losses to all crops [3]. Nematodes are also of significant medical and veterinary importance [4], and free-living nematodes are crucial to nutrient recycling in the environment. Therefore, accurate identification is of paramount significance to understand nematode diversity and design efficient control and management strategies. Traditionally, identification is based on characteristics such as body length, morphology of sexual organs, mouth and tail parts, and other physical characters. This morphology-based classification can prove inadequate due to lack of clear variation among closely related taxa and the need for highly skilled taxonomists, whose number is on the decline [5]. Morphology-based identification is also a demanding endeavor, especially when large numbers of samples are involved. Various sub-organismal (protein- and DNA-based) methods have been employed to supplement or circumvent the limitations associated with morphology-based classification of nematodes. The highly influential work of Blaxter et al. [6] employed sequencing of nematode ribosomal DNA (rDNA) and led to improved understanding of nematode evolutionary relationships and identification. We will not spend time discussing the evolution of nematodes and phylogenetic relationships, but it is important to understand the significance of correct nematode identification and, more to the point, how we define a nematode species. As pointed out by Adams [7] there is a trade-off between an operational species definition and that with a strong philosophical integrity. While there is a justified need to place species within the correct evolutionary lineage, more often, nematode identification techniques are driven by an operational definition of species to assess potential threats to animal and plant health. Here, we review current methods and their progenitors in nematode taxonomic techniques and suggest potential advances.

## 2. Morphological and Image-Based Analyses

### 2.1. Classical Morphological Identification

Classic identification of nematodes is based on morphological and anatomical differences using microscopic image analysis. Morphological identification is among the cheaper identification methods and helps relate morphology with possible function [5]. While most effective for nematodes that have distinct differences, nematodes that share subtle morphological and morphometric differences like body length, presence, and shape of a stylet, the shape of the tail, etc., are difficult to distinguish morphologically. For example, root-knot nematodes (RKN; *Meloidogyne* spp.) were previously diagnosed based on adult female perineal patterns [8,9], i.e., posterior region comprising the vulva-anus area (perineum), tail terminus, phasmids, lateral lines and surrounding cuticular striae; a set of characters that was originally proposed to distinguish among *Meloidogyne incognita*, *M. javanica*, *M. arenaria* and *M. hapla* [10]. With the discovery of new species, however, perineal patterns became inadequate because perineal patterns (and other morphometric characters; [11]) overlapped between species [12,13]. Currently, RKN species are identified using a combination of morphological and molecular characteristics (e.g., [14,15]).

Another example is in cyst nematodes (*Heterodera* spp. and *Globodera* spp.), which are among the major pathogenic plant parasitic nematodes with worldwide distribution [16]. *Heterodera* and *Globodera* can be distinguished from each other by the morphology of their cysts: lemon shaped in the former and round in the latter [17]. Species identification within *Heterodera* is based on few morphological traits including vulval cone [18], cone top [19], vaginal [20] and lip [21,22] structures. Taxonomic distinction within *Globodera* is mainly based on morphology of cyst and second stage juveniles [23]. Host plant association may also be indicative of the cyst nematode species, though this may be misleading at times as is the case with the cereal cyst nematode group of *Heterodera* [17]. Morphological identification of cyst nematodes requires taxonomic expertise and can be challenging if samples contain mixed species. Moreover, both genera include species complexes whose members are difficult to distinguish based on morphology alone [24,25].

Important morphological identification characters in nematodes include shape of head, number of annules, body length, length of stylet, shape of stylet knob, structure of lateral fields, presence/absence and shape of spermatheca, shape of female tail terminus, shape and length of spicule and gubernaculum [26]. Measurements of these characteristics and processing of samples for this purpose requires skilled taxonomists, whose number is on the decline [5]. Morphology may also be altered due to variation in geographic location, host plant, nutrition, and other environmental factors as is observed among some free-living and plant parasitic nematodes. Concisely, it can be difficult for non-specialists to identify a nematode species with a high level of confidence based on morphology alone [27], and an integration of sub-organismal data such as DNA sequence can be required for accurate identification. However, recent advances in high performance computing may augment human image analyses.

### 2.2. Machine Learning

Advances in machine learning, also referred to as deep learning or artificial intelligence (AI), have opened a new avenue for nematode identification and quantification based on image analysis. The technique is especially suitable for handling large numbers of samples as well as detecting rare and microscopic objects, such as nematode eggs in complex backgrounds.

Machine learning for automated detection of phenotypes takes place in multiple stages. First, a large number of images (of nematodes, their eggs, or cysts) is taken and independently annotated (labeled) by a group of experts to reduce subjectivity. These are then used to build an algorithm that learns (captures) the salient features of the objects from the images in a layer-wise hierarchy while masking (rejecting) the noise in the background. The pattern of interest in the in-put images is then reconstructed using a network model with a supervised learning scheme. Using this technique, Akintayo et al. [28] designed a novel end-to-end Convolutional Selective Autoencoder (CSAE) to identify soybean cyst nematode (SCN) eggs in different backgrounds to cover for variations in background noise across samples from different sources. The authors trained the CSAE to identify SCN eggs using many labeled image segments (patches) that were smaller than the entire image. Information from multiple overlapping local patches was then combined to reconstruct a complete image and determine the existence of an egg in a particular patch. The model correlates pixel intensity values to reconstructed images to show the degree of confidence in predicting the object in the image is indeed an SCN egg. Tests done using two sets of samples collected from regions with different soil properties showed that egg counts done by trained personnel and using this AI technique were comparable at the 95% confidence level.

Another AI technique developed by Hakim et al. [29] using *Caenorhabditis elegans* combines the capabilities of different image processing programs for a fully automated and simultaneous processing of informative phenotypic features in a single platform called WorMachine. The image processor in this platform binarizes, identifies and crops individual worms from still in-put images taken using the bright-field with or without overlapping fluorescent acquisitions. Morphological and fluorescent features are then extracted from the cropped worm masks and analyzed individually by the feature extractor, which also allows labeling of different worms. Based on the features and labels obtained, the machine learner algorithm then conducts a binary classification or scoring of complex phenotypes using principal component analysis (PCA) and *t*-distributed stochastic neighbor embedding (*t*-SNE). The authors distinguished between males (XO), hermaphrodites (XX) and a range of phenotypes in between using fluorescent reporters for sex-specific expression patterns in mutant *C. elegans*. To demonstrate that WorMachine can be used to quantify continuous morphological phenotypes, they used strain CB5362 that is mutated in the sex-determination genes, and quantified intersex phenotypes in worms grown at different temperatures. For each worm, they determined the degree of masculinization from measurements of tail shape, gonad width (larger mid-width in egg-bearing worms), body length and area (males being smaller), brightness of head and tail (darker tails in males in bright-field), analyzed using PCA and *t*-SNE. They reported that the results agreed with those from previous studies, which showed increased masculinity at higher temperatures.

These studies show that AI can play a big role in the detection, quantification as well as classification of nematodes. As such, it will help address some of the limitations associated with the traditional morphology-based classification including the dwindling number of taxonomists, subjective decision making, and provide fast and accurate identification. Ironically, however, generating sufficient training data may present a bottleneck in developing AI due to the declining number of taxonomists. Limitations arising from shared morphological features between taxa would likely remain, but there is the possibility that machine learning will be able to elucidate unique characters discriminating nematodes that have been undetected even by the trained human eye.

### 2.3. Autoflorescence

A potential supplement to traditional light microscopy is the utilization of natural autofluorescence of microorganisms. Bhatta et al. [30] demonstrated that the emission and excitation spectra of the bacterial genera *Lactobacillus* and *Saccharomyces* were distinct. They also reported on the potential of these spectroscopic fingerprints to discriminate between different fungal species within the genus *Saccharomyces* without the need for fluorescent staining. Qazi et al. [31] built on this and demonstrated that eggs of different helminths revealed characteristic florescence when illuminated at different wavelengths ranging from white light to infrared. They also showed that differences in florescence lifetime values (decay in florescence intensity) were diagnostic of the species considered, *Ascaris lumbricoides* and *A. suum*. Qazi et al. [31] concluded that spectroscopic features and lifetime value measurements of autofluorescence in nematodes are promising tools in the taxonomy of these organisms.

## 3. DNA-Based Methods

Many forms of DNA-based methods have been developed for the identification of nematodes (e.g., [32–37]). These can be broadly categorized into fingerprint- and nucleotide-based methods. Fingerprint-based methods may include Restriction Fragment Length Polymorphism (RFLP), Amplified Fragment Length Polymorphism (AFLP), Random Amplification of Polymorphic DNA (RAPD) and the use of species-specific primers, which relies on the presence/absence of a PCR amplification product. Except for RFLP, where PCR may not be needed, all fingerprint-based methods involve PCR followed by electrophoresis. The resulting DNA fingerprint, i.e., the pattern of resolution of the DNA fragments, is used for identification and/or phylogenetic analyses of the nematode taxa considered. On the other hand, nucleotide-based methods involve PCR amplification, specific probe hybridizations and sequencing of a region(s) of the DNA, which is then used in phylogenetic analyses. Each of these methods has its own advantages and/or disadvantages compared to other nematode identification methods, DNA-based or otherwise. However, it is notable that nematode sequences have greatly altered our understanding of the evolutionary relationships between taxa [6].

### 3.1. Fingerprint-Based Methods

RFLP analyses can be made using fingerprints generated from genomic DNA (gDNA) digested with one or more endonucleases. Alternatively, fingerprints may be generated from PCR-amplicons (PCR-RFLPs) (e.g., [37–39]). gDNA-RFLPs tend to be complex, but potentially reveal more polymorphisms owing to the size of the gDNA template. Also, gDNA-RFLPs do not require knowledge of sequence information a priori, which is not the case with PCR-RFLPs. In both cases, however, care must be taken to let restriction digestions go to completion since incomplete digestions may lead to non-reproducible fingerprints.

The AFLP technique improves upon gDNA-RFLP by selectively amplifying fewer restriction products and producing less-complex fingerprints (e.g., [32,40]). gDNA is digested with two restriction enzymes that produce sticky ends, to which are ligated adaptors. A subset of these adaptor-ligated fragments is then selectively amplified using primer sets that recognize sequences of the adaptors, the sticky ends, and one to three nucleotides inside the restriction sites. As with gDNA-RFLPs, AFLPs do not require prior knowledge of sequence information, and completion of restriction digestions is crucial for reproducible fingerprints.

RAPD involves PCR amplification of gDNA fragments using short (usually 10 bp) primers of arbitrary sequences (e.g., [34,41]). The primers bind to several regions on the DNA, and amplification results if two primers bind on opposite strands of the DNA with their 3′-ends facing each other at a distance that can be traversed by the polymerase. Consequently, fragments of various sizes may be generated, with sizes of the larger fragments dependent on efficiency of the polymerase used. The use of large, intact gDNA template is important for this reason. Because RAPDs are done at lower temperatures, which create lower stringency for primer annealing, reproducibility especially between laboratories also poses a limitation. One advantage of this method is that it does not require prior knowledge of sequence information about the template DNA.

The use of primer sets that amplify a PCR product only in a taxon of interest is commonplace nowadays (e.g., [42–44]). Such primer sets can be designed based on fragments that uniquely identify the taxon in fingerprint analyses or based on taxon-dependent nucleotide sequence differences in aligned sequence data. In either case, care must be taken to include as much of the genetic variation within the taxon of interest as well as that of its close phylogenetic relations to ensure specificity of the primer sets. The diagnostic value of species-specific primers is based on amplification of a product only in the species for which they are designed. Therefore, it is necessary to have an internal control for a successful PCR and avoid false negatives by multiplexing the reaction with a second set of primers that amplify a product nonspecifically; after electrophoresis, two bands would be diagnostic of the species of interest while single bands corresponding to the internal control indicate otherwise.

### 3.2. Microarrays and Probe-Based Methods

DNA microarray is a collection of pico-moles of microscopic DNA fragments fixed at defined positions on a solid surface such as a glass slide. For nematode identification, these DNA fragments can be generated from sequence characterized amplified regions (SCARs) and are used as probes to which test samples containing florescent-labeled PCR products or gDNA are made to hybridize in high-throughput diagnostics. Data from hybridized slides are acquired using an array scanner at the emission wavelengths of the florescent dyes used. François et al. [45] investigated the suitability of DNA microarray technique for identification of nematodes using *M. chitwoodi*-specific oligonucleotides as probes. The probes were designed based on nucleotide sequences internal to binding sites of the primer sets used to amplify SCAR and satellite DNA fragments in *M. chitwoodi*, but not in *M. arenaria*, *M. javanica*, *M. fallax* and *M. hapla*. In agreement with the specificity of the primer sets in standard PCRs, both SCAR- and satellite DNA-based probes detected *M. chitwoodi* irrespective of the geographical origin of the nematode. However, cross-hybridization with *M. chitwoodi* targets was observed when satellite DNA-based probes designed from the pMfFd satellite DNA family of *M. fallax*, a closely related species, was used. This shows that careful selection of probes is important. This is the only study that we came across where DNA microarray technology was used in nematode diagnostics.

TaqMan qPCR also employs labelled DNA probe(s) for the detection and quantification of nematodes. At the start of TaqMan qPCR, the labelled probe binds to the template DNA within the site circumscribed by the primers. As the reaction progresses and the polymerase reaches the probe, its endogenous 5′ nuclease activity cleaves the probe, separating the dye from the quencher at the 3′-end of the probe. With each PCR cycle, more dye molecules are released, resulting in an increase in fluorescence intensity proportional to the amount of amplicon synthesized. The inclusion of probe(s) makes the technique more specific than standard PCRs and the amount of florescence detected can be used to quantify the number of nematodes in the sample. Primers and probes may be designed from aligned sequence data (e.g., [46]) as described above for species-specific primers, or from SCARs (e.g., [47]). Using this technique, Sapkota et al. [46], for example, developed a real-time PCR assay for the detection of M. hapla in soil and in root galls. They were able to differentiate *M. hapla* DNA from among those of 14 other *Meloidogyne* spp. included in their study except for *M. minor*. Based on aligned sequences from the 14 species, the authors concluded that the *M. minor* DNA must have been contaminated with that of *M. hapla* for amplification to result using these primer sets and probe. The authors reported *M. hapla* DNA extracted from 250 mg of soil (containing the equivalent of a third of an egg) could be detected by this technique. Similar studies have been carried out for other nematode taxa as well, which reported on the suitability of TaqMan qPCR for detection and quantification of nematode taxa (e.g., [47,48]).

### 3.3. Sequence-Based Methods

Sequence-based methods may involve analyses of nucleotide sequence information from specific segment(s) of the nuclear DNA, mitochondrial DNA (mtDNA), or the whole genome (for examples of gene regions and the corresponding primer sets, see: [42,49–54]). The rDNA and mitochondrial cytochrome c oxidase subunit I (*COX1*) genes are preferred by most studies (e.g., [54–58]) for diagnostic purposes because they have variable regions circumscribed by conserved ones. The higher level of sequence diversity in the variable region makes *COX1* preferable for resolution at lower taxonomic levels such as species and subspecies groups (e.g., [59]), while the higher level of sequence conservation in the flanking regions, which allows for 'universal' primers to be designed [56], has made the rDNA more suitable for use in wider taxonomic levels. The bulk of the sequence variability in the rDNA is harbored in the internal transcribed spacer (ITS), which is interrupted by the 5.8S coding region in the rDNA cistron into ITS1 and ITS2 [60], making the ITS useful in molecular systematics of closely related nematode species (e.g., [61–63]). ITS2 alone has been used for species diagnosis in *Caenorhabditis* [64] involving genetic crosses of newly collected isolates with known biological species, though the authors do not advocate for the use of ITS2 as an absolute criterion for species diagnosis because of the potential that distinct species may share identical ITS2 sequences. An added advantage of *COX1* and rDNA is that both genes occur in multiple copies in nematode genomes enabling PCR amplifications form small amounts of DNA templates such as that can be obtained from single nematodes. Sequence information generated is then used in character-based or phylogenetic analyses to resolve and/or identify the taxa involved; the latter analysis allows for evolutionary inferences.

The rDNA encompasses conserved coding regions (28S, 18S, and 5.8S subunits) and variable non-coding regions (ITS and ETS; the external-transcribed region) organized as tandem repeats, with intergenic spacers separating the repeating units [60]. As mentioned above, the rDNA provides phylogenetic resolution at a wide range of taxonomic levels and allows 'universal' primers to be designed for use in these taxa. This has led researchers to propose different regions of the rDNA for use as DNA barcode in different organisms; unique nucleotide sequences that can potentially be used to identify each species. Proposed DNA regions include ITS for fungi [65], 16S for bacteria [66], and 18S for nematodes [67,68]. The barcode region used in animals is the *COX1* region [69]. As such, DNA barcodes use sequence information from defined regions of the DNA to identify species using primers that are applicable for the broadest possible taxonomic group. Intraspecific variations should be smaller than interspecific variations in the barcode region.

Floyd et al. [67] used sequence information from the 18S (small subunit; SSU) to group soil nematodes into molecular operational taxonomic units (MOTUs). Each of these MOTUs was comprised of a cluster of sequences that differed from one another by less than three bases over aligned sequence data. The aligned data contained 349 to 396 nucleotides after removal of gaps, ambiguous characters and unresolved base calls from 450–500 nucleotide-long raw sequences generated using primer SSU94 [6]. MOTU content was predicted from neighbor joining trees generated using absolute character differences as a measure of distance. The authors reported that MOTUs largely corresponded with morphologically defined species or genera. Powers et al. [70] also studied a region of the 18S as a potential barcode for nematodes of suborder *Criconematina*. This region does not overlap with that used by Floyd et al. [67] and lies closer to ITS1. The authors used both phylogenetic and character state differences to define MOTUs. Among the 132 polymorphic sites in the aligned dataset, 56 were singletons and defined 56 MOTUs, each consisting of identical sequences. Most clades did not have strong statistical support, and morphologically identified species did not correspond with phylogenetically supported clades except for Clade B. Apart from a single MOTU, Clade B exclusively consisted of *Discocriconemella limitanea*, represented by 11 MOTUs, which may be cryptic species according to the authors. Conversely, some individual MOTUs identified a complex of species. For example, MOTU 76 corresponded to *Ogma* spp. that have scales arranged singularly in longitudinal rows along the length of the body, or arranged in rows consisting of clusters of 4–6 scales, or with scales densely packed on the annules forming a continuous elongated fringe.

The value of a barcode is directly related to the taxonomic rank it can effectively be applied to. The regions of SSU tested for their potential as barcodes by Powers et al. [70] and Floyd et al. [67] resolved the respective soil nematodes into named taxa and/or MOTUs. However, it is evident that MOTUs cannot be compared between the two studies because they were established based on incongruent sequence information; a phylum-wide barcode would be more powerful, but possible only if taxa representing the whole phylum were analyzed for the same DNA region. The sequence heterogeneity in individual nematodes that was reported by Powers et al. [70] is also suggestive of sequence variation among different copies of the SSU in the rDNA tandem repeat. Though, Dorris et al. [71] and Floyd et al. [67] stated that there is no evidence in nematodes of one species carrying more than one very distinct sequence variant. Bik et al. [72], however, have demonstrated that there exists intragenomic rRNA polymorphism and copy number variation in nematodes, and that the existence of minor variant gene copies in the rRNA repeats presents substantial challenges for biodiversity estimates and the analysis of marker-based datasets. Care must be taken to exclude such variable sites during analyses if the variation is greater than the cut-off value (see below). Another issue that needs to be addressed is how to interpret the barcode. DeSalle et al. [73] contend that a non-tree-based population aggregation analysis (PAA; [74]) is the most appropriate approach because tree-building approaches are flawed for many reasons. Firstly, morphology-based methods are character-based rendering the union of classical methods and distance-based DNA barcoding difficult. Secondly, tree-building methods are hierarchical while the underlying system consisting of individuals and populations is not. Thirdly, cut-off values are rather subjective; there is no objective set of criteria to delineate taxa when using distances. DeSalle et al. [73] emphasize that the best approach is to look for diagnostic characters in the aligned sequences themselves.

A great advantage of sequence-based methods is that sequence information is stored in publicly available databases such as GenBank (ncbi.nlm.nih.gov) and NEMBASE (nematodes.org). This facilitates identification of nematodes based on sequence information through comparison with that available in these databases. Accuracy of identification, however, depends on the quality of sequences deposited in the databases and the authenticity of the taxa the sequences originated from.

Most journals require that sequences be submitted to open-access databases as part of the publication process. But there is no such requirement for alignments. Unavailability in these databases of aligned sequence datasets may affect identification, especially that based on character states. This is because though alignments are generated using software, they invariably need manual editing particularly when larger datasets containing ambiguous sites are involved, which may introduce variations in alignments.

While gene-specific sequence information is commonly employed at lower taxonomic levels, there is a growing effort to include whole mitochondrial or whole genome sequence information at all taxonomic levels now that sequencing has become more affordable. Comparative genomics enables retrieval of additional information such as synteny and gene order for the investigation of underlying evolutionary mechanisms like inversion, translocation, fusion, and so on, in addition to aiding a more advanced understanding of nematode biology. After the call to sequence 959 nematode genomes by Kumar et al. [75], progressively larger number of nematode genomes have been sequenced. It would be advantageous if whole genome sequencing projects involve morphological type specimens where possible as the availability of sequence information from type specimens in the databases would help improve the accuracy of sequence-based identification of nematode samples.

## 4. Protein-Based Methods

Like DNA-based methods, protein sequences, mass-to-charge ratios, and immunological techniques focus on using unique protein composition and structures to delineate nematode species. Proteins provide a reduced vocabulary compared to DNA due to redundancy of the genetic code; however, the alphabet used is vastly more complex, utilizing 20 plus characters compared to the four DNA bases. Additionally, protein structure and post-translational modifications increase the potential diversity available to define nematode species and facilitate identification. Nonetheless, the requisite specialization in protein-based techniques is often a significant deterrent.

### 4.1. Isozyme Analyses

Enzyme phenotypes were among the first non-morphology-based methods used for the identification of nematodes. Briefly, this technique involves the extraction of soluble proteins from whole nematodes in buffer solutions, resolving the resulting extracts by starch or polyacrylamide gel electrophoresis followed by staining for specific enzymes. This electrophoretic method, also known as Multi-locus Enzyme Electrophoresis (MEE), relies on the migration patterns of isozymes, owing to differences in electrical charge, molecular weight, and conformation stemming from slight variations in amino acid compositions. The most commonly utilized enzymes were esterases [76], though malate dehydorgenase, superoxide dismutase, and glutamate-oxaloacetate transaminase have also been employed to various degrees [76,77]. This technique supplemented morphological methods and shed light in the phylogenetic relationships, especially among the major species in the genus *Meloidogyne*. However, the method was still cumbersome and time consuming; and the need to include known samples for reference purposes are among its limitations [76].

### 4.2. Two-Dimensional Gel Analyses

Two-dimensional gel electrophoresis (2-DGE) has been employed in taxonomic studies of nematodes. The technique allows resolution of complex protein mixtures by charge using isoelectric focusing in one-dimension followed by mass-based resolution in a dimension perpendicular to the first. The resolution pattern is then compared among isolates to determine similarities/differences, which can be scored as presence/absence for phenetic and/or cladistic analyses of the resulting data matrix. Navas et al. [78] used 2-DGE to show proteomic variations among 18 root-knot nematodes representing four species. They demonstrated that some of these variations were species-specific, while other variations revealed evolutionary relationships among the different species.

The technique has a number of pros and cons as applied to nematode taxonomy. One of the pros of 2-DGE is that it allows evolutionary inferences to be made about the taxa considered. Species-specific polypeptides can also be excised and analyzed using mass spectrometry (see below) allowing inferences to be made about the encoding genes. The cons include that the number of polypeptides resolved, and the polymorphism observed depend on the procedure used and the number of samples analyzed. For example, the number of polypeptides Navas et al. [78] observed among the 18 isolates ranged from 73–203. The authors stated that scoring the spots was difficult at times because it was hard to assess if some of the observed differences were real or due to deformations in the gel. For this reason, they scored only 95 spots that were consistently expressed in the two replicates they used for each nematode. Thirty-seven of these spots were monomorphic and thus uninformative. Considering that two of the species in their study were represented by single isolates only, it can be concluded that both the total number as well as the number of informative spots would have been different from what they observed had they used larger number of isolates.

### 4.3. Mass Spectral Analyses

Matrix-assisted laser desorption/ionization (MALDI) is an ionization technique, which uses laser energy-absorbing matrix to generate gaseous ions from large molecules in solid state. Embedded in a suitable matrix, the sample is applied onto a plate and irradiated with pulsed laser resulting in vaporization of the sample and the matrix material. Molecules are ionized by loss/gain of proton(s) in the hot plume of ablated gases and accelerated into a mass spectrometer for detection. Time of flight mass spectrometer (ToF-MS) measures the time taken by these ions to reach the detector as determined by the mass/charge ($m/z$) values, with smaller and/or more charged ions travelling faster. Since MALDI results in minimum fragmentation, the ions generated are predominantly non-fragmented and single-charged, which makes it easy to determine parental ion masses from mass spectra [79].

The basis of taxonomic identification using MALDI-ToF-MS is the ability to detect protein/peptide ions or protein profiles that are diagnostic to the taxa being considered. Perera et al. [80] used intact second stage juveniles (J2s) and/or proteins extracted from these using various organic solvents and discriminated between *Anguina tritici, A. funesta* and *M. javanica* based on unique peaks in their spectra and/or the spectral profiles. However, the authors advised that care must be taken when selecting the solvent for protein extraction and the matrix material for the MALDI as reproducibility and quality of the spectra vary with the material used. Ahmad et al. [81] built on this study to show that single *M. incognita* nematodes (an adult female or a J2) washed or unwashed, crushed or intact, can be used for diagnostics using MALDI-ToF-MS. Their study revealed that protein profiles differed between adults and J2s, each with its own diagnostic peaks; more masses and stronger peaks were also observed when washed and/or crushed samples were used. Both studies reported that careful optimization of instrument settings is also crucial.

Navas et al. [78] generated MALDI-ToF-MS spectral profiles for species-specific proteins obtained from excised 2-DGE gels to identify the proteins for use as biomarker molecules. Their attempt to identify the proteins using similarity matches, however, returned no hit for lack of sufficient information in the databases at the time. A similar study involving 2-DGE and MALDI-ToF-MS analyses of proteomes of two nematomorph species, *Paragordius tricuspidatus* and *Spinochordodes tellinii*, was carried out by Biron et al. [82]. Biron et al. [82] reported that while 36.2% of total protein spots on the 2-DGE analyses were shared between the two hairworm species, 38.0% were specific to *P. tricuspidatus* and 25.8% to *S. tellinii*; a genetic distance of 0.47 separated the two species confirming the distant relationship reported previously for these species. Unlike Navas et al. [78], Biron et al. [82] were successful in identifying MS fingerprints of proteins obtained from excised gel plugs using similarity searches in the databases.

These studies [78,80–82] have demonstrated that 2-DGE coupled with MALDI-ToF-MS provide a powerful tool in nematode taxonomy. The methods allow for inferences to be made regarding evolutionary relationships among taxa as well as for development of species-specific markers. Results, however, can be affected by a number of factors including the protein extraction method, the quality of 2-DGE runs, and setup of instrument. Protein expression profiles are also known to differ depending on the developmental stage of nematodes and the growth conditions.

### 4.4. Serological Analyses

Since Bird [83] first reported on the possibility of generating antisera against nematodes, the application of poly- and monoclonal antibodies (mAbs) has been explored by several researchers with mixed results (for a summary see [84]). For example, Lee [85] reported that antiserum raised against *M. incognita* did not produce the trademark arc-shaped precipitation band when paired with antigens from another species within the same genera, *M. hapla,* in the Ouchterlony double diffusion assay, indicating a lack of cross-reactivity. However, it was noted that the apparent specificity may be due to the small number of nematodes used in the assay. Further studies [86–88] also confirmed a lack of specificity in reactivity of antisera from *Meloidogyne* spp. Similar mixed results were also observed among cyst nematode *Heterodera* and *Globodera* species (summarized in [84]). Cross-reactivity of polyclonal antisera raised against whole macerated nematodes, including the associated microbiome and metabolites, in their bodies is to be expected.

The development of the hybridoma technique by Kohler and Milstein [89] raised the hope of the nematology community to develop mAbs for diagnostic purposes. The technique involved isolating mature B-cells from animals immunized with nematode antigens, fusing these B-cells with mouse lymphoid tumor cells to produce hybridomas that can be maintained indefinitely in vitro for continuous production of the antibodies. mAbs provide more specificity depending on the immunogen the antibodies were raised against. mAbs were raised against a variety of agriculturally important nematodes including *H. glycines* [90], *M. incognita* [91], *G. rostochiensis* and *G. pallida* [84] using the hybridoma technique. Schots et al. [84] reported that some mAbs differentiated between *G. rostochiensis* and *G. pallida* isolates. The authors also showed that these mAbs were so sensitive that protein equivalents of less than one egg were detected using immunoassays. The hybridoma technique becomes cumbersome with increasing number of nematode samples. The low proportion of successful fusions obtained between B- and tumor cells also presents a handicap. Next-generation sequencing technologies may prove to revitalize this line of nematode identification techniques as single B-cell receptor sequencing (scBCR-seq) can reconstruct antigen binding site sequences for comparative studies [92].

## 5. Conclusions

The purpose of taxonomy is to understand biodiversity, categorize organisms, and aid the communication of biological information. Scientific naming is a prerequisite for communication in taxonomy, and valid naming is only possible with type specimens and corresponding morphological information. However, this is not always possible, particularly when dealing with environmental samples (eDNA). Furthermore, it is now generally accepted that there are insufficient morphological features to describe biological diversity, and the use of molecular information to supplement and/or circumvent this limitation is commonplace. Nonetheless, a taxon is more meaningful if its members possess unique biological features, rather than the taxon only representing a group of individuals sharing similar morphological or molecular features.

Morphology-based classification forms the foundation of taxonomy. It has benefited from recent advances in image analysis. AI helps circumvent limitations associated with the scarcity of highly qualified taxonomists and enables objective decision making, coupled with fast and accurate identification. Spectroscopic features and lifetime value measurements of autofluorescence also provide additional traits that can be exploited for identification purposes.

The relative ease of molecular methods (Table 1) has led to the recognition of many new taxa; some, based on sequence information alone. These taxa would have been impossible to describe morphologically not only for lack of taxonomists and sufficient morphological differences, but also because members of these taxa are difficult to culture. Taxa identified using different molecular approaches, however, are not always congruent; for example, when sequence information from different regions of the DNA is used in different studies, or when sequence data generated from the same DNA region are analyzed differently between studies. Likewise, taxa based on morphological features do not always correspond with those based on molecular information and vice versa. Consequently, no single method by itself provides all the answers all the time; and the choice of method(s) depends on the question asked, the nature of the samples and the availability of resources.

**Table 1.** Comparison of different nematode identification methods.

| Method | Expertise | Cost | Resolution |
|---|---|---|---|
| Morphological and Image-Based | | | |
| Classical Morphometrics | High | Low | Medium |
| Machine Learning | High | Low | Medium |
| Autoflorescence | High | Low | Medium |
| DNA-Based | | | |
| Fingerprint | Medium | Medium | Medium |
| Microarray / Probe-Based | Medium | Low | Medium |
| Sequencing | Medium | High | High |
| Protein-Based | | | |
| Isozyme Analyses | Medium | Medium | Medium |
| 2-D Gel Analyses | Medium | Low | Medium |
| Mass Spectrometry | Medium | Medium | Medium |
| Serological Analyses | High | High | Medium |

If the question at hand is identification of a nematode sample, the most direct approach would be to examine the sample microscopically and assign the nematode to the lowest taxonomic rank possible. The source of the sample may also provide a clue in this regard. However, this may require some level of taxonomic expertise. Based on this information, a molecular technique may then be employed to identity the nematode to species or even subspecies level. If the question has to do with quarantine, molecular methods that are specific to the quarantined nematode species may be employed to ascertain whether the nematode at hand is quarantined. If the objective is assessment of diversity in a field population(s), any of the fingerprinting techniques and/or sequence analyses based on one or a few genes may do. High-throughput sequencing using second or third generation technologies and the appropriate bioinformatic techniques are useful to study the diversity of nematodes in an environmental sample (eDNA).

**Author Contributions:** Drafting and writing of the manuscript, M.B.; drafting parts of the manuscript, A.B.; conceptualizing of the review and editing of the manuscript, P.D. All authors have read and agreed to the published version of the manuscript.

**Funding:** This research received no external funding.

**Conflicts of Interest:** The authors declare no conflict of interest.

## References

1.  Abad, P.; Gouzy, J.; Aury, J.M.; Castagnone-Sereno, P.; Danchin, E.G.J.; Deleury, E.; Perfus-Barbeoch, L.; Anthouard, V.; Artiguenave, F.; Blok, V.C.; et al. Genome sequence of the metazoan plant-parasitic nematode *Meloidogyne incognita*. *Nat. Biotechnol.* **2008**, *26*, 909–915. [CrossRef] [PubMed]
2.  Abebe, E.; Mekete, T.; Thomas, W.K. A critique of current methods in nematode taxonomy. *Afr. J. Biotechnol.* **2011**, *10*, 312–323. [CrossRef]
3.  Jones, J.T.; Haegeman, A.; Danchin, E.G.J.; Gaur, H.S.; Helder, J.; Jones, M.G.K.; Kikuchi, T.; Manzanilla-López, R.; Palomares-Rius, J.E.; Wesemael, W.M.L.; et al. Top 10 plant-parasitic nematodes in molecular plant pathology. *Mol. Plant Pathol.* **2013**, *14*, 946–961. [CrossRef] [PubMed]
4.  Blaxter, M. Nematodes: The worm and its relatives. *PLoS Biol.* **2011**, *9*. [CrossRef]
5.  De Oliveira, C.M.G.; Monteiro, A.R.; Blok, V.C. Morphological and molecular diagnostics for plant-parasitic nematodes: Working together to get the identification done. *Trop. Plant Pathol.* **2011**, *36*, 65–73. [CrossRef]
6.  Blaxter, M.L.; De Ley, P.; Garey, J.R.; Liu, L.X.; Scheldeman, P.; Vierstraete, A.; Vanfleteren, A.; Vanfleteren, J.R.; Mackey, L.Y.; Dorris, M.; et al. A molecular evolutionary framework for the phylum Nematoda. *Nature* **1998**, *392*, 71–75. [CrossRef]
7.  Adams, B.J. The species delimitation uncertainty principle. *J. Nematol.* **2001**, *33*, 153–160.
8.  Karssen, G.; Van Aelst, A.C. Root-knot nematode perineal pattern development: A reconsideration. *Nematology* **2001**, *3*, 95–111. [CrossRef]
9.  Eisenback, J.D.; Hunt, D.J. General Morphology. In *Root Knot Nematodes*; Perry, R.N., Moens, M., Starr, J.L., Eds.; CABI: Wallingford, CT, USA, 2009; pp. 18–54, ISBN 9781845934927.
10. Chitwood, B.G. "Root-knot nematodes"—Part I. A revision of the genus *Meloidogyne* Goeldi, 1887. *Proc. Helminthol. Soc. Wash.* **1949**, *15*, 90–104.
11. Eisenback, J.D.; Hirschmann, H.; Triantaphyllou, A.C. Morphological comparison of *Meloidogyne* female head structures, perineal patterns, and stylets. *J. Nematol.* **1980**, *12*, 300–313.
12. Brito, J.; Powers, T.O.; Mullin, P.G.; Inserra, R.N.; Dickson, D.W. Morphological and molecular characterization of *Meloidogyne mayaguensis* isolates from Florida. *J. Nematol.* **2004**, *36*, 232–240. [PubMed]
13. Villar-Luna, E.; Goméz-Rodriguez, O.; Rojas-Martínez, R.I.; Zavaleta-Mejía, E. Presence of *Meloidogyne enterolobii* on jalapeño pepper (*Capsicum annuum* L.) in Sinaloa, Mexico. *Helminthologia* **2016**, *53*, 155–160. [CrossRef]
14. Ye, W.; Robbins, R.T.; Kirkpatrick, T. Molecular characterization of root-knot nematodes (*Meloidogyne* spp.) from Arkansas, USA. *Sci. Rep.* **2019**, *9*, 1–21. [CrossRef] [PubMed]
15. Da Cunha, T.G.; Visôtto, L.E.; Lopes, E.A.; Oliveira, C.M.G.; God, P.I.V.G. Diagnostic methods for identification of root-knot nematode species from Brazil. *Ciência Rural* **2018**, *48*, 1–11. [CrossRef]
16. Turner, S.J.; Subbotin, S.A. Cyst Nematodes. In *Plant Nematology*; Perry, R.N., Moens, M., Eds.; CABI: Wallingford, CT, USA, 2006; pp. 109–143, ISBN 9781845930561.
17. Cook, R.; Noel, G.R. Cyst Nematodes: *Globodera* and *Heterodera* species. In *Plant Resistance to Parasitic Nematodes*; Starr, J.L., Cook, R., Bridge, J., Eds.; CABI: Wallingford, CT, USA, 2002; pp. 71–105, ISBN 9780851994666.
18. Mathews, H.J.P. Morphology of the Nettle Cyst Nematode *Heterodera urticae* Cooper, 1955. *Nematologica* **1970**, *16*, 503–510. [CrossRef]
19. Mulvey, R.H. Identification of *Heterodera* cysts by terminal and cone top structures. *Can. J. Zool.* **1972**, *50*, 1277–1292. [CrossRef]

20. Green, C.D. The Vulval Cone and Associated Structures of Some Cyst Nematodes (Genus *Heterodera*). *Nematologica* **1975**, *21*, 134–144. [CrossRef]

21. Stone, A.R. Cyst nematodes—Most successful parasites. *New Sci.* **1977**, *10*, 355–356.

22. Stone, A.R. Recent developments and some problems in the taxonomy of cyst nematodes, with a classification of the Heterodeoidea. *Nematologica* **1977**, *23*, 273–288. [CrossRef]

23. Stone, A.R. Three Approaches to the Status of a Species Complex, with A Revision of Some Species of Globodera (Nematoda: Heteroderidae). In *Concepts in Nematode Systematics, Systematics Association Special Volume*; Stone, A.R., Platt, H.M., Khalil, L.F., Eds.; Academic Press: London, UK, 1983; pp. 221–223.

24. Kumari, S. Morphological and molecular characterizations of cereal cyst nematode *Heterodera avenae* Wollenweber, 1924 from the Czech Republic. *J. Integr. Agric.* **2017**, *16*, 532–539. [CrossRef]

25. Thevenoux, R.; Folcher, L.; Esquibet, M.; Fouville, D.; Montarry, J.; Grenier, E. The hidden diversity of the potato cyst nematode *Globodera pallida* in the south of Peru. *Evol. Appl.* **2020**, *13*, 727–737. [CrossRef] [PubMed]

26. Handoo, Z.A.; Carta, L.K.; Skantar, A.M. Taxonomy, Morphology and Phylogenetics of Coffee-Associated Root-Lesion Nematodes, *Pratylenchus* spp. In *Plant-Parasitic Nematodes of Coffee*; Souza, R.M., Ed.; Springer: Dordrecht, The Netherlands, 2008.

27. Carneiro RM, D.G.; de Oliveira Lima, F.S.; Correia, V.R. Methods and Tools Currently Used for the Identification of Plant Parasitic Nematodes. In *Nematology—Concepts, Diagnosis and Control*; Shah, M., Mahamood, M., Eds.; IntechOpen: Rijeka, Croatia, 2017; pp. 19–35, ISBN 978-953-51-3416-9.

28. Akintayo, A.; Tylka, G.L.; Singh, A.K.; Ganapathysubramanian, B.; Singh, A.; Sarkar, S. A deep learning framework to discern and count microscopic nematode eggs. *Sci. Rep.* **2018**, *8*, 1–11. [CrossRef] [PubMed]

29. Hakim, A.; Mor, Y.; Toker, I.A.; Levine, A.; Neuhof, M.; Markovitz, Y.; Rechavi, O. WorMachine: Machine learning-based phenotypic analysis tool for worms. *BMC Biol.* **2018**, *16*, 1–11. [CrossRef] [PubMed]

30. Bhatta, H.; Goldys, E.M.; Learmonth, R.P. Use of fluorescence spectroscopy to differentiate yeast and bacterial cells. *Appl. Microbiol. Biotechnol.* **2006**, *71*, 121–126. [CrossRef]

31. Qazi, F.; Khalid, A.; Poddar, A.; Tetienne, J.P.; Nadarajah, A.; Aburto-Medina, A.; Shahsavari, E.; Shukla, R.; Prawer, S.; Ball, A.S.; et al. Real-time detection and identification of nematode eggs genus and species through optical imaging. *Sci. Rep.* **2020**, *10*, 1–12. [CrossRef] [PubMed]

32. Semblat, J.P.; Wajnberg, E.; Dalmasso, A.; Abad, P.; Castagnone-sereno, P. High-resolution DNA fingerprinting of parthenogenetic root-knot nematodes using AFLP analysis. *Mol. Ecol.* **1998**, *7*, 119–125. [CrossRef]

33. Randig, O.; Leroy, F.; Castagnone-Sereno, P. RAPD characterization of single females of the root-knot nematodes, *Meloidogyne* spp. *Eur. J. Plant Pathol.* **2001**, *107*, 639–643. [CrossRef]

34. Abd ElAzim, A.M.; Khashaba, E.H.K.; Ibrahim, S.A.M. Genetic polymorphism among seven entomopathogenic nematode species (Steinernematidae) revealed by RAPD and SRAP analyses. *Egypt. J. Biol. Pest Control* **2019**, *29*. [CrossRef]

35. Han, H.; Cho, M.R.; Jeon, H.Y.; Lim, C.K.; Jang, H.I. PCR-RFLP Identification of Three Major *Meloidogyne* Species in Korea. *J. Asia. Pac. Entomol.* **2004**, *7*, 171–175. [CrossRef]

36. Correa, V.R.; dos Santos, M.F.A.; Almeida, M.R.A.; Peixoto, J.R.; Castagnone-Sereno, P.; Carneiro, R.M.D.G. Species-specific DNA markers for identification of two root-knot nematodes of coffee: *Meloidogyne arabicida* and *M. izalcoensis*. *Eur. J. Plant Pathol.* **2013**, *137*, 305–313. [CrossRef]

37. Smith, T.; Brito, J.A.; Han, H.; Kaur, R.; Cetintas, R.; Dickson, D.W. Identification of the peach root-knot nematode, *Meloidogyne floridensis*, using mtDNA PCR-RFLP. *Nematropica* **2015**, *45*, 138–143.

38. Širca, S.; Stare Geric, B.; Strajnar, P.; Urek, G. PCR-RFLP diagnostic method for identifying *Globodera* species in Slovenia. *Phytopathol. Mediterr.* **2010**, *49*, 361–369. [CrossRef]

39. Handoo, Z.A.; Skantar, A.M.; Hafez, S.L.; Kantor, M.R.; Hult, M.N.; Rogers, S.A. Molecular and morphological characterization of the alfalfa cyst nematode, *Heterodera medicaginis*, from Utah. *J. Nematol.* **2020**, *52*, 58–66. [CrossRef] [PubMed]

40. Marché, L.; Valette, S.; Grenier, E.; Mugniéry, D. Intra-species DNA polymorphism in the tobacco cyst-nematode complex (*Globodera tabacum*) using AFLP. *Genome* **2001**, *44*, 941–946. [CrossRef]

41. Naz, I.; Rius, J.E.P.; Blok, V. Species Identification of Root Knot Nematodes in Pakistan By Random Amplified Polymorphic DNA (RAPD-PCR). *Sarhad J. Agric.* **2013**, *29*, 71–78.

42. Toumi, F.; Waeyenberge, L.; Viaene, N.; Dababat, A.; Nicol, J.M.; Ogbonnaya, F.; Moens, M. Development of two species-specific primer sets to detect the cereal cyst nematodes *Heterodera avenae* and *Heterodera filipjevi*. *Eur. J. Plant Pathol.* **2013**, *136*, 613–624. [CrossRef]

43. Amarante, M.R.V.; Bassetto, C.C.; Neves, J.H.; Amarante, A.F.T. Species-specific PCR for the identification of *Cooperia curticei* (Nematoda: Trichostrongylidae) in sheep. *J. Helminthol.* **2014**, *88*, 447–452. [CrossRef]

44. Dong, K.; Dean, R.A.; Fortnum, B.A.; Lewis, S.A. Development of PCR primers to identify species of root-knot nematodes: *Meloidogyne arenaria, M. hapla, M. incognita* and *M. javanica*. *Nematropica* **2001**, *31*, 271–280.

45. François, C.; Kebdani, N.; Barker, I.; Tomlinson, J.; Boonham, N.; Castagnone-Sereno, P. Towards specific diagnosis of plant-parasitic nematodes using DNA oligonucleotide microarray technology: A case study with the quarantine species *Meloidogyne chitwoodi*. *Mol. Cell. Probes* **2006**, *20*, 64–69. [CrossRef]

46. Sapkota, R.; Skantar, A.M.; Nicolaisen, M. A TaqMan real-time PCR assay for detection of *Meloidogyne hapla* in root galls and in soil. *Nematology* **2016**, *18*, 147–154. [CrossRef]

47. François, C.; Castagnone, C.; Boonham, N.; Tomlinson, J.; Lawson, R.; Hockland, S.; Quill, J.; Vieira, P.; Mota, M.; Castagnone-Sereno, P. Satellite DNA as a target for TaqMan real-time PCR detection of the pinewood nematode, *Bursaphelenchus xylophilus*. *Mol. Plant Pathol.* **2007**, *8*, 803–809. [CrossRef] [PubMed]

48. Huang, D.; Yan, G.; Gudmestad, N.; Skantar, A. Quantification of *Paratrichodorus allius* in DNA extracted from soil using TaqMan Probe and SYBR Green real-time PCR assays. *Nematology* **2017**, *19*, 987–1001. [CrossRef]

49. Folmer, O.; Black, M.; Hoeh, W.; Lutz, R.; Vrijenhoek, R. DNA primers for amplification of mitochondrial cytochrome c oxidase subunit I from diverse metazoan invertebrates. *Mol. Mar. Biol. Biotechnol.* **1994**, *3*, 294–299. [CrossRef] [PubMed]

50. Avó, A.P.; Daniell, T.J.; Neilson, R.; Oliveira, S.; Branco, J.; Adão, H. DNA barcoding and morphological identification of benthic nematodes assemblages of estuarine intertidal sediments: Advances in molecular tools for biodiversity assessment. *Front. Mar. Sci.* **2017**, *4*, 66. [CrossRef]

51. Alanio, A.; Desnos-Ollivier, M.; Garcia-Hermoso, D.; Bretagne, S. Investigating clinical issues by genotyping of medically important fungi: Why and how? *Clin. Microbiol. Rev.* **2017**, *30*, 671–707. [CrossRef]

52. Ye, W.; Zeng, Y.; Kerns, J. Molecular characterisation and diagnosis of root-knot nematodes (*Meloidogyne* spp.) from turfgrasses in North Carolina, USA. *PLoS ONE* **2015**, *10*, 1–16. [CrossRef]

53. Holterman, M.; Van Der Wurff, A.; Van Den Elsen, S.; Van Megen, H.; Bongers, T.; Holovachov, O.; Bakker, J.; Helder, J. Phylum-wide analysis of SSU rDNA reveals deep phylogenetic relationships among nematodes and accelerated evolution toward crown clades. *Mol. Biol. Evol.* **2006**, *23*, 1792–1800. [CrossRef]

54. Van Megen, H.; Van Den Elsen, S.; Holterman, M.; Karssen, G.; Mooyman, P.; Bongers, T.; Holovachov, O.; Bakker, J.; Helder, J. A phylogenetic tree of nematodes based on about 1200 full-length small subunit ribosomal DNA sequences. *Nematology* **2009**, *11*, 927–950. [CrossRef]

55. Donn, S.; Neilson, R.; Griffiths, B.S.; Daniell, T.J. Greater coverage of the phylum Nematoda in SSU rDNA studies. *Biol. Fertil. Soils* **2011**, *47*, 333–339. [CrossRef]

56. Hadziavdic, K.; Lekang, K.; Lanzen, A.; Jonassen, I.; Thompson, E.M.; Troedsson, C. Characterization of the 18s rRNA gene for designing universal eukaryote specific primers. *PLoS ONE* **2014**, *9*. [CrossRef]

57. Blouin, M.S.; Yowell, C.A.; Courtney, C.H.; Dame, J.B. Substitution bias, rapid saturation, and the use of mtDNA for nematode systematics. *Mol. Biol. Evol.* **1998**, *15*, 1719–1727. [CrossRef] [PubMed]

58. Derycke, S.; Vanaverbeke, J.; Rigaux, A.; Backeljau, T.; Moens, T. Exploring the use of cytochrome oxidase c subunit 1 (COI) for DNA barcoding of free-living marine nematodes. *PLoS ONE* **2010**, *5*. [CrossRef] [PubMed]

59. Morise, H.; Miyazaki, E.; Yoshimitsu, S.; Eki, T. Profiling Nematode Communities in Unmanaged Flowerbed and Agricultural Field Soils in Japan by DNA Barcode Sequencing. *PLoS ONE* **2012**, *7*. [CrossRef] [PubMed]

60. Long, E.O.; Dawid, I.B. Repeated genes in eukaryotes. *Annu. Rev. Biochem.* **1980**, *49*, 727–764. [CrossRef] [PubMed]

61. Powers, T.O.; Todd, T.C.; Burnell, A.M.; Murray, P.C.B.; Fleming, C.C.; Szalanski, A.L.; Adams, B.A.; Harris, T.S. The rDNA internal transcribed spacer region as a taxonomic marker for nematodes. *J. Nematol.* **1997**, *29*, 441–450.

62. Hung, G.C.; Chilton, N.B.; Beveridge, I.; Gasser, R.B. A molecular systematic framework for equine strongyles based on ribosomal DNA sequence data. *Int. J. Parasitol.* **2000**, *30*, 95–103. [CrossRef]

63. Bu, Y.; Niu, H.; Zhang, L. Phylogenetic analysis of the genus *Cylicocyclus* (Nematoda: Strongylidae) based on nuclear ribosomal sequence data. *Acta Parasitol.* **2013**, *58*, 167–173. [CrossRef] [PubMed]

64. Félix, M.A.; Braendle, C.; Cutter, A.D. A streamlined system for species diagnosis in *Caenorhabditis* (Nematoda: Rhabditidae) with name designations for 15 distinct biological species. *PLoS ONE* **2014**, *9*. [CrossRef]

65. Schoch, C.L.; Seifert, K.A.; Huhndorf, S.; Robert, V.; Spouge, J.L.; Levesque, C.A.; Chen, W.; Bolchacova, E.; Voigt, K.; Crous, P.W.; et al. Nuclear ribosomal internal transcribed spacer (ITS) region as a universal DNA barcode marker for Fungi. *Proc. Natl. Acad. Sci. USA* **2012**, *109*, 6241–6246. [CrossRef]

66. Hugenholtz, P.; Goebel, B.M.; Pace, N.R. Impact of culture-independent studies on the emerging phylogenetic view of bacterial diversity. *J. Bacteriol.* **1998**, *180*, 4765–4774. [CrossRef]

67. Floyd, R.; Abebe, E.; Papert, A.; Blaxter, M. Molecular barcodes for soil nematode identification. *Mol. Ecol.* **2002**, *11*, 839–850. [CrossRef] [PubMed]

68. Blaxter, M.; Mann, J.; Chapman, T.; Thomas, F.; Whitton, C.; Floyd, R.; Abebe, E. Defining operational taxonomic units using DNA barcode data. *Philos. Trans. R. Soc. B Biol. Sci.* **2005**, *360*, 1935–1943. [CrossRef] [PubMed]

69. Hebert, P.D.N.; Ratnasingham, S.; DeWaard, J.R. Barcoding animal life: Cytochrome c oxidase subunit 1 divergences among closely related species. *Proc. R. Soc. B Biol. Sci.* **2003**, *270*, 96–99. [CrossRef] [PubMed]

70. Powers, T.; Harris, T.; Higgins, R.; Mullin, P.; Sutton, L.; Powers, K. MOTUs, morphology, and biodiversity estimation: A case study using nematodes of the suborder criconematina and a conserved 18S DNA Barcode. *J. Nematol.* **2011**, *43*, 35–48. [PubMed]

71. Dorris, M.; De Ley, P.; Blaxter, M. Molecular analysis of nematode diversity and the evolution of parasitism. *Parasitol. Today* **1999**, *15*, 188–193. [CrossRef]

72. Bik, H.M.; Fournier, D.; Sung, W.; Bergeron, R.D.; Thomas, W.K. Intra-Genomic Variation in the Ribosomal Repeats of Nematodes. *PLoS ONE* **2013**, *8*, 1–8. [CrossRef]

73. DeSalle, R.; Egan, M.G.; Siddall, M. The unholy trinity: Taxonomy, species delimitation and DNA barcoding. *Philos. Trans. R. Soc. B Biol. Sci.* **2005**, *360*, 1905–1916. [CrossRef]

74. Davis, J.I.; Nixon, K.C. Populations, genetic variation, and the delimitation of phylogenetic species. *Syst. Biol.* **1992**, *41*, 421–435. [CrossRef]

75. Kumar, S.; Schiffer, P.H.; Blaxter, M. 959 Nematode genomes: A semantic wiki for coordinating sequencing projects. *Nucleic Acids Res.* **2012**, *40*, 1295–1300. [CrossRef]

76. Esbenshade, P.R.; Triantaphyllou, A.C. Isozyme phenotypes for the identification of *Meloidogyne* species. *J. Nematol.* **1990**, *22*, 10–15.

77. Esbenshade, P.R.; Triantaphyllou, A.C. Use of enzyme phenotypes for identification of *Meloidogyne* species. *J. Nematol.* **1985**, *17*, 6–20. [CrossRef] [PubMed]

78. Navas, A.; López, J.A.; Espárrago, G.; Camafeita, E.; Albar, J.P. Protein variability in *Meloidogyne* spp. (Nematoda: Meloidogynidae) revealed by two-dimensional gel electrophoresis and mass spectrometry. *J. Proteome Res.* **2002**, *1*, 421–427. [CrossRef] [PubMed]

79. Ahmad, F.; Babalola, O.O.; Tak, H.I. Potential of MALDI-ToF mass spectrometry as a rapid detection technique in plant pathology: Identification of plant-associated microorganisms. *Anal. Bioanal. Chem.* **2012**, *404*, 1247–1255. [CrossRef] [PubMed]

80. Perera, M.R.; Vanstone, V.A.; Jones, M.G.K. A novel approach to identify plant parasitic nematodes using matrix-assisted laser desorption/ionization time-of-flight mass spectrometry. *Rapid Commun. Mass Spectrom.* **2005**, *19*, 1454–1460. [CrossRef] [PubMed]

81. Ahmad, F.; Gopal, J.; Wu, H.F. Rapid and highly sensitive detection of single nematode via direct MALDI Mass Spectrometry. *Talanta* **2012**, *93*, 182–185. [CrossRef] [PubMed]

82. Biron, D.G.; Joly, C.; Marché, L.; Galéotti, N.; Calcagno, V.; Schmidt-Rhaesa, A.; Renault, L.; Thomas, F. First analysis of the proteome in two nematomorph species, *Paragordius tricuspidatus* (Chordodidae) and *Spinochordodes tellinii* (Spinochordodidae). *Infect. Genet. Evol.* **2005**, *5*, 167–175. [CrossRef] [PubMed]

83. Bird, A.F. Serological studies on the plant parasitic nematode, *Meloidogyne javanica*. *Exp. Parasitol.* **1964**, *15*, 350–360. [CrossRef]

84. Schots, A.; Hermsen, T.; Schouten, S.; Gommers, F.J.; Egberts, E. Serological differentiation of the potato-cyst nematodes *Globdera pallida* and *G. rostochiensis*: II. Preparation and characterization of species specific monoclonal antibodies. *Hybridoma* **1989**, *8*, 401–413. [CrossRef]

85. Lee, S.H. Attempts to use immunodiffusion for species identification of *Meloidogyne* (Abstr.). *Nematologica* **1965**, *11*, 41.

86. Hussey, R.S. Serological Relationship of *Meloidogyne incognita* and *M. arenaria*. *J. Nematol.* **1972**, *4*, 101–104.

87. Hussey, R.S.; Sasser, J.N.; Huisingh, D. Disc-Electrophoretic Studies of Soluble Proteins and Enzymes of *Meloidogyne incognita* and *M. arenaria*. *J. Nematol.* **1972**, *4*, 183–189. [PubMed]

88. Misaghi, I.; McClure, M.A. Antigenic Relationship of *Meloidogyne incognita, M. javanica,* and *M. arenaria*. *Phytopathology* **1974**, *64*, 698–701. [CrossRef]

89. Köhler, G.; Milstein, C. Continuous cultures of fused cells secreting antibody of predefined specificity. *Nature* **1975**, *256*, 495–497. [CrossRef] [PubMed]

90. Atkinson, H.J.; Harris, P.D.; Halk, E.J.; Novitski, C.; Leighton-Sands, J.; Nolan, P.; Fox, P.C. Monoclonal antibodies to the soya bean cyst nematode, *Heterodera glycines*. *Ann. Appl. Biol.* **1988**. [CrossRef]

91. Hussey, R.S. Monoclonal antibodies to secretory granules in esophageal glands of *Meloidogyne* species. *J. Nematol.* **1989**, *21*, 392–398.

92. Goldstein, L.D.; Chen, Y.J.J.; Wu, J.; Chaudhuri, S.; Hsiao, Y.C.; Schneider, K.; Hoi, K.H.; Lin, Z.; Guerrero, S.; Jaiswal, B.S.; et al. Massively parallel single-cell B-cell receptor sequencing enables rapid discovery of diverse antigen-reactive antibodies. *Commun. Biol.* **2019**, *2*. [CrossRef]

MDPI
St. Alban-Anlage 66
4052 Basel
Switzerland
Tel. +41 61 683 77 34
Fax +41 61 302 89 18
www.mdpi.com

*Plants* Editorial Office
E-mail: plants@mdpi.com
www.mdpi.com/journal/plants